ROUTLEDGE LIBRARY EDITIONS: HISTORY AND PHILOSOPHY OF SCIENCE

BRITISH SCIENTISTS OF THE TWENTIETH CENTURY

BRITISH SCIENTISTS OF THE TWENTIETH CENTURY

By

J. G. CROWTHER

Volume 9

Routledge
Taylor & Francis Group
LONDON AND NEW YORK

First published in 1952

This edition first published in 2009 by
Routledge
2 Park Square, Milton Park, Abingdon, Oxon, OX14 4RN

Simultaneously published in the USA and Canada
by Routledge
270 Madison Avenue, New York, NY 10016

Routledge is an imprint of the Taylor & Francis Group, an informa business

British Library Cataloguing in Publication Data
A catalogue record for this book is available from the British Library

Library of Congress Cataloging in Publication Data
A catalog record for this book has been requested

ISBN 10: 0-415-42029-6 (Set)
ISBN 10: 0-415-47453-1 (Volume 9)

ISBN 13: 978-0-415-42029-7 (Set)
ISBN 13: 978-0-415-47453-5 (Volume 9)

Publisher's Note

The Publisher has gone to great lengths to ensure the quality of this reprint but points out that some imperfections in the original copies may be apparent.

Disclaimer

The Publishers have made every effort to trace copyright holders and would welcome correspondence from those they have been unable to contact.

1. Lord Rutherford

frontispiece

BRITISH SCIENTISTS OF THE TWENTIETH CENTURY

by

J. G. CROWTHER

ROUTLEDGE & KEGAN PAUL LIMITED
Broadway House, 68–74 Carter Lane
London

First published in 1952
by Routledge & Kegan Paul Limited
Broadway House, 68-74 Carter Lane
London E.C.4
Printed in Great Britain
by William Clowes & Sons Limited
London and Beccles

CONTENTS

		page
	INTRODUCTION	ix
	ACKNOWLEDGEMENTS	xiii
I	JOSEPH JOHN THOMSON, 1856–1940	1
II	ERNEST RUTHERFORD, 1871–1937	43
III	JAMES HOPWOOD JEANS, 1877–1946	93
IV	ARTHUR STANLEY EDDINGTON, 1882–1944	140
V	FREDERICK GOWLAND HOPKINS, 1861–1947	197
VI	WILLIAM BATESON, 1861–1926	248
	SELECT BIBLIOGRAPHY	311
	INDEX	313

LIST OF ILLUSTRATIONS

LORD RUTHERFORD	*frontispiece*
	FACING PAGE
J. J. THOMSON AS A CHILD	1
SIR J. J. THOMSON	1
RUTHERFORD'S PARENTS	48
SIR JAMES JEANS	104
SIR ARTHUR EDDINGTON	144
SIR FREDERICK GOWLAND HOPKINS	234
WILLIAM BATESON	288
ACADEMICIAN T. D. LYSENKO	304

INTRODUCTION

HISTORY is advancing at a tremendous pace. Within the last fifty years there have been two world wars, the Russian Revolution, an immense growth of American industrial capitalism, and the achievement of independence by the great Chinese nation with its ancient and profound culture.

The most important change that has occurred to Britain in this period is her decline from world pre-eminence. This has produced a change in the assumptions and attitudes of British culture of which British scientific activity is a part.

One of the aims of this book is to look at the life and work of great British scientists of the immediate past from the perspective engendered by the new British situation.

The six eminent scientists discussed have only recently died. Their work and personality have not yet receded into the same depth of perspective as their predecessors of the nineteenth century, but the historic changes that have occurred have already cut them off sharply from our own generation. We can see them as scientists of another social period, and yet as men whom many of us knew personally.

It is concluded that they were deeply influenced by various phases of British capitalist development and imperialism. J. J. Thomson belonged to the rising Liberal phase, Rutherford to the imperialist phase, Jeans and Eddington to capitalist science in decline, and Hopkins to the socialist science of the future. Bateson was a striking misfit, who was frustrated in his advancement of biology, especially by the medieval traditions of an ancient university set in a capitalist industrial society.

Rutherford appears indeed as a mighty scientist, but an unconsciously tragic social figure. His discoveries were exploited, after the end of his optimistic life, by decadent social forces for anti-human purposes.

Of the six, Hopkins emerges with the most enhanced stature. From his youth he steadily resisted scientific and social reaction, and he continued the struggle through his later years of the highest conventional eminence, right to the end of his very long working life. By so doing, he aligned himself with developments passing beyond those which dominated his own times.

The six scientists discussed were all Cambridge men, and five of them belonged to Trinity College. This indicates a considerable degree of uniformity in the cultural environment of the leadership of British science, which, from the aspect of world-civilization, has necessarily had a certain narrowing tendency.

This is reflected in J. J. Thomson and Rutherford, and especially in Jeans and his younger twin-star, Eddington. Both of the latter appear in their solid scientific work as highly conventional applied mathematicians of the typically British school. Their philosophical and expository writings are on the whole a cultural reflection of imperial decline.

Hopkins was the least influenced by the Cambridge environment, as he did not begin to work in it until he was over forty years of age, and entirely mature.

The book has been written on the following plan. The main facts of the life of each scientist have been given. These have been followed by a fairly complete outline of his chief discoveries, to enable the reader to appreciate the foundation of his scientific reputation. The salient features of his ideas and style have been illustrated as far as practicable by quotations of his own words, especially in those cases where expression through writing was a major part of his activity.

J. J. Thomson and Rutherford were experimental physicists. Their primary work consisted of experimenting and organizing laboratory research. They were not particularly concerned with exposition, and their books were secondary to their main activity.

Jeans and Eddington were theorists, also concerned with philosophy and general exposition. Their books were a primary activity. They recorded their discoveries in their learned scientific books, and they pursued their philosophical and expository work in their books for the non-technical reader.

Eddington was greater than Jeans in fundamental discovery, and as an expositor almost as widely read. Jeans was a prose writer. At his best he had a majestic literary style, which was related to his love of music, and especially of counterpoint and fugue. Eddington's expository style was essentially poetic.

Hopkins' scientific reputation rests in the first place on his experimental discoveries. But the study of his works as a whole shows how indefatigably he struggled with the elaboration of the philosophy of biochemistry and the careful literary expression of the scientific principles of his science. Of the six scientists discussed he had the finest literary style and was the profoundest philosopher of science. He also had the deepest insight into the relations of science with social affairs. Considerable attention has accordingly been given to his views and writings on this subject.

Bateson was a scientist of great abilities who finally capitulated to Morgan against his innermost judgment. Among his pupils was N. I. Vavilov, who developed Mendel-Morgan studies in the Soviet Union on a big scale. The results were judged unrewarding, and have led to a return from the Mendel-Morgan tendency to explain genetics in terms of microscopy, chemistry and mathematics to a re-examination of the fundamental biological principle of heredity in living organisms.

The cause of Bateson's frustration is attributed to the crippling of biology by industrialism in British capitalist society. The re-examination of the biological principle of heredity started by T. D. Lysenko is a result of the role attained by biology in the scientific socialized agriculture of the Soviet Union, and in the vast projects for the transformation of nature in whole regions of that country.

Bateson was concerned with the great significance of genetics for humanity, but the difficulties imposed on him by his social and cultural situation prevented him from finding the proper way of exploring the question. In consequence, he recklessly misapplied biological conceptions to social problems, contributing to the development of racialist theories the results of which would have horrified him if he had lived to see them.

In reviewing Bateson's life and work, special attention has been given to his writings on social affairs.

The parallelism of trends in biology and physics is noticeable, how the Mendel-Morgan genetics and the prostitution of Rutherford's nuclear physics are seen to flourish in the same place, side by side; and, in contrast, how sounder biological theories, together with beneficent applications of atomic energy, are flourishing elsewhere, in great constructive works for the dignity and happiness of man.

Hopkins alone of this group had a deep insight into the social role of science. The works of the other five had vast significance for science itself, but not for the use of science. In this they are now out of date, for the question is not merely the advance of science but also its use for the benefit of man.

The scientists of today, and still more those of tomorrow, cannot, for better or for worse, have the same conception of science and of its role in human activity. The advance of science now depends primarily on the proper social use of science. If science is not properly used, it will destroy society, and with the ending of humanity will end itself.

The main aim of this book is to give a deeper understanding of these great scientists who are so near and yet so far from us, so that we can appreciate their genius and celebrate their contributions to civilization, and at the same time see how and why they have been limited and frustrated.

This helps us to realize what must be done by their successors, and by all of us, to provide the conditions which will enable scientists to give the fullest aid to man in his struggle for the future, and ensure that life shall not end through the misuse of their discoveries.

J. G. CROWTHER.

ACKNOWLEDGEMENTS

THANKS are due to the following ladies and gentlemen, publishers and institutions for permission to reproduce photographs or quote from works.

Miss W. Eddington for a photograph of Sir Arthur Eddington; the Cambridge University Press for a photograph of J. J. Thomson as a child, and one of Rutherford's parents; Professor Sir Lawrence Bragg, F.R.S., for the portrait of Sir J. J. Thomson by Arthur Hacker; Dr. Joseph Needham, F.R.S., for a photograph of Sir F. G. Hopkins, and to quote from *Hopkins and Biochemistry*; Messrs. Ramsey & Muspratt for a photograph of Lord Rutherford; the Planet Agency for a photograph of Sir James Jeans; the National Portrait Gallery for a drawing of W. Bateson by W. Arnold Forster; Mrs. Joan Charnock and Messrs G. Bell & Sons, Ltd., to quote from Sir J. J. Thomson's *Recollections and Reflections*; Mr. Maurice Cornforth and Messrs. Lawrence & Wishart, Ltd., to quote from Mr. Cornforth's *Science versus Idealism*; The Cambridge University Press to quote from *The Life of Sir J. J. Thomson* by Lord Rayleigh; *Applications of Dynamics to Physics and Chemistry*, by J. J. Thomson; *Rutherford*, by A. S. Eve; *Problems of Cosmogony and Stellar Dynamics*, by J. H. Jeans; *The Universe Around Us*, by J. H. Jeans; *The Mysterious Universe*, by J. H. Jeans; *Space Time and Gravitation*, by A. S. Eddington; *The Mathematical Theory of Relativity*, by A. S. Eddington; *The Internal Constitution of the Stars*, by A. S. Eddington; *The Nature of the Physical World*, by A. S. Eddington; *The Philosophy of Physical Science*, by A. S. Eddington; *Reflections on the Philosophy of Sir A. Eddington*, by A. D. Ritchie; *William Bateson: Essays and Addresses, and Life*, by Beatrice Bateson; the Clarendon Press to quote from *Recent Researches in Electricity and Magnetism*, by J. J. Thomson; *Stars and Atoms*, by A. S. Eddington; Messrs. George Allen & Unwin, Ltd., to quote from *Science and the Unseen World*, by A. S. Eddington; Messrs. John Lane & Co., Ltd., to quote from *The Crisis in Physics*, by Christopher Caudwell; Messrs. Macmillan & Co., Ltd., to quote from *Stellar Movements and the Structure of the Universe*, by A. S. Eddington, and from *Materials for the Study of Variation*, by W. Bateson; Messrs. Methuen, Ltd., to quote from *Letters from the Steppe*, by W. Bateson; the Physical Society of London to quote from *Report on Radiation and the Quantum Theory*, by J. H. Jeans.

The main sources of information on which the book is based are given in the bibliographies at the end of each section. Among those which have been of particular help are the obituary notices in the *Proceedings* of the Royal Society and the Physical Society and in *Nature*.

I am glad to record here my indebtedness to all of the many authorities on which I have depended.

Finally, I owe special thanks to Dr. Nora Wooster for helpful criticisms, and reading the proofs.

J. G. C.

2. Joseph John Thomson as a child

Sir J. J. Thomson, 1902, from the painting by Arthur Hacker in the Cavendish Laboratory

face p. 1

JOSEPH JOHN THOMSON

1856–1940

I

J. J. THOMSON was the first to give conclusive proof of the existence of the electron. He determined its main properties, and immediately saw its significance for the constitution of matter and the structure of atoms. He proposed an atomic model, based on the scattering of X-rays by matter, which gave a qualitative explanation of the periodic table of the elements, and of the difference between physics and chemistry. He invented the method of atomic analysis by which isotopes and any atomic particles of differing mass might be separated. In addition to these he made many other discoveries, some of them little less important.

Besides gaining the Nobel Prize himself, eight of his pupils also received this award. Twenty-seven were elected Fellows of the Royal Society. Seventy-nine professorships were held by them. He trained hundreds of scientists who subsequently achieved distinction.

Thomson held the offices usual to a man of his achievement. He was President of the Royal Society from 1915–1920. He was Chairman of the Government Committee on the Position of Natural Science in the Educational System of Great Britain. He was the scientific chairman of the Board of Invention and Research in the First World War. He had an influential part in the creation of the Department of Scientific and Industrial Research in 1916. After being Cavendish Professor of Experimental Physics at Cambridge for thirty-four years, from 1884 until 1918, and performing all his duties of teaching and administration besides research, he became Master of Trinity College.

He took the chair at the first meeting of the newly formed National Union of Scientific Workers in 1918.

He was the son of a moderately prosperous Manchester bookseller who died at the age of thirty-nine. Thomson was then sixteen years old, and had to pursue his education with the aid of scholarships and such support as his widowed mother could provide. His early means were very modest, yet when he died he left a fortune of £82,000.

The enormous output of discoveries by the Cavendish Laboratory was accomplished by an expenditure on special research equipment which never exceeded £550 in any of the thirty-four years of his directorate, though in some years there were forty research workers in the Laboratory.

J. J. Thomson was indeed a major scientist. In assessing his contribution to science in England, and in the world, it is necessary, however, to consider the things he and his school did not discover.

He was appointed in 1884 to the Cavendish chair first held by Clerk Maxwell, and was in a degree his scientific heir. But it was Hertz who proved the existence of Clerk Maxwell's electromagnetic waves, in 1887.

Though he had been investigating electric discharges in vacuum tubes for sixteen years, it was Röntgen who discovered X-rays.

It was Planck who invented the quantum theory, which, again, was a derivation from the statistical mechanics created by Clerk Maxwell to explain the properties of gases.

It was Einstein who derived the theory of relativity from Clerk Maxwell's equations for describing the behaviour of waves of light and electricity.

It was Becquerel who discovered radioactivity, and the Curies who discovered radium.

J. J. Thomson's mind, and that of his school, was evidently selective: it seized on certain classes of primary truths and passed others by. What did his mind select, and why? How far was his achievement due to intrinsic mental characteristics, and how far to training and intellectual environment? His personality, circumstances and achievement are the data for the answers to these questions.

From his Cambridge student days he was known as 'J.J.' Born has remarked that even his son, G. P. Thomson, addressed him as 'J.J.', in 1925. It is significant of his popularity and human qualities.

II. YOUTH

Joseph John Thomson was born in Cheetham in Manchester on December 18th, 1856, and was of pure Scottish descent. He had an ancestry of Scottish hardheadedness and a background of Lancashire individualism. These were softened by the interests of a bookseller's family. His parents wanted him to be an engineer. They could not apprentice him until he was sixteen, and were advised to send him to Owens College to gain some preliminary scientific training. Consequently, he entered Owens College at the age of fourteen. His father died two years afterwards.

Thomson has recorded that his father had introduced him to Joule, and said that one day he would be proud to say that he had met that gentleman. And indeed he subsequently was. A cousin of J.J.'s mother was at one time secretary of the famous Manchester Literary and Philosophical Society. The family was not out of touch with the atmosphere of science.

Mrs. Ryan, a cousin of J.J., described Mrs. Thomson as a small lady with bright dark eyes, beaming with kindness. She wore her hair in clusters of ringlets in an old-fashioned style seen in the illustrations of Dickens' novels. In later years, she used to give her nephews and nieces an annual pantomime outing and party. After sitting in the middle of the front row in the dress circle, they were driven home in ancient four-wheeled horse cabs, or 'growlers', and given a sumptuous Lancashire high tea. J.J. used to join these parties when he was from twenty-five to thirty-five years of age. He was one of the liveliest members, and evidently had a great liking for and understanding of children.

Mrs. Thomson used to give each of the children a new florin and a large bar of chocolate cream when they left for home.

J.J. and his brother always spent their summer holidays with their mother. He was a keen reader of the 'yellow-back' novels which preceded the modern detective story. He had a large repertoire of music-hall ditties, of which 'My Maria's a fairy queen' was a favourite. He used to sing these about the house.

How Joe Thomson appeared at the age of eleven to a little girl aged nine has been recorded by Miss Gertrude Mellor, a relative of the family by marriage. She saw him 'one afternoon, a small boy in a little grey suit and a blue silk tie, sitting on rather a high chair, with his legs dangling—very silent—and no doubt shy . . .' Some years later she saw him again. 'He remarked that when he grew up, he intended to go in for "Original Research", and I remember my brother-in-law laughed and tapped him on his head, and said: "Don't be such a little prig, Joe".' Years later still, she met him in Oxford Road near Owens College. 'He was wearing a nice brown suit, with a brown tie (I think it had white spots) the bow of which was under one ear, a fact of which I daresay he was quite unaware, and was possibly indifferent, and I longed to rectify. . . .'

This Child was the Father of the Man who is seen in Hacker's portrait, with his tie awry.

He entered Owens in 1871, and his father died in 1873. After Mr. Thomson's death, the family could no longer expect to afford the high fees for apprenticeship to an engineering firm. J.J.'s mother was advised to keep him at Owens until he had completed his course.

He attended the College for five years, and graduated in engineering at the age of nineteen. During the first two years the College was still housed in Cobden's old house in Quay Street. The rooms were extremely rough and overcrowded. This brought the students into close relations with the professorial staff, which was exceptionally brilliant. Osborne Reynolds was professor of engineering, Balfour Stewart of physics, W. C. Williamson of natural history, Stanley Jevons of logic and political economy, Adolphus Ward of history, and Bryce, who subsequently became Lord Bryce and Ambassador at Washington, was professor of law. H. E. Roscoe was professor of chemistry, and in 1874 Owens had created the first chair of organic chemistry in England, which was filled by Schorlemmer, Engels' friend.

Roscoe was one of the most energetic advocates of the cultivation of research and the application of science. He campaigned for these ends for forty years. He was the author of very widely used textbooks and the organizer of popular lectures. During the unemployment in Manchester caused by the American Civil War, he started popular scientific lectures which drew attendances of four thousand weekly.

Osborne Reynolds, the discoverer of the 'Reynolds number', which today has such a governing effect on the possibility of supersonic flight, was one of the most original and independent of men. He never did anything, or expressed himself, like others. He taught his own son to do the contrary of what he was told. When Reynolds said to him 'Come here', the boy would go away, and Reynolds would be delighted. He was a very obscure but stimulating lecturer.

Balfour Stewart had announced in 1858 the law of the reciprocity of radiation and absorption, that is, a body absorbs the same kind of radiation, such as light, as it emits, and vice versa. This was more than a year before Kirchhoff, to whom the discovery is usually assigned.

The Owens teacher who most influenced J.J. was Thomas Barker, the professor of mathematics. When Thomson's chance of becoming an engineer had faded, Barker advised him to compete for a scholarship at Trinity College in Cambridge. Barker was himself an old Trinity man and a very good mathematical teacher. His research passion was, however, in botany. He retired early, having acquired a fortune by successful investment, and devoted himself to the study of mosses. He bequeathed his fortune to Owens College to found the Barker Chair of Cryptogamic Botany (the botany of mosses).

The brilliance of Owens College at this time was a consequence of the vigour of Manchester capitalist enterprise, free trade and liberalism. Manchester was at the height of its prosperity, and had already a substantial scientific tradition in which Dalton and Joule were the chief stars. It was a confident atmosphere, in which middle-class men could hope to make money by skilful investment, and were accustomed to expect that reasonable efforts would receive a satisfactory reward. At the same time, world trade created a need for understanding men of other traditions and nations as a prerequisite to business, and acted as one of several spurs to the development of education.

The Dalton Chemical Scholarship, established at Owens in 1856, was the first research fellowship for science founded in England.

At Owens, Thomson was one of Balfour Stewart's half-dozen pupils.

He said that he first gained a liking for physics from him. His lectures 'were so clear that, child as I was, I could understand them. No one since Faraday has ever combined such scientific genius and such deep piety.'

Balfour Stewart, who had a weakness for trying experiments beyond the sensitiveness of his instruments, attempted to discover changes in mass during chemical change. He heated mixtures of mercury and iodine to this end. J.J. assisted him, and on one occasion the tube burst and seriously hurt J.J.'s eyes, nearly blinding him.

As might be expected from his discoveries on radiation, Balfour Stewart had a profound understanding of the conservation of energy and of its scope in physical phenomena. J.J. acquired from him an appreciation of comprehensive theorizing and the stimulus that led him to the theme of his first important book.

His career at Owens was very successful. He gained an engineering scholarship in 1873 and the Dalton Senior Mathematical Scholarship in 1874. Before he had left the College, he had published his first paper, 'On Contact Electricity of Insulators', in the *Proceedings* of the Royal Society. He was 'a rather pallid, boney youth with the air of a serious but happy student, unassuming and modest without diffidence, very approachable and friendly.'

III. STUDENT

J. J. Thomson competed for a scholarship at Trinity College, Cambridge, in 1875, and was unsuccessful. He tried again in the next year, and was awarded a minor scholarship. This was not a very remarkable beginning.

He went up to Cambridge in October 1876, and found comfortable lodgings. He stayed in these for four years, and his landlady was so attentive that in all that time he never had to make up the fire himself. He found this very helpful, as he could never remember to do it while he was working, and when his room had got cold he could not work properly.

The incident reveals the extent of the social privilege enjoyed by young men at Cambridge in the middle of the nineteenth century, and the favourable conditions for intellectual work for those who desired to study. The change for a winner of a scholarship from the lower-middle-class atmosphere of a Manchester bookseller to the gentlemanly atmosphere of Cambridge was profound. J.J. has remarked that no man ever owed more than he to scholarships. He has commented on the enormous difference between the world of his later years and that of his youth, when there were no bicycles, motor-cars, aeroplanes, electric lights, telephones, wireless, gramophones, electrical engineering, X-rays, cinemas, or germs, 'at least none recognized by the doctors'. The spirit of revolutionary change in the contemporary world, on which J.J. commented, did not penetrate his private life. Apart from physics, this was remarkably quiet and uneventful. It had the assured stability of British nineteenth-century society.

As a scholar, J.J. set out on the usual course to secure a high degree. At that time the conventional avenue to natural philosophy was through mathematics. Modern physics had only just been given appropriate status at Cambridge by Clerk Maxwell, who had recently re-introduced questions on heat and electricity in the mathematics examination. The degree examination, the Mathematical Tripos, had evolved from medieval beginnings, and had been strongly impressed by Newtonian thought and method, in which the application of mathematics was dominant. This derived from the Newtonian interest in astronomy, and indeed from the astronomical interests of British mercantile society, conceived in world commerce and navigation, before the industrial revolution.

The training in Newtonian mathematics at Cambridge had in the course of a hundred and fifty years been reduced to a peculiar exercise. Thomson entered into it when it was at the height of formal perfection, but already decaying at heart.

Newton was primarily interested in mathematics as a tool to solve astronomical problems. He was less interested in the development of mathematics as a consistent and elegant system, in the study of mathematics for its own sake, or pure mathematics. Problems lend themselves very much better than expositions of mathematical theory, then called bookwork at Cambridge, to competitive examinations. Competition in examinations had been strengthened by several tendencies at the beginning of the nineteenth century. There was the ancient rivalry among students to excel. This was coloured by the spirit of competition from private enterprise and from the sportive competition of the leisured gentlemen.

The Cambridge Mathematical Tripos examination required a training comparable to that for a horse-race at the neighbouring course of Newmarket. The hotels in Cambridge were as likely to accommodate leisured backers of horses as relatives of competitors in mathematical examinations.

Thomson placed himself under the direction of E. J. Routh, the unparalleled trainer for the mathematical competition, who in his career taught no less than twenty-seven Senior Wranglers, or winners of the first place in the examination. Routh had worked out to the finest pitch just what a candidate should learn, and how to present it to the maximum effect in an examination consisting of a number of three-hourly intervals. The winners had to acquire judgment in selecting the nature and order of questions, so that they could answer the maximum number in the minimum time. The system put a premium on facility of thought, and even of writing. It was particularly suited for training lawyers, and the greatest triumph ever won in the examination was by a man who subsequently became a famous lawyer. This was Fletcher Moulton, who more than doubled the second man's marks. He became a fellow of the Royal Society with similar facility, succeeding in a few months, while reading for the bar, in making useful experimental discoveries concerning the discharge of electricity through gases.

Thomson was deeply absorbed for about three years in training for the examination, which he took in 1880. It was held in January in the unheated Senate House: a gruelling experience. He was second to Joseph Larmor, who subsequently was elected to the Lucasian professorship, the chair held in the seventeenth century by Newton.

It was notable that Thomson and Larmor never had a great deal to do with each other in later life, in spite of being attached to neighbouring colleges for more than sixty years.

J.J. was not a particularly good examinee. He could not write quickly (he was 'pathetically helpless with his hands', which made his subsequent achievements as an experimental physicist all the more remarkable), and he had not the flair to judge when he had spent the proper time on a question and should begin to write out the answer. The consummate examinee would attack the easiest questions first, and judge nicely how much time to spend on finding the solution and how much to spend on giving it polished expression.

Thomson used to go to a barber's in the interval for lunch, and have a shampoo, like a boxer being massaged in his corner between the rounds.

The extraordinary effects of the old Mathematics Tripos system are seen in the remarkable passage of Isaac Todhunter, the famous mathematical teacher, and historian of the theory of probability, in which he deplored experiments as undermining the authority of the teacher and unsettling the mind of the student. 'It may be said', he wrote, 'that the fact makes a stronger impression on the boy through the medium of his sight, that he believes it the more confidently. I say that this ought not to be the case. If he does not believe the statements of his tutor—probably a clergyman of mature knowledge, recognized ability and blameless character—his suspicion is irrational, and manifests a want of power of appreciating evidence, a want fatal to his success in that branch of science which he is supposed to be cultivating.'

Todhunter was still teaching in Cambridge when Clerk Maxwell became Cavendish professor. The survival of such scholastic authoritarianism through the medium of mathematics showed how serious was the need for the introduction of systematic experimental research into Cambridge science.

One day, Clerk Maxwell, who had with much trouble ground out a plate of doubly refracting crystal, asked Todhunter whether he would like to see the phenomenon of conical refraction demonstrated with it.

'No ', he replied, 'I have been teaching it all my life, and I do not want to have my ideas on it upset.'

The mathematician A. R. Forsyth, who was a fellow student of Thomson at Cambridge, has written that he was already respectfully and affectionately known as 'J.J.' As Alfred Lyttelton and A. G. Steel, the contemporary cricket heroes, were worshipped by the sporting, so 'J.J.' was worshipped by the reading students.

As J.J. studied mathematics, he did not work under Clerk Maxwell. He did not enter the Cavendish Laboratory until one month after Clerk Maxwell had gone.

J.J.'s first research in the Cavendish Laboratory, under the direction of the third Lord Rayleigh, was to assist in the determination of the ratio of the electromagnetic and electrostatic units of electricity—roughly speaking, of electricity in motion and of electricity at rest. The two units could be measured in the laboratory with suitable arrangements of batteries, condensers, resistances, galvanometers, etc.

According to Clerk Maxwell's theory, the ratio should be a velocity, and this should be equal to that of light. The deduction of the velocity of light from laboratory apparatus in a little room seemed very extraordinary, as indeed it was, for the first determination of the velocity of light had been made by Rømer in the seventeenth century by making the vast solar system his laboratory. He deduced the speed of light from the time it took to cross the solar system, as revealed by the eclipses of Jupiter's moons.

J.J.'s results were not very good, and the experiments were repeated with far greater success years later with the collaboration of G. F. C. Searle.

Thomson applied for the new chair of Applied Mechanics founded at Manchester in 1881, but Schuster, who was five years his senior, was preferred. He was already lecturing at Owens, and had in fact been one of J.J.'s teachers.

J.J. was elected to a fellowship of his college, which provided him with the prospect of leisure for research for the next seven years. He looked for a topic on which he could exercise his highly tuned mathematical muscles. William Crookes had published in 1879 very stimulating experimental researches on the rays emitted by the cathode in electric discharges through gases at low pressures. Most of Crookes' observations had already been noted by Hittorf in 1869, but Crookes' research was independent and presented in a very interesting style. E. Goldstein, who named these rays the 'cathode rays' and worked in Helmholtz's laboratory in Berlin, published notable papers in the English *Philosophical Magazine* in 1880 on the same subject.

Crookes was convinced that the cathode rays consisted of electrified particles. Goldstein was convinced that they were waves, because they threw sharp shadows of objects placed in their way. Both views were determined largely by habits of mind and traditions of thought. All of the British physicists believed that the rays were particles, and nearly all of the Germans, with the exception of Helmholtz, who was in closer touch with British scientific thought than most Germans, believed that the rays were waves.

In 1874 Goldstein had observed that parallel beams of cathode rays repel each other; Crookes observed a similar phenomenon in 1879. Helmholtz, in whose laboratory Goldstein was working, was immediately convinced that the rays must consist of particles. Schuster has reported that he 'was rather sore that the idea did not originate in his own laboratory '. 'Of course', said Helmholtz to Schuster some years later, 'as soon as Stokes became acquainted with Crookes' experiment he guided him into the right path.'

A short note was published by Riecke in 1881, showing that the properties of the cathode rays could be explained on the supposition that they were particles, but it did not influence general Continental opinion. The German tendency towards wave explanations was strengthened by Hertz's demonstration in 1887 of the existence of electromagnetic waves, or radio waves. Hertz investigated the cathode rays and found in 1892 that while they could be deflected by magnets, they appeared not to be deflectable by electric fields. He concluded, following Hittorf, that the rays must be flexible currents in the ether, which could be deflected by magnets in the same way as a loose wire carrying an electric current can be deflected by a magnet.

Returning to the situation in 1880, the explanation of the properties of the cathode rays was one of the central problems of physics. Thomson decided to try to calculate what effect charged particles would have on each other, to see whether an explanation of the behaviour of the cathode rays could be deduced on the assumption that they consisted of electrically charged particles and obeyed the laws of Clerk Maxwell's theory of electricity and magnetism.

The paper giving his results was published in 1881, and is in some ways the most interesting single piece of work Thomson ever did. He concluded that a moving electric charge must have a mass and that this increases with velocity. It was the first explicit step towards the great modern discovery that mass and energy are equivalent, as finally expressed by Einstein. In his paper J.J. calculates how much the mass of the earth is increased owing to carrying with it in its journey through space an electric charge; it is about 650 tons, a quantity quite negligible in comparison with the ordinary rest mass of the earth. J.J. pointed out that a moving body charged with electricity produces the effect of an electric current. The transference of electricity by the cathode rays, which resembled an electric current, the deflection by magnets, and the mutual repulsion of beams of cathode rays could all be adequately explained by the hypothesis that they consisted of electrified particles.

Thomson's forecast that the mass of the particles would increase with speed was confirmed experimentally by Kaufmann twenty years later.

Thomson's remarkable paper of 1881, written when he was twenty-four years old, was like a perfect answer to a super-Tripos question. It was extremely clear, concise, powerful and elegant. The mathematical technique, though superbly used, was entirely subordinated to the solution of a physical question. It exhibited the genius and the limitations of the Tripos mathematical athleticism.

IV. PROFESSOR

Clerk Maxwell had died prematurely in 1879, and Rayleigh had accepted the Cavendish professorship for five years, to give the University time to find a suitable successor in an unexpected vacancy.

The electors had the courage to choose Thomson, who became Cavendish professor in December 1884, the month in which he reached his twenty-eighth year.

He had done little research in experimental physics, and that had not been particularly good. Several much more experienced experimentalists and theorists senior to him were candidates. These included G. F. Fitzgerald, W. Garnett, R. T. Glazebrook, Osborne Reynolds and Schuster, all of whom, except the first, had been his teachers. Owens had preferred Schuster to J.J., but now Cambridge preferred J.J. to Schuster.

Schuster's mind was less powerful than J.J.'s, but he had wider interests. His reputation has suffered an undeserved decline, and will undoubtedly rise as more consideration is given to insight into the historical evolution of science. Schuster's comments on the physics of his time, considered apart from the weight of personal discovery, are often more illuminating than J.J.'s. His three papers on the discharge of electricity through gases, the last published in 1890, contributed a striking advance in the conception and understanding of the phenomenon. Several unsuccessful competitors were a good deal nettled at being passed over in favour of a young man of twenty-eight who had been their student. J.J.'s great ability was patent, but the prestige of the peculiar Cambridge type of Tripos mathematical athlete probably influenced the selection: a man who was master of this technique was qualified to undertake anything, from governing India to research in experimental physics. It was the mark of the English intellectual gentleman.

His Adams Prize Essay on the subject of Vortex Rings was published in 1883. It was a contribution to the theory of atoms conceived as vortices, which had become fashionable, and was his first paper on the problem of atomic structure.

J.J.'s next outstanding publication after his famous paper of 1881 was his book on *Applications of Dynamics to Physics and Chemistry* in 1888.

His aim was to extend the idea of the equality of action and reaction to the equality of phenomena; from considering 'two bodies, attracting each other' to 'two phenomena which mutually influence each other'. The mathematical method of attacking these problems was suggested to him by Clerk Maxwell's paper on the Electromagnetic Field, which had so charmed him as a boy that he had copied it out by hand, and was equivalent to similar methods worked out by Massieu and Willard Gibbs. He applied the method to the effect of temperature on bodies, the electrical forces produced by temperature differences, evaporation, the properties of dilute solutions, the dissociation of molecules, chemical equilibrium, the change of state from solid to liquid, and electric forces associated with chemical change.

For instance, if the magnetization of an iron bar alters its length, then one can calculate the change in magnetization of the bar if its length is altered.

Thomson incidentally and independently discovered Le Chatelier's principle that if one of the factors of any system in equilibrium is changed, the effect produced tends to restore the factor to its original value. He rediscovered some of the results of Willard Gibbs, such as the adsorption equation. It is conceivable that if J.J. had continued in this field he might have become the dominant physical chemist of his time. Though he worked on chemical dynamics for only a short time, he became one of the founders of the subject.

His discussions on the dissociation of molecules of gases and of evaporation were of particular importance for his future work. He showed that if a water-drop were electrified, its evaporation would be retarded because a diminution of its size would increase the concentration of electricity on its surface and thus tend to prevent any more atoms leaving it.

It seems very probable that his intense concentration on the theory of dissociation in these early days prepared his mind for the idea of atomic dissociation, or ionization, in which electrons fly off and on atoms. This was one of the reasons why he found it so much easier than others to accept the notion of atoms losing small parts and then catching them again. The notion of chemical dissociation made the idea of the splitting of the atom so much the more acceptable.

As Maxwell attained his spectacular success in research by trying to work out a consistent theory of Faraday's electrical experiments, applying a conception of wave-motions derived from one of William Thomson's early papers, so J. J. Thomson set out to attain his by trying to work out a consistent theory of Crookes' electrical experiments, applying a conception of physical action derived from one of Clerk Maxwell's papers.

Thomson conceived his action-and-reaction of phenomena idea, the theme of his *Applications of Dynamics*, in the form of the transference, rather than the transformation, of energy while still a student at Manchester under Balfour Stewart, but ten years passed before he could work it out.

Though Clerk Maxwell and Thomson entered research in very similar ways, the outcomes were strikingly different. Maxwell's attempt to discover a general theory of electricity was successful, and formed one of the grand syntheses of science.

J. J. Thomson's attempt to discover a general theory of comparable power did not succeed, but it had the indirect effect of preparing his mind for an advance of comparable importance. It was filled with ideas of particles splitting and recombining, attracting and repelling each other, of droplets evaporating, and of a multitude of phenomena of the kind that were subsequently found to occur in the discharge of electricity through gases. His mathematical ability and his Tripos training inclined him to conceive each of these phenomena as a Tripos question, which he analysed mathematically with clarity, concision, and power. This ability and training gave him a more exact picture and grasp than any other investigator of what was happening in the discharge tube, and enabled him to question nature by experiments which could give conclusive answers. In spite of his mathematical ability and training, he was not really interested in mathematics. As Max Born has noted, he never extended his fundamental mathematical equipment after his Tripos days. He had no mathematical inspiration. His Cambridge training merely enabled him to visualize what was happening in the vacuum tube more clearly and quantitatively than others.

Besides not being an imaginative theorist like Maxwell, he was not a natural experimenter such as Faraday or Crookes. He had not the restless observation which seizes on peculiar facts without definite guidance from theory, and worries new principles of nature out of them. He made experiments not for manipulative pleasure, but to secure answers to mathematical conceptions of material action, which he had formulated in his head. He was clumsy with his hands, and a bad manipulator of apparatus. His genius was not for imaginative theoretical syntheses, nor for the discovery of raw new facts of nature, but in the intermediate field of semi-manufactured scientific goods. He took the raw new facts discovered in general by others and shaped them down with his Tripos mathematics until they became susceptible to conclusive experimental questioning, and then built the results together to form the foundation for a new conception of matter. His achievement was as great as that of the purest theorists and the greatest pure experimentalists, but it was not of either kind. It was in an intermediate field of its own. It marks a stage in the evolution of the technique of physical research, corresponding to the evolution of the production of intermediates, semi-finished goods, in industry.

Thomson began experimental researches on the discharge of electricity through gases in 1885, just after his appointment as professor. His choice of topic was influenced by Clerk Maxwell's belief in its importance as a means to an understanding of the nature of electricity, besides the interest raised by the recent experimental researches of Crookes, Goldstein, and others. His experiments were designed to test whether, when a gas conducts electricity, some of its molecules might be split, by the electric forces acting on the gas, into positively and negatively electrified atoms. These would then conduct the current like the ions which transport currents through liquids.

In 1893 Thomson published his *Recent Researches in Electricity and Magnetism*. It was intended to be a sequel to Clerk Maxwell's Treatise on Electricity and Magnetism. It has the same characteristic qualities as his *Applications of Dynamics* of 1888, but it shows a larger preoccupation with experimental work. He mentions that students who commence the subject after a long course of mathematical training have a tendency to regard Maxwell's theory as the solution of certain differential equations. They do not attempt to form a mental picture of the physical processes they are investigating.

He uses Faraday's conception of 'tubes of force' to correct the tendency to use mathematics as the modern equivalent 'of the Philosopher's Machine in the Grand Academy of Lagado, and to regard as the normal process of investigation the manipulation of a large number of symbols in the hope that every now and then some valuable result may happen to drop out. . . . Analysis is undoubtedly the greatest thought-saving machine ever invented . . .' but it is not 'necessary . . . to prevent students from thinking too much. . . . It frequently happens that more thought is required, and a more vivid idea of the essentials of a problem gained, by a rough solution by a general method, than by a complete soluton arrived at by the most recent improvements in the higher analysis.'

These views, expressed in 1893, help to explain why Thomson did not discover the quantum theory, relativity theory, or the wave-theory of matter. His mind was not inclined to theoretical and mathematical speculation. In spite of his mathematical ability, he was not primarily a mathematical physicist, such as Planck, Einstein, or de Broglie. He was a mental, as distinguished from a manual, experimental physicist. F. W. Aston, who had been one of J.J.'s scientific assistants and one of the most gifted of scientific manipulators, subsequently wrote that Thomson's intuitive ability to understand a piece of apparatus, perceive why it would not work, and suggest what should be done to cure it, all without handling it himself, seemed almost miraculous.

Though Thomson was in Cambridge during the last three years of Clerk Maxwell's life and belonged to the same college, he appears not to have known him. Clerk Maxwell was absorbed in editing Henry Cavendish's papers, and then died rather suddenly, and was probably not easily available. Yet it is strange that the young man who had been influenced so much by Clerk Maxwell's work should not have sought him out. It may have been due to accident or to a combination of shyness and independence, but perhaps it may also have been due to a kind of indifference owing to an unconscious conviction that he was not at heart a Maxwellian theorist.

The *Recent Researches*, published in 1893, contain the first summary in English of the results of world investigation into electric discharge in gases, which by that time had become very extensive. The cathode rays discovered by Plücker in 1858, and more clearly recognized and explored by Hittorf in 1869, after the invention in 1851 of the induction coil by Ruhmkorff for producing sparks, were the centre of the phenomena, but thirty-five years of experiment and calculation had passed and their exact nature had not yet been elucidated. It was not even yet known whether they were waves or particles. If they were particles, what was their mass? With what speed were they moving? Were they electrically charged and, if so, were the charges positive or negative? Scores of fascinating experiments had been performed, but the answers were incomplete. Schuster in 1890 had measured the deflection of the cathode rays by a magnetic field of known strength. He assumed that the rays consisted of particles which, through many collisions, had the same average energy or velocity as the molecules of gas with which the tube was filled. He assumed further that the charge on the particles was the same as that on the ions which transport electric currents through liquids. Combining his assumed velocity and the assumed charge with the magnitude of his magnetic deflection, he deduced that the particles were of atomic mass; indeed, he supposed that in a tube filled with nitrogen, they would be negatively electrified nitrogen atoms.

This and other admirable experiments of Schuster's had the weakness typical of so much research on the discharge of electricity in gases up to that date. It had excellent points, but the conclusions had to be helped out by various more or less plausible assumptions. His assumption that the particles had acquired the same average energy as the ordinary atoms of gas in the tube was incorrect. It was incompatible, in J.J.'s view, with the sharp outline of a beam of cathode rays. Would not the edges be diffuse and frayed if there were a general sharing-out of energy by collisions?

The English workers assumed almost without question that the rays consisted of particles. Yet in 1892 Hertz, two years before he died at the age of thirty-seven, of tuberculosis, probably caught from the walls of his laboratory, which consisted of a disused tuberculosis hospital, showed that the cathode rays could shoot right through gold and aluminium foil. Not only that, they emerged with an electric charge and could be deflected after emergence by a magnetic field. It seemed incredible that electrified particles could go right through solid metal foils and retain their charge. In 1894, however, J.J. had shown by a rotating mirror that certain cathode rays had a velocity only about one two-thousandth that of light. This was a strong indication that the cathode rays were not waves akin to light.

By the beginning of 1895 a great deal of work had been done on the discharge of electricity through gases. Many interesting results had been obtained, but the picture was confused. J. J. Thomson was then thirty-eight years old. He had been Cavendish professor for ten years. He had organized the routine of the Laboratory; he had pursued experimental researches on the electrical discharge in gases for ten years. He had published two particularly notable books, which included summaries of many of his own papers, and had submitted many aspects of the phenomena of electrical discharges to powerful mathematical analysis.

He had done much impressive work, but had not as yet accomplished a major achievement. He was, however, by combination of theoretical and experimental experience, the best-prepared expert in the world on the discharge of electricity in gases.

V. DISCOVERER

There were several reasons why the subject of the discharge of electricity in gases was in confusion. One was the inadequacy of experimental technique. Discharges in gases were obtained either by very strong spark discharges at voltages of about 30,000 or by heating the gases in flames. Both of these methods were difficult to control and gave capricious results.

In 1895 Röntgen discovered X-rays. Their startling nature gave the strongest impetus to research. J.J. immediately tried their effect on his particular interests. He found that they would discharge an electroscope. Like several other workers at the time, including Röntgen himself, he discovered that they rendered gases electrically conducting. Exact investigation showed that this occurred at the low voltage of 300. It made reliable, controlled, exact measurements of what was happening in discharge tubes possible.

Another event of a different kind, but also of critical importance, occurred at Cambridge in 1895. The University introduced the award of degrees to graduates of other universities who came to Cambridge and performed an approved piece of research. Formerly, research students could come to Cambridge, but they could not secure the recognition of their work by a degree unless they went through the whole of the normal three-year course for undergraduates first. The new regulation opened the University to graduates from others. The first student who arrived to take advantage of the new system was Rutherford, from New Zealand. Within the following hour Townsend arrived from Ireland. In 1897 Langevin arrived from France. These, and many others, were research workers, with no teaching duties. J.J., who already had a staff of seven recognized teachers, including Glazebrook and Searle, and certain holders of Cambridge research studentships, such as C. T. R. Wilson, was suddenly reinforced by this brilliant group from overseas with nothing to do except research. It was a pure intellectual bonus for a Laboratory system which was already completely functioning.

J.J. had founded the Cavendish Physical Society in 1893, which provided the opportunity for group-discussion of research problems. In 1895 he had the problem, the laboratory system, the experience, and the men. Had he the mental ability to command the intellectual attack, the physical strength to direct the laboratory, and the personality to lead the men?

He invited Rutherford to join with him in analysing the effect of X-rays on gases. They showed that the rays released charged ions in the gas, which moved under the application of electric forces. With Rutherford's collaboration, the discharge of electricity through gases was transformed from a qualitative to a quantitative science, whose results were expressed in metrical terms.

He found that Hertz had failed in his experiments of 1883 to detect the deflection of cathode rays by electric fields because he had not used a sufficiently low pressure. J.J. noticed in his early experiments that there was a flicker of movement in the rays when the electric field was switched on. Investigation showed that the deflection died out if too many charged ions were present, as they masked the rays from the field.

The improvement of vacuum pumps, stimulated by the needs of the electric-lamp industry, and Dewar's introduction of charcoal cooled with liquid air for absorbing residual gases, made many decisive experiments possible and had a profound influence on the future of physics. Thomson was now able to combine the application of magnetic and electric fields to the deflection of the rays, so that, assuming that they were particles, he could measure the ratio of their charge to their mass. He found that, assuming the charge was the same as that on the hydrogen atom in electrolysis, the mass of the cathode ray particles must be only about one-thousandth of that of the hydrogen atom.

J.J. had first suspected that the cathode rays consisted of particles other than charged atoms when he found the magnetic deflection so large: it was far greater than could be expected with hydrogen atoms, the lightest then known. He was not the first to publish this result. Wiechert of Königsberg published a beautiful paper in January, 1897, showing that cathode-ray particles in some cases had a velocity of about one-tenth of that of light and their mass was about two to four thousand times less than that of the hydrogen atom, assuming that the charges of the particle and of the atom were the same. A preliminary note of J.J.'s result was published in May, 1897 and a full account in October, 1897. Kaufmann published a similar result in July, 1897.

J.J.'s discovery that the cathode rays consist of a new kind of particle, now called electrons, about a thousand times lighter than the lightest atom, was not clear-cut. He was, however, the first to make it conclusive and completely free from assumptions. Hertz had tried in 1883 to detect directly whether the cathode rays had a charge by catching them in a metal cylinder, but had failed to get any result. Perrin in 1895 succeeded in performing this experiment with the aid of a better experimental arrangement and better vacua, and was the first to prove that the cathode rays had a negative charge. He found that the cylinder became negatively charged.

J.J. varied Perrin's experiment and placed a thermopile inside the cylinder, so measuring the amount of heat produced by the particles. This gave their energy, and hence average velocity, directly without any assumptions. Reduction of these data also indicated that the cathode particles were a thousand times lighter than the hydrogen atom.

Thomson, like other observers, had noted that the cathode rays appeared to be of the same nature, whatever their origin, whether they came from one kind of metal cathode or another. The cathode rays inside discharge tubes were subject to various electric forces not fully understood. J.J. therefore investigated rays of a similar kind which are emitted when zinc plates are illuminated with ultra-violet light. These particles, now called photo-electric electrons, are free from the kind of electrical disturbances present inside discharge tubes. J.J. found that they also had masses of the order of one-thousandth of that of a hydrogen atom. This showed that the results obtained from discharge tubes were not peculiar to them and dependent on some obscure masking effect. He obtained the same results with the particles emitted from hot carbon filaments, the thermionic effect upon which radio-valves and talking-films depend.

It was necessary, too, to measure the electric charge on the particles directly and prove without any assumption that it was equal to the charge found on the hydrogen ion in liquids carrying a current. This was first done by Townsend. J.J., who had a decade ago discussed the effect of electrification on the evaporation of water drops, had in his Laboratory C. T. R. Wilson, who, following Aitken, had begun his famous work on the role of ions in cloud formation. Aitken had discovered that air saturated with water vapour would not produce cloud when expanded if it was completely free from dust. Wilson showed that if the air were electrified with X-rays, cloud would be formed when it was expanded, even if no dust was present. Evidently the electrified particles, the ions, were able to replace the particles of dust as nuclei on which cloud-formation could begin. Further, the expansion required to produce condensation on the negatively charged ions was less than on the positively charged ions. Wilson's discovery enabled the negatively charged particles, that is, particles of the cathode-ray type, to be separated from the ordinary atoms and positively charged particles.

The next task was to find how many of the particles were produced, so that the total amount of electricity on them could be divided by their number, giving the size of the charge on each. Townsend showed how to calculate the number of drops in a cloud formed by an expansion of a given amount. The quantity of water in the cloud would be distributed among the drops, which would be visible and equal in number to the negatively charged particles.

Townsend had been well trained in hydrodynamics when a student at Dublin, and saw that the size of the drops, settling like parachutists, in Rayleigh's phrase, could be calculated from their rate of fall in the air after they had been formed. Hence the number of drops could be calculated by dividing the total amount of water known to be formed by the average amount in a drop. As each drop contained a negatively charged particle, the number of particles was known.

The total electric charge on the drops and hence on the particles was measured by allowing the drops to fall on a suitable electrical receiver. The division of this quantity by the known number of particles gave the size of the electric charge on each particle.

J.J., from his own work and that of his school, thus produced conclusive proof, sound at every point, that a negative electrical particle existed which was more than a thousand times lighter than the hydrogen atom and could be obtained from every kind of matter. He was aware, from the moment of realization that the electron existed, of its full significance for the theory of the structure of atoms and the constitution of matter. He explained that the atom, hitherto thought eternal and indivisible, had been split, since small pieces could be removed from it.

He saw that electrons must be the prime material, or one of the prime materials, out of which all matter must be built. He expressed his views to the British Association at Dover in 1899, when the meeting was attended by many foreign scientists, and his conclusions were beginning to be accepted: 'I regard the atom as containing a large number of smaller bodies which I will call corpuscles; these corpuscles are equal to each other; the mass of the corpuscle is the mass of the negative ion in a gas at low pressure, i.e. about 3×10^{-26} of a gramme. In the normal atom, this assemblage of corpuscles forms a system which is electrically neutral. Though the individual corpuscles behave like negative ions, yet when they are assembled in a neutral atom the negative effect is balanced by something which causes the space through which the corpuscles are spread to act as if it had a charge of positive electricity equal in amount to the sum of the negative charges on the corpuscles. Electrification of a gas I regard as due to the splitting up of some of the atoms of the gas, resulting in the detachment of a corpuscle from some of the atoms. The detached corpuscles behave like negative ions, each carrying a constant negative charge which we shall call for brevity the unit charge, while the part of the atom left behind behaves like a positive ion with the unit charge and a mass large compared with that of the negative ion. On this view, electrification essentially involves the splitting up of the atom, a part of the mass of the atom getting free and becoming detached from the original atom.'

J.J. immediately conceived the electrical conductivity of metals as due to clouds of mobile electrons which could freely move inside. He applied the kinetic theory of gases to these clouds of conducting electrons, and suggested in 1900 that as the temperature was raised some of these would evaporate, or jump out of the metal. He forecast the phenomenon of the emission of electrons with increase of temperature, which explained the phenomenon of leakage of electricity from hot carbon filaments, first noticed by Edison.

O. W. Richardson, who began research in the Laboratory in 1900, independently arrived at the same idea, which he called thermionic emission, investigated it thoroughly, both experimentally and theoretically, thus providing the immediate basis for the radio valve, broadcasting, and talking films.

J.J.'s final greatness was seen in the way that he immediately advanced, without pause, from the proof of the existence of the electron to its application to the explanation of atomic structure, the constitution of matter, and the Periodic Table of the elements. As J. A. Crowther has well said, 'a gateway had been suddenly opened into new territories of vast, though unknown, potentialities, and through this gateway Thomson marched to take possession', accompanied by his research students.

Thomson had explained at Dover that the existence of the electron indicated that all matter might be made out of this primary stuff, on the lines of the hypothesis proposed by Prout in 1815. The atom had a structure, the electrons being arranged symmetrically inside according to their mutual interactions. He had, many years earlier, shown Mayer's old experiment of 1878 in his lectures. In this, bunches of floating magnets are seen to arrange themselves in concentric circles under the influence of an external magnetic field. J.J. had suggested that the phenomenon had a bearing on the Periodic Table of the elements, long before the discovery of the electron, and indeed some of his students were known to have dismissed the idea as fanciful.

Mayer's experiment led Thomson to suggest that the electrons would arrange themselves in layers or shells inside the atom, and he was prompted to develop Kelvin's conception of the atom as a sphere of positive electrification containing a number of electrons. He mentioned at Dover that regularities in the structure of atoms, now attributed to electrons, would help to explain the Periodic Table.

He expressed all these views while the majority of physicists were still doubtful of the existence of the electron and many chemists were hostile to the idea. When he had described his work on the electron to the Royal Institution in 1897, some of the learned audience thought he was perpetrating a hoax. Here his independence, which sometimes hindered him from taking sufficient notice of the work of others, enabled him to express inspiring suggestions that have been found by all subsequent work to contain major truths. It is to his lasting credit that he pushed constructive speculation to the uttermost, with complete indifference to his position and prestige.

During the years 1896–1900 one hundred and four original research papers were published from the Cavendish Laboratory. Thomson reported the results of this brilliant period of research in the *Conduction of Electricity through Gases*, published in 1903. In this great book he introduced to the world at large the new spirit of the experimental research worker of the twentieth century. The nineteenth-century mathematical tripos manner of his earlier books had disappeared. Mathematics was completely assimilated to experiment.

Thomson was now forty-seven years old. It was not to be expected that his future work would have the unerring greatness of the period that had just ended. Many of his brilliant young men had departed and were exploring the new field, with fresh minds and youthful energies, in their own laboratories. It was only natural that the advantage in making the deepest discoveries of the next generation should lie with them.

Nevertheless, Thomson had a wonderful Indian summer of research. He perceived that the scattering of X-rays by matter should help to reveal how the electrons were distributed inside the atoms. The X-rays could be used as a kind of probe. In 1904 his school explored the atom by these means, and Barkla, who had since gone to Liverpool, showed that in several gases the number of electrons contained by their atoms was about half the atomic weight.

This marked the limit of the main contributions of Thomson and his personal school to the creation of the modern theory of the atom. New fundamental ideas were required for the advances of the next generation. Yet his major contributions to science were not finished.

The investigation of the negatively charged rays in discharge tubes had led to the discovery of the electron. What was the nature of the positively charged rays, first noted by Goldstein in 1886? W. Wien had the foremost part in answering this question. J.J. projected a narrow beam of the positive rays on to a photographic plate. He subjected the beam to two forces acting at right angles to each other: one electric, the other magnetic. All the particles of the same mass in the beam hit the photographic plate at some point, depending on their velocity, on a particular curve: a parabola. The consequence is that particles of a particular mass form a particular parabola, which can be used to identify them. If the beam consists of particles of various masses, a series of parabolas is formed on the plate. Hence a beam of mixed positive rays can be analysed and the masses of the various kinds of constituent particles identified.

Though the apparatus is beautifully simple in principle, Thomson had much difficulty in making it work. This was due to the slight penetrating power of positively charged particles, thousands of times heavier than electrons. They could not pass through the gas in the tube and reach the photographic plate unless the pressure was very low, and it was difficult to make the electric discharge, which propels the rays, pass through the tube. After overcoming these difficulties Thomson secured wonderful photographs, in which the presence in the tube of atoms and molecules of such substances as carbon, nitrogen, and oxygen could be revealed simultaneously. He directly detected chemical radicles, such as hydroxyl, the combination of one atom of hydrogen with one of oxygen, whose existence had hitherto been only inferred. As the particles are registered on the photographic plate in less than a millionth of a second, the presence of very short-lived substances, such as occur in flames and explosions, could be detected.

He analysed for the first time not a chemical compound, but one of the natural elements. He found that neon, the gas with which the well-known red advertisement lights are filled, consists of a mixture of two kinds of atoms, one of atomic weight of 20 and the other of 22. He had accomplished the first separation of isotopes, that is, atoms of the same chemical properties and therefore inseparable by chemical means, which nevertheless have different atomic masses. Soddy had deduced the existence of isotopes in 1911 and had given them their name. Thomson analysed neon in 1912. He was then fifty-six, an advanced age for discovery of the highest class.

The outbreak of the First World War in 1914 dispersed many of his colleagues and stopped research in the Cavendish Laboratory. Thomson's career as a personal leader of investigation virtually ended, and his main influence was subsequently exerted in the direction of national scientific policy.

VI. THE CAVENDISH LABORATORY

J.J. was not only a teacher and discoverer, he was a creator in the method and organization of research. As J. A. Crowther has remarked, during his direction of the Cavendish Laboratory research changed from a 'personal idiosyncrasy' to the 'normal completion of a University training'. One might say that research had changed from an amateur into a professional occupation.

The great Cambridge physicists of the past, such as Newton and

Stokes, had not created schools of experimental research. Neither did Clerk Maxwell nor the third Lord Rayleigh, though they built the Cavendish Laboratory and set it going.

Clerk Maxwell rarely had more than two or three students at his lectures. When he visited his handful of experimental researchers, he was usually accompanied by his dog Tobi, with whom he discussed the difficulties in monosyllabic undertones.

By 1890 there were twenty research students in the Laboratory. J.J. visited them every morning. The daily study of their varied problems and suggestions for solution of difficulties had become a hard and systematic labour, and a very important part of the Laboratory method. There was a change of character from casual artistic labour to the systematic production of results.

The creation of the modern physical research laboratory depended on many factors. The fourth Lord Rayleigh suggested that the first of these is a stimulating leader. One can entirely agree with him that it is an essential factor; it is not, however, the first. There have been many stimulating leaders in the past who failed to create schools because the other necessary factors were not present. Rayleigh's outline of the qualities needed in a leader of a modern school of research are excellent: he must be 'one who is not only abounding in energy and ideas, but also one who can without too great an effort throw himself into the difficulties of others. This requires a peculiar kind of versatility not always easily combined with great powers of concentration on any one line of thought.' Rayleigh also points out very truly that 'it is necessary to have a productive line of investigation opening up', and 'it is necessary to have the right kind of pupils. These must be men of the not very common kind of ability which makes a scientific investigator: they must not be too young: and they must be provided with the means of subsistence while the work goes on.'

In earlier days this presented a serious difficulty at Cambridge, as the only persons available were fellows of colleges and those with private means. The introduction in 1895 of the post-graduate research system, formulated mainly by a theologian, J. Armitage Robinson, afterwards Dean of Westminster, mitigated this difficulty.

At the same time, the Commission of the Exhibition of 1851 changed the rules governing the scholarships awarded to holders nominated by Universities in the British Commonwealth. Formerly, scholars could spend a year at the University which nominated them. From 1896 scholars were required to go to a University other than their own. Consequently young scientists in Australia, Canada, Ireland, etc., were under much greater compulsion to go to research abroad, and naturally tended to choose England.

Up to 1895, twenty-nine physicists were given Exhibition of 1851 scholarships. Only two of these went to Cambridge, both in 1895. From

1896 to 1921 one hundred and three were awarded to physicists, of whom sixty went to Cambridge.

It is evident that Cambridge was changing in 1895 from an institution serving the needs of England into one serving the British Empire. It was changing from a national into an imperial institution; this was another of the factors which contributed to the creation of the modern Cavendish Laboratory.

This world-wide movement of research students to Cambridge was stimulated by the need for science teaching. Most of J.J.'s pupils became professors and lecturers who taught physics to the increasing numbers of science students in the Universities as a whole, at home and overseas. The general increase in the number of science students was due to the growing demands for science in industry, government, and medicine.

In 1885, J.J.'s first year as Cavendish professor, there were already ninety-seven students taking courses in practical physics. By 1890 the numbers had increased to 220, mainly owing to the demand for courses in physics from medical students and as a training for posts in industry and government.

The direct effect of industrial methods on Laboratory organization had already begun. J.J.'s old pupil and great friend Richard Threlfall, who had become influential in increasing the place of science in Government activity, and was the first to suggest the use of helium for filling airships, wrote to him from New York in 1889, after visits to Edison: he 'is a capital fellow himself . . . he has about 100 assistants and manages . . . to keep them working all night . . . very often. His Laboratory is a splendid building like a factory . . . and his store room is filled with everything that has ever been manufactured in the way of material.'

While imperial needs and the methods and conceptions of industrial production were affecting research, the results of research were affecting industry, none more than the discoveries in the Cavendish Laboratory concerning the electron. Research and social development react on each other, and the outcome is a result of this interaction, forth and back.

In the Cavendish Laboratory in 1896, the twenty research workers were scattered in twos or threes in the larger rooms. Each was working on his own problem, which he had consciously or unconsciously chosen from the multitude in the air or had adopted at the professor's suggestion. Rutherford's choice was typical. He brought with him from New Zealand an electromagnetic detector for radio waves, which had a great success and for a time held the long-distance radio communication record. He dropped the development of this brilliant original idea and accepted J.J.'s invitation to collaborate on the investigation of the mobility of ions in gases.

Nearly all of the gifted young men were working on some aspect of the grand scheme of research into the constitution of matter and electricity which existed in J.J.'s head. The fourth Rayleigh, who was working in the Laboratory at the time, says there was no team work. But does not this appear to be team work, even if only unconsciously organized?

J.J. regularly visited his research students in the morning. They asked him many difficult questions, and he nearly always had an immediate answer, though his suggestions were sometimes very impractical. He did not resent objections and criticisms, but he always had the last word, even when he was certainly wrong. In such cases, when he came round on the next morning, he accepted the correction tacitly.

The combined attack on the secrets of nature by unconscious and conscious team work stimulated the need for group discussion of research problems. The Cavendish Physical Society, founded in 1893, was the first British colloquium, or modern research discussion group. The idea of the colloquium seems to have been started by Jacobi at Königsberg in 1834. The Cavendish Physical Society provided the centre where all new discoveries could be immediately and informally discussed, and in which anyone could take part.

The professor's wife used to preside at a tea before the meetings of the Society. This increased the pleasure and social homogeneity of the group. The presence of ladies in University activities was much rarer at that time than now.

When J.J. came to Cambridge, the ancient medieval rule that fellows of colleges must have taken holy orders had only recently been abolished, and the rule that they must remain unmarried was not rescinded until 1882. The marks of the monastery were still very heavily impressed on University society. Fellows generally postponed their marriages until their later years, when they were already assured of a good appointment outside Cambridge.

J.J. himself did not marry until he was thirty-three. Very appropriately, he found his wife in the Cavendish Laboratory. He first met his future wife shortly after he had been appointed professor. She was Miss Rose Paget, the daughter of Sir George Paget, the noted professor of medicine in the University. Paget had studied mathematics. He had been 8th Wrangler in the Tripos, and had been awarded a fellowship at Caius College. One of the terms governing its award was that the holder should study medicine, so for this reason Paget became a doctor, and subsequently achieved great success in his profession.

Miss Paget had some of her father's scientific tastes. This led her to attend classes in the Cavendish Laboratory. In 1889 she undertook research on soap films thrown into vibration under the influence of sound waves. J.J. used to look in from time to time to see how she was getting on. He and Searle were collaborating on the re-measurement of the ratio of the electrostatic and electromagnetic units, which has the same value as the speed of light. J.J. used to come in at about 4 p.m., and then say he must run upstairs for a few minutes. These visits used to expand into periods of an hour. One day he came downstairs in very high spirits and Miss Paget left the Laboratory with a lively colour. They were married in 1890.

Owing to the survival almost up to these times of the rules on holy orders and marriage, there were still very few ladies in Cambridge University society—only about sixty families at the time of J.J.'s marriage. The burden of entertainment on such few ladies as there were was very heavy, for parties were preponderantly male.

Girton and Newnham Colleges began on a very modest scale about 1870, bringing a few women students. Even today the number of women students is limited. It was not until 1948 that a woman received a full degree at Cambridge; the first recipient was the Queen.

J.J. had a son and a daughter. His son, George Paget Thomson, was born in 1892. He also studied physics, and began research in the Cavendish Laboratory. In 1927 G. P. Thomson independently found experimental proof of de Broglie's theory of the wave-nature of the electron. Together with Davisson, he shared the Nobel Prize in 1937 for this discovery.

Born has commented on the extraordinary fact that the father should have given conclusive proof of the particulate aspects of the electron and the son conclusive proof of its wave aspects. J.J. was intensely proud of his son's achievement, and became more sympathetic than he had been before to the physical theories that had evolved in his later years. Father and son collaborated in the revision of the *Conduction of Electricity in Gases*, which was published in two large volumes, one in 1928 and the other in 1933.

The daily Laboratory tea was started in 1896. It was quite different from the Physical Society teas. Rutherford has said that they were started at his suggestion. It became the most stimulating and enjoyable part of the Laboratory life. Rayleigh describes the crockery and biscuits as of the plainest kind, but the conversation, led by J.J. himself, was sparkling. Everything was discussed except physics: politics, novels, sport, theatre, personalities. J.J. told stories well, but was also a good listener, and drew out the shiest men. He had a very infectious laugh. At his best, he paced the room like a caged lion, criticizing the British generals in the Boer War or selecting an English Test team to beat the Australians. On one day when the Boat Race was being discussed, the American physicist Bumstead mentioned that in America railways ran alongside the river where the race between Harvard and Yale was rowed, and that spectators engaged trains so that they could follow the race from start to finish. 'Oh, is that how fast your trains go?' said J.J. 'No,' replied Bumstead; 'that's how fast our crews row.' J.J. laughed heartily at this hit.

In later years J.J. regarded the Laboratory tea as one of his chief engagements, and did his best to make his other engagements fit in with it. There were men and women from many different countries and of different races and religions. The exchange of ideas and points of view between them was of great mutual value, and strengthened the international solidarity of science and scientists.

J.J.'s research students Townsend, McClelland, Langevin, Zeleny, and others initiated the first Cavendish Dinner, which was held in the Christmas Vacation of 1897. It was in the Prince of Wales Hotel. J.J. remarked on the following day that he had no idea that the Laboratory held such a nest of singing birds. W. Craig Henderson has written that J.J. was as happy as a sandboy. The uproar was such that it attracted the attention of the University Proctors, who asked the landlord of the hotel what was happening. It is said that the Proctors withdrew on learning that the Cavendish professor was presiding.

The Cavendish Dinner became an occasion for the singing of entertaining parodies of popular songs, referring to the life of the Laboratory. Many were written by A. A. Robb. In one of them, set to the tune of 'Father O'Flynn', there is a description of J.J.:

When the professor has solved a new riddle,
Or found a fresh fact, he's fit as a fiddle,
He goes to the tea-room and sits in the middle
And jokes about everything under the sun.
Then if you try to look grave at his jest,
You'll burst off your buttons which fasten your vest,
  For when he starts chaffing,
  Your tea you be quaffing,
  You cannot help laughing
    Along with the rest.

If Robb's verses are compared with Clerk Maxwell's, a reflection of the difference in atmosphere during the first and the third Cavendish professorships will be seen. Thomson looked outward, healthily and happily playing with nature. Clerk Maxwell looked inward as well as outward, lightly expressing the soul as well as the outward play.

After Rutherford left for Canada in 1898, H. A. Wilson succeeded to his place in Thomson's room. Rayleigh also joined them, as he had been suffering from the miserably inadequate central heating upstairs, due to the narrow bore of the copper pipes, originally put in by Clerk Maxwell for freedom from magnetic disturbances. Copper pipes of the proper size would probably have cost more than could have been afforded.

Clerk Maxwell and the third Rayleigh had both used the inmost, quietest room in the Laboratory. J.J. moved out of it and chose the busiest, through which a stream of Laboratory traffic surged.

The research student was left to acquire his technique by trial and error. He had to make most of his apparatus. J. A. Crowther has remarked that the method was painfully slow, and that a larger supply of ready-made equipment and more continuous direction would have doubled the output of results. But, in Rayleigh's view, J.J. all through his professorship carefully fostered individualism in research. He believed that this was educationally valuable, besides being the best way of promoting discovery. He believed that research strengthened the mind, in addition to training it.

When J.J. had important results, he liked to publish them quickly. For this reason, he preferred the *Philosophical Magazine* to the *Proceedings of the Royal Society*. The professional importance of priority was growing.

The demand from research students for places in the Laboratory rapidly increased. By 1906 it had become very much overcrowded. There was competition for space even on a single table.

The third Rayleigh assigned £5000 of his Nobel Prize, received in 1904, for an extension to the Laboratory.

J.J. was now fifty years old, and some of his most brilliant birds had flown and were building their own nests elsewhere. His former pupils, in other Universities and in Cambridge, were diverging more along their own lines. A separation of fields of research began. He concentrated on the analysis of positive rays; while others, now senior, in some cases created new subjects.

The Nobel Prize for physics was awarded to J.J. in 1906 for 'the great merits of his theoretical and experimental researches on the discharge of electricity through gases'; not specifically for the discovery of the electron. He was knighted in 1908.

J.J. thought that H. L. Callendar, of all his pupils, had the most interesting career. Callendar gained a first-class degree in classics in 1894 and a first in mathematics in 1895. Then, in that year, he entered the Laboratory. He had never done any practical work before. It was at once evident that he was a beautiful and exact manipulator. J.J. recommended him to investigate the causes of the failure of the platinum resistance thermometer, introduced by Siemens. Callendar soon showed that if the platinum were free from contamination and strain the thermometer could be very accurate, and within eight months he revolutionized the science of the measurement of temperature.

Callendar reviewed the resistance of metals at very low temperatures, and pointed out in 1899 that his measurements suggested that in certain metals electrical resistance vanishes above absolute zero.

Unique in the galaxy of J.J.'s collaborators was C. T. R. Wilson. This most admired and Scottish of Scotsmen graduated with a first-class degree in botany at Manchester University, and then became a meteorologist, studying the wonderful sky and cloud phenomena which could be seen from the Observatory on top of Ben Nevis, the highest mountain in Scotland.

Wilson became interested in Aitken's work on the role of dust in producing clouds. He was elected to the Clerk Maxwell studentship at Cambridge in 1895. After his discovery of the possibility of separating positive and negative ions by cloud expansion, a tactical result of great importance in the discovery of the electron, he steadily pursued his studies of the formation of ions and clouds. Wilson's observation, published in 1900, that a few ions are always being produced in the air, was one of the most pregnant ever made, and an outstanding incident in Cavendish history. He found that if the base of the insulating support bearing the gold leaves in an electroscope was kept at a slightly higher voltage than the leaves, so that it was impossible for electricity to leak from the leaves down the support, nevertheless a leakage occurred. It could be due only to leakage through the air surrounding the leaves. Elster and Geitel independently made a similar observation in the same year. Wilson even suggested that the leakage might be due to ionizing radiations coming from outside the earth. It was the first well-founded suggestion of the existence of cosmic rays.

In the 1928 edition of the *Conduction of Electricity through Gases* J. J. Thomson and G. P. Thomson remarked that 'It would be one of the romances of science if these obscure and prosaic minute leakages of electricity from well-insulated bodies should be the means by which the most fundamental problems in the evolution of the cosmos came to be investigated.'

C. F. Powell has pointed out that recent research on cosmic rays, which has shown that they consist of atomic nuclei moving at speeds approaching that of light, seems to be confirming the Thomsons' predictions. These nuclei are shattered into lighter atomic fragments when they collide with the atoms of the air, and the flying fragments as they pass through the air towards the earth produce mesons.

The study of these collisions and disintegrations, leading to the discovery of new particles and modes of interaction, promises for the physics of the future discoveries comparable with those made by J. J. Thomson at the birth of the present century.

Wilson's long and steadily continued researches on the formation of cloud culminated in his invention, in 1911, of the cloud expansion chamber, which would reveal the tracks of atoms. This instrument has had an outstanding role in the experimental investigation of atoms and the elucidation of cosmic rays.

His work was the complete foil to J.J.'s. It exhibited superb manipulative skill and elegance. He was most Scottish in his caution and thoroughness, but perhaps more akin to Clerk Maxwell than most of his countrymen in his sense of the aesthetic in science. No one has ever performed more beautiful experiments.

Near the beginning of the century, Wilson made calculations on the condensation of water on charged particles akin to those published by J.J. in 1888, in his studies of condensation on ions made by X-rays. He deduced a good value of the ionic charge by this method. He showed the work to Kelvin, who approved his observations, but disapproved his theoretical calculations, so characteristically he never published his results. Kelvin, especially in his later years, was enthusiastic about other people's experiments, but nearly always disapproved of their theories.

Wilson always demanded extreme perfection in his work. His written accounts, as exquisite as his experiments, cost him agonies of labour.

Wilson was appointed University Lecturer in 1900, and did not retire until 1925, when he was appointed Jacksonian Professor. The influence of the presence and teaching of such an experimental genius for a quarter of a century is great. It establishes a tradition of quality which cannot be created in any other way.

H. A. Wilson, who succeeded in 1898 to Rutherford's place in J.J.'s room, suggested in 1903, after reading one of Townsend's papers, that the size of the charge on the electron might be measured directly by observing the rate of fall of drops containing electrons, with and without an electric field. He observed the rate of subsidence of the top of the cloud under these conditions. Subsequently, R. A. Millikan improved the method for single drops and made very accurate measurements of the electronic charge: one of the fundamental constants of nature.

C. G. Barkla, who left the Cavendish Laboratory in 1902, nevertheless continued his line of work on the scattering of X-rays by matter. He discovered the characteristic X-ray spectra produced by the scattering of the X-rays by the various shells of electrons within the atoms of the bombarded matter. His experiments provided J.J. with the data for his calculations of the number of electrons inside an atom from the degree to which X-rays were scattered by the atom. J.J.'s formula became the basis for all future determinations of atomic structure by scattering effects.

Towards the end of J.J.'s direction, entirely new branches of physics were born in the Laboratory, such as W. L. Bragg's discovery of the reflection method of analysis of crystal structure by X-rays.

A recapitulation of some of the most famous of J.J.'s pupils and workers gives a final impression of the Laboratory's greatness under his direction: Appleton, Aston, Barkla, Born, W. H. Bragg, W. L. Bragg, Callendar, Eddington, Langevin, Lodge, McLennan, Papalexi, Przibram, the fourth Rayleigh, O. W. Richardson, Rosenhain, Rutherford, Smoluchovski, G. I. Taylor, G. P. Thomson, Townsend, C. T. R. Wilson, H. A. Wilson, and Zaviska.

Outstanding among the teachers was G. F. C. Searle. He began work in the Laboratory in 1888, and did not retire until 1935, and even returned to service during the Second World War. He was inexhaustibly fertile in devising original experiments for practical laboratory classes, and the exceptional teaching of practical physics given by him, which showed the beginner the meaning of elegance, originality, and exactitude, was one of the foundations for the high quality of research done subsequently by so many of his pupils.

Searle also made considerable original discoveries, but he was mainly interested in teaching. The presence of men of his kind of aptitude was essential to the Laboratory's success.

The finance of the Laboratory is one of the most remarkable of its features. J.J. gave an account of it in his *Recollections*. The University paid the salary of the professor, and two university lecturers, £50 each; two University Demonstrators, £100 each; and two Assistant Demonstrators, £60 each. Supplementary teachers cost a further £1,500 a year. The Laboratory received £270 from the Museum and Lecture Rooms Syndicate.

The wages of the workshop staff and apparatus and supplies for teaching and research had to be found by the Laboratory. Cavendish staff had to eke out their salaries with college fellowships and coaching fees. It has been said that one reason why Cambridge physicists discovered so much in those days was that they were too poor to be able to indulge in distractions. They could not afford to do anything except work.

The Laboratory income came from students' fees and was administered by the professor. This system induced the lectures and demonstrations to be as attractive as possible. Thomson found that the 'most useful instrument a laboratory can have is a good balance in the bank.' A new wing was built in 1896 as a result of twelve years' saving.

When the Laboratory was at the height of achievement expenditure still had to be argued with him, and he was fertile in suggesting alternatives, generally at the expense of students' time. Cavendish men struggled for days with crude apparatus such as antiquated electrometers. Good ones were purchasable, but as they were expensive the Laboratory possessed only one.

W. G. Pye, who was head of the workshops, had to plead earnestly for a milling machine. He later became the founder of a well-known firm of radio manufacturers which, in the Second World War, made the first proximity fuses. When new accumulators were not sanctioned, harassed conspirators arranged that the electrical current for the Professor's own experiments should fail. J.J. was doubtful whether he could afford five pounds for a new electroscope designed by Curie.

The extreme stringency created severe competition for such apparatus as was available. It was said that the investigator assembling apparatus for a research had to carry a drawn sword in his right hand, and his apparatus, or rather somebody's else's apparatus, in his left. A few of those who could afford to buy their own apparatus did so.

In 1912–13 the extra expenses incurred by research were less than £550, and in no earlier year of J.J.'s directorship had they ever been more. The apparatus required for researches such as those of Röntgen, the Curies, and C. T. R. Wilson had cost virtually nothing. The developments from their basic discoveries become more expensive.

By 1936 the Austin donation of £250,000 was needed to extend and re-equip the Cavendish Laboratory to suit contemporary requirements. The organization of fundamental research had evolved to a new stage and required a new order of expenditure.

VII. THE MAN

Rutherford's first impression of J.J. is recorded in a letter to his fiancée, the future Lady Rutherford: 'You ask whether J.J. is an old man. He is just forty and looks quite young, small, rather straggling moustache, short, wears his hair (black) rather long, but has a very clever-looking face, and a very fine forehead and a radiating smile, or grin as some call it when he is scoring off anyone.'

J.J. has recorded that he could not remember a day in sixty years when his work had been interrupted by illness. He said that the climate of Cambridge suited him and that he felt well and vigorous when he was there. His capacity for refreshing sleep and his strong nerves are shown by an incident in the First World War. During an air raid while he was staying in London, he was awakened by a prodigious noise which was due to a bomb that fell nearby. He soon fell asleep again. When he went down for breakfast in the morning, he learned that everybody except himself spent the night in the basement.

He was physically very active. He frequently went for long walks, and he would vault over gates and stiles. He was very fond of watching games, especially of Rugby football. He described a good passing movement as the most thrilling in any game. He used to rush up and down the sidelines, though sometimes, when he became conscious of being watched by his students, he would pause in momentary embarrassment.

J.J.'s ardour as a spectator of sports may have had some relation to his mental processes in experimental research. As an experimenter, he was also a spectator. His capable manipulative assistant, E. Everett, made the apparatus and performed the experiments, after J.J. had invented and performed the experiment in his head. No doubt he played Rugby football in the same way in his head, and thus found recreative activity for that special faculty with which he accomplished his greatest achievements.

One may ask whether the natural tendency of Thomson, and British physicists in general, to be inclined to interpret nature in terms of particles rather than waves has something to do with the concrete, empirical outlook, and extrovert activity characteristic of British ideas and conceptions. Did the German bias towards wave-interpretations of the cathode rays, and of nature generally, spring from the same roots as their passion for music, while the British bias for particles was related to their love of and skill in games?

Did the British boy grow up with an insight into the way a particle would behave, which he had gained from experience in the cricket or football field? Some of the scientists from other countries who have made discoveries in the particulate aspect of nature, such as Niels Bohr and Frederic Joliot, were also enthusiastic footballers in their youth.

Righi remarked in 1913 that J.J., from the commencement of his researches, was unconsciously preparing himself for the grand discovery of the electron. One could see that the discoverer was aided not by blind fortune but the possession of the special attitude of mind and the appropriate scientific preparation.

In a sense Thomson was chosen from the company of his fellow-physicists by the advancing movement of physics to make this discovery, because he was particularly suited to the task. The electron discovered Thomson just as much as he discovered it.

J.J. was remarkably insensitive to interruption. When Rutherford was his chief collaborator and working in the same room, they talked on and off at all times. The rooms of the laboratory opened into each other, and workers wandered through them continually. He used his telephone under these conditions, discussing engagements and personalities in a loud voice, apparently oblivious of involuntary hearers.

He replied one day to the fourth Rayleigh, his pupil and future friend and biographer, who had been commenting on the difficulty of some work: 'Yes, that is why there is so much credit in doing anything.' He had a great respect for grit and perseverence, and said that he had known far more men fail through lack of them than from lack of ability.

He was never defeatist and always inclined to consider what a theory or experiment might prove, and not what it could not prove. He had no superior airs, and was rarely out of temper or difficult to approach. He was accessible, and willing to talk on any scientific or political matter of interest, even in casual meetings in corridors. He was friendly with his staff, from demonstrators down to Laboratory boys. He hated persons whom he regarded as toadies of the great.

He was a shrewd judge of scientists. He remarked in 1934 that he had known the third Lord Rayleigh intimately and admired his fine judgment and his ideas, which were in front of his time: 'His collected works still go on selling.'

He had found Kelvin different from all others. He was a bad lecturer. He remembered that Kelvin was once due to give an Evening Discourse to the British Association. He arrived unprepared and spoke for one and three quarter hours. Kelvin had more imagination than Rayleigh. His pupils had become professors everywhere, especially in the colonies. He believed that Kelvin had influenced industry, but industry had not influenced Kelvin. His achievements before the age of forty had been wonderful. He had great drive, and made the Atlantic cable a success. Except for the mirror galvanometer, his many instruments were not very practical, though much time passed before better ones were introduced.

J.J. said that Kelvin never read a book through in thirty years. His mind was full of ideas. He always interrupted people who explained things, because he thought of developments of the ideas that were explained. He worked chiefly in the train between London and Glasgow. He had an engineering imagination, and always thought things out for himself from first principles. He did not read what had been written on a subject before he began to think of developing it.

J.J. believed in this, and was inclined to do it himself. He said that students should, within reason, attempt to arrive at their own methods of solving problems and not be biased by the approach of previous workers.

He was in favour of the engineering training. He had studied for three years under Osborne Reynolds, whose reputation continues to increase.

J.J. remembered that one day Kelvin was in his Laboratory and J.J. was explaining to him the current research on atoms. Kelvin said he didn't believe in atoms. Stokes, who happened to be there, explained for a quarter of an hour, in a beautiful impromptu little lecture, why the atomic theory must be accepted. This was the only time J.J. ever heard Kelvin hear out an explanation to the end: Stokes had a compelling personality.

He remarked that Helmholtz, who was pro-English in his experimental ideas and methods, became the dictator of German physics. He was once asked to support a manifesto affirming the superiority of the metaphysical approach in scientific research. He declined to sign it, and stated he could not subscribe to a method which was contrary to that of Faraday.

J.J. considered that Helmholtz's book on *Tones* showed an enormous intellectual power.

As the director of his Laboratory, J.J. had sometimes to deal with unpleasant matters. He never shrank from them, though he did not tell 'home truths', unless it was necessary. He was generally aware of the existence of friction and personal difficulties.

He had intense enthusiasm. Rayleigh has described how he met J.J. in King's Parade, a Cambridge street, in the summer of 1897. He was highly excited, saying that the cathode rays had turned out to be particles and quite different from atoms. He walked past St. John's College, round into Sidney Street, and after continuing talking to his pupil for several minutes, went on. When he was fired by an idea, he found delays and obstacles very trying.

He had enormous fertility of ideas. This made him sometimes unreceptive to the ideas of others. He also liked to think things out for himself, even when he could have saved time and energy by asking colleagues to explain. He was not thoroughly at home with any new thing unless he had rediscovered it for himself. This self-sufficiency caused him occasionally to be ignorant of prior researches which he should have known, such as those of Schuster on the magnetic deflection of cathode rays, which were published in the Bakerian Lecture of 1890. Apparently he had not read this through, and consequently failed to give an adequate account of Schuster's work in his *Recent Researches* in 1893.

The spate of ideas in his mind may have prevented him from always clearly recognizing which were his own and which his pupils'. It has been said that Lenard's anti-British diatribes were due partly to J.J.'s insufficient appreciation of his work.

He did not always remember how ideas had come into his mind. His gifted young colleagues made many suggestions, and sometimes Thomson used to hear these and come back days later with the same suggestion, under the impression that it was his own. This caused a considerable amount of tension, for the raw new idea was the young man's research ore, and Thomson was apt to return with a piece of finished metal. The unconsciousness of its origin combined with the more perfected form was peculiarly trying.

It is said that the focusing mass-spectrograph was developed in some privacy, in order to avoid any misunderstanding as to its origin. It is notable that when Aston's results on the analysis of the elements, and the establishment of the whole-number rule were published in 1921, J.J. was remarkably inappreciative of them.

In his speech at the opening of the Royal Society Mond Laboratory in 1933, he jocularly remarked that forty years ago there were two kinds of physicists in Cambridge: those who made discoveries and those who got the credit for them. Perhaps there was some unconscious memory here of earlier arguments.

J.J. had a strong acquisitive instinct. His extremely careful collecting of fees, his very small expenditure on apparatus and the running of the Laboratory seem to have exceeded necessity. He was inclined to save money, and did not apparently make any special effort to collect it for the Laboratory.

The acquisitive tendency was clearly seen in the amassing of his large private fortune from modest beginnings. His fortune, £82,000, was more than eight times as much as was spent on research apparatus in the whole thirty-four years of his directorship of the Laboratory.

It is a striking illustration of the values of the time and society in which he lived that the fruits of the management of investments in a scientist's spare time should be so much greater than the cost of researches that have changed civilization.

Thomson preferred mathematical methods that were simple and well worn, rather than elegant but more difficult. He found the older mathematical methods more manageable and efficient. As Born has said, he was a combination of revolutionary and conservative in his scientific work: he took nothing for granted, but he clung to classical ideas. He was primarily interested in getting his result. Mentally he was a North Countryman, engaged in 'getting there', without frills.

He rated pioneering experiments more highly than refined measurements, and he was happier in rough new regions, where the facts were simple and important, and the chief requirement was originality and independence in appreciating them, than in the subtle development of old ideas.

His openness of mind was exhibited in his investigations of spiritualism. Like Crookes, the third Rayleigh, and other notable physicists of his period, he believed that these phenomena should be carefully investigated, as, if they contained even a grain of reality, they would be exceedingly important.

J.J. had a characteristic kind of humour. He was very fond of telling stories, sometimes twice on the same occasion. He did this even when at the height of his vigour.

He liked to relate how J. G. Frazer, the famous author of the *Golden Bough*, offered to resign his fellowship at Trinity College because someone had detected a misquotation in his great work. Thomson was quite without this kind of sensitiveness, and his temperament was in complete contrast with that of Helmholtz, who had been discovered one day by Roscoe complaining that his whole scientific career was endangered because someone had queried one of his results. Schuster remarked that Helmholtz bore his title of 'Excellency' with great dignity. Perhaps this helped him to suppress some internal lack of confidence. How different from J.J.'s serenity and affectionate nickname.

J.J. was much amused by the action of one of his Laboratory boys who found a girl student from Newnham who had fainted in the Laboratory. The boy, proud of his promptitude, had turned the fire hose at her.

He liked to illustrate the need for awareness in research by a story of his bedmaker at Trinity. He remarked to her one day that in Oxford the college servants were male. She said at once that in that case the staircases would be dirty. She went to Oxford by the next excursion, and then reported to J.J. on her return that she was right about the staircases. He asked her what she thought of Oxford, but she had not been interested in the city—'she had proved her theory, but she had missed Oxford.'

He was fond of jokes relating to games and sports. While on the way to America in 1923, he heard from the radio that Oxford had won the boat race. In 1935 he wrote 'this was not expected and has not occurred again.' He did not always score, however, as when Bumstead caught him on the speed of American boat crews. When he visited Yale, he saw in the house of one of the professors a dog which had been trained to pretend to be sick when he heard the name Harvard.

J.J. remembered how Stokes, who was a very illegible writer, and an early and enthusiastic user of the typewriter, recommended J. C. Adams to get one. 'Why should I?' said Adams. 'People can read *my* writing.' He appreciated the saying attributed to W. H. Thompson, the Master of Trinity in his student days: 'We are none of us infallible, not even the youngest.'

After he became Master of Trinity himself, he had to preside at meetings for the election of new fellows. At one of these, an ardent sponsor argued in favour of his candidate that he had really said the last word on his subject. J.J. commented that what the College required were fellows who would say the first word on their subjects, not the last. He related that if all of the forecasts of future greatness by sponsors of candidates at all of the elections he had attended had proved correct, then the College would have added to its rolls four more Newtons and three more Bentleys.

In 1930 he sent to a friend 'a little book on History, which contains only two dates, one of which, 1066, I knew.' He remarked that there was a great craze for people to read what they do not understand. It reminded him of the Browning fashion of fifty years ago, but now it was for the scientific speculations of Jeans and Eddington.

The outbreak of the First World War in 1914 and his election to the presidency of the Royal Society in 1915 brought J.J. from absorption in the atmosphere of Cambridge and the Cavendish Laboratory into public affairs.

Fisher had asked for the establishment of a Board of Invention and Research for the application of science to defence. The Board was appointed when Arthur Balfour was the First Lord of the Admiralty. J.J. had had connections with the Balfour family for many years. The wife of his Cavendish predecessor, the third Lord Rayleigh, and mother of his pupil, the fourth, was a sister of Arthur Balfour. Mrs. Sidgwick, with whom he had collaborated in his first research in the Laboratory, was another of Balfour's sisters. In early days, Balfour himself used to wander into the Laboratory sometimes, and was made to help in taking observations.

J.J. usually took the chair at meetings of the Board of Invention and Research. It was especially active in submarine detection. At Thomson's request, Rutherford began the work, and W. H. Bragg became the director of Laboratory research in 1916. The most important development came from France, where Thomson's old pupil Langevin invented the quartz plate method for producing strong beams of ultrasonic waves. The French scientific liaison officer was the Duc de Broglie.

The First World War exposed the backwardness of Britain in applied science. The shortage of explosives, dyes and drugs, the lack of optical scientific instruments, the absence of equipment for detecting submarines brought the country into extreme danger. The Government was forced to undertake the fostering of science for the national benefit.

Formerly, scientific research in Britain had been maintained almost entirely by private and amateur effort. A fundamental change now occurred very quickly. Thomson, as President of the Royal Society, led a deputation of scientific bodies to the Lord President of the Council, pressing for appropriate action. The Government had announced the foundation of a £1,000,000 fund for the encouragement of cooperative research. In his speech he made one of the best statements of the method and value of pure research. He instanced the discovery of X-rays. 'It was not the result of a research in applied science starting to find an improved method of locating bullet wounds. This might have led to improved probes, but we cannot imagine it leading to the discovery of X-rays. No, this method is due to an investigation in pure science, made with the object of discovering what is the nature of electricity. . . . In fact, research in applied science leads to reforms, research in pure science leads to revolutions, and revolutions, whether political or industrial, are exceedingly profitable things if you are on the winning side.'

The Government created the Department of Scientific and Industrial Research, which has since had a deep influence on the fortunes of the country. Two illustrations may be quoted from many: the provision of means for the study of the ionosphere, which enabled E. V. Appleton and R. A. Watson Watt to pursue their researches, and again the knowledge and skill from which radar could be invented; and the endowment of studentships for research, one of the first holders of which was P. M. S. Blackett, who was thus enabled to adopt the profession of research.

Thomson had no patience with the industrialists who preferred to buy the results of research rather than do any themselves. He said in 1911 that he considered such a course undignified, unsportsmanlike, and quite unworthy of the great traditions of British engineering, and he doubted even whether it was good business.

Thomson was greatly assisted in his promotion of Government support for research by his old friend Richard Threlfall. This remarkable man and experimenter had collaborated with J.J. in 1885, and in 1886 was appointed Professor of Physics in Sydney. When he arrived, he found there was no Laboratory equipment and no provision for it. He ordered £2,000 worth of instruments on his own responsibility. In the meantime he became friendly with the Prime Minister, who was very fond of mechanical hobbies. Threlfall gave him a few tips on the lathe. Presently the Prime Minister told him the cost of his new equipment had now been put on the estimates, and Threlfall's Laboratory became the best in the British Commonwealth at the time.

Threlfall returned to England to join a firm of industrial chemists manufacturing phosphorus. He was asked to join the Board of Invention and Research. This brought him into cooperation with Governmental and military science. He developed the use of phosphorus for smoke screens, tracer bullets and bombs, with great success. The resources of his works Laboratory enabled him to make large-scale experiments without delay.

He became one of the foundation members of the Advisory Council of the Department of Scientific and Industrial Research, and remained on it until 1926. H. T. Tizard described him as 'the hardest-worked of any who had no direct responsibilities for the work' of the Department's research stations. He regarded him as the 'most informed scientific man actively engaged in industry' in England in his day.

Threlfall was a physically big man, a native of Lancashire, who had been educated at Clifton College, and had shared a room with Douglas Haig. He was a magnificent footballer, and claimed to have interested J.J. in the game. Threlfall was a scientist, a man of business, and a country squire; a man after J.J.'s own heart.

J.J. wrote to a friend in 1915 that he did not expect the kind of world they had known ever to return. The war and age had brought a change in the continuity of his attention. The death of Montagu Butler caused the Mastership of Trinity College to be vacant, and Thomson was appointed to it by the Crown, in whose gift it is.

When J.J. accepted the Mastership in 1918, he had hoped to continue as Cavendish professor, in spite of the new administrative duties, by retiring from advisory committees, etc. It soon became clear that this hope was impractical, so he resigned from the Cavendish chair in 1919. The obvious successor was Rutherford. J.J. wished, however, to retain some place in the Laboratory. This created a delicate situation, which was successfully solved by the frankness and tact of both men, before Rutherford accepted the succession. A special new professorship was created for J.J. while he lived, and some rooms were placed at his disposal.

J.J. came in almost daily until he became infirm. His period of creative discovery was, according to nature, completed, but his interest in physics continued unabated. He still went into the Cavendish Laboratory to research even after Rutherford had come, and gone. His small bent figure, seen disappearing through the Laboratory doors into his research rooms, after Rutherford had been buried in Westminster Abbey, seemed to have a legendary quality, a manifestation of the permanence of his spirit and works.

J.J. undoubtedly had great natural aptitude for business. This was proved by his large fortune, made by skilful Stock Exchange operations. He liked to talk of how fortunes had been made in unusual ways, and he had a keen nose for any swindle or hoax.

As Master of Trinity he had to examine the proposals of the professional advisers on the management of the College's funds. He criticized them with the peculiar shrewdness of a North-country business man. If the magnitude of his fortune, revealed by his will, had previously been known, perhaps more attention might have been paid to his advice on investments. Apart from the Nobel Prize, then worth about £7,000, he never received any large gift or fee.

J.J. had few cultural interests apart from science. He did not like classical concerts. He used to say that he had had to go to the Hallé concerts too often in his youth. He was very fond of the Gilbert and Sullivan operas. He went to the first nights of several of them in London.

He liked Scott and Dickens, Trollope and Thackeray, Jane Austen and George Eliot. He was delighted with J. B. Priestley's *Good Companions*, but liked his later books less. He read a large number of thrillers. He was interested in botany and wildflowers all his life. He bought herbals from the secondhand bookstalls, and his collection sold for more than it had cost. His biographer says that he could not remember J.J. ever using a classical or literary allusion.

J.J. was progressive in University business. He supported the admission of women to degrees, and he opposed compulsory examinations of all students in Greek. He considered the time he had spent on it had been entirely wasted.

As Master of Trinity, J.J. was in notable contrast with Montagu Butler, his predecessor. Butler was large and stately; J.J. was small and quick, and without concern for dignity. Their relations with the College were very different. Butler had become Master in 1886, when a young man. As the years passed by, he became the embodiment of the College. He was familiar with every detail of its affairs. He wrote personal letters to the humblest undergraduate entering the College.

J.J. became Master at the age of sixty-two, his greatness had been achieved in science, not in the guidance of the College. Trinity College became Butler's life, whereas the electrical theory of matter was J.J.'s.

J.J. preferred personal discussion and gossip to philosophical conversation. He probably preferred the company of the ordinary student to that of the intellectual. He was more interested in their personal idiosyncrasies and amusements than in their work. Though he owed so much to scholarships himself, he did not like to see his Laboratory assistants attempting to become University graduates and did not believe in making the educational ladder easy to climb.

Rayleigh records that J.J. wrote in September 1938, during the Munich Crisis, when he was eighty-two years old, that 'The Prime Minister is magnificent.' Rayleigh compares J.J. with Neville Chamberlain, both of whom he knew well. He thought there was a certain affinity of character between them, in the simple and straightforward point of view taken by each, and in their grasp of what was within the bounds of practicability, in their readiness for reasonable compromises, and their willingness to exchange ideas with anyone who had something to say.

J.J. died on August 30th, 1940, aged eighty-three. His ashes were interred near those of Newton, Kelvin and Rutherford in Westminster Abbey.

One of the measures of a great man is the volume of important work that he accomplishes. J.J. published some two hundred and fifty works, consisting of 230 papers and 11 books, one in collaboration with his son, G. P. Thomson. His number of personal publications was not quite so stupendous as that of Kelvin (661) and the third Lord Rayleigh (447), but he accomplished far more than they as a leader of organized research. His total achievement was of the same order.

The contribution of the Cavendish Laboratory to the creation of the modern age of electronics was greater than that from any other single institution. The radio valve, the electron microscope, the electronic calculating machine—a hundred new devices and industries have come from the electron. The expenditure by Britain alone on radio and radar devices in the Second World War has been estimated by R. A. Watson Watt at £500,000,000.

Yet the total cost of research apparatus in the Cavendish Laboratory during the whole of J.J.'s thirty-four years of directorship was less than £10,000. The contrast is startling, though on analysis this impression requires some modification. The conditions of research at Cambridge, in some ways so attractive and fertile, were created at the cost of centuries of accumulation of wealth and learning. It may be true that J.J.'s discoveries were made with a pittance for apparatus, but the cost to mankind as a whole of a leisured life for Newton, Clerk Maxwell, and the scores of geniuses who created the subtlety and depth of the Cambridge tradition was great. From the comprehensive social perspective, J.J.'s fortune of £82,000 was, in a degree, part of his payment by society. The cost of providing a kind of society in which it was possible for J.J. to acquire a large fortune in his spare time must be entered as part of the social cost of the Cavendish discoveries. They were, in fact, far more costly in absolute historical expense than appears at first sight.

Nevertheless, after all allowances, the contrast between the economic value of the applications with the cost of the fundamental research is sufficiently striking.

For sixty-four years, Cambridge was J.J.'s home. His research life was spent entirely in England, and within England, in Cambridge. Yet he was a Lancashire boy by origin and his native characteristics were never transformed. His careful saving habits, his complete division of his own money affairs from the finances of the Laboratory, the sharp separation between personal prosperity and institutional poverty, his homely humour and jokes and gossip, his interest in sport, his self-sufficiency in research method, his religious piety (in which he resembled

Faraday, Maxwell, and Kelvin), were rooted in his origins. He was an individualist from the Lancashire lower middle class who rose into Cambridge and higher English society through his ability.

He started life with Free-Trade and Liberal ideas of the Gladstonian kind, and like Gladstone was not much interested in social reform. The pursuit of physics led him to transcend his origins. The individualist in research did more than any other to create the research team.

Though J.J. was born and grew up in the England of Gladstone and Queen Victoria, he did more than any other English scientist to found the scientific outlook of the twentieth century. His Dover Address of 1899 on the constitution of atoms was the manifesto for twentieth-century research in experimental atomic physics.

It is impossible to overlook the adolescent, uncultivated, unintellectual aspects of his mind and school. J.J. was a very extrovert personality. Here again he was like the shrewd self-made Lancashire business man, and in social contrast with the subtle, imaginative, philosophical Clerk Maxwell, the Scottish laird and leisured country gentleman, who, in spite of his origins, was just as sociable as J.J., and who actively worked at social reform.

During J.J.'s directorate, the Cavendish Laboratory greatly advanced the professionalization of research. The new scientists seemed, however, to show signs of the cultural narrowness and juvenility often associated with specialization. J.J. and the Cavendish Laboratory made a tremendous contribution to discovery, to the creation of the modern outlook in physics, and to research method. Much, however, remained to be done in bringing the results of this mighty achievement into balanced relation with other cultural and human activities and the rest of life.

BIBLIOGRAPHY

Recollections and Reflections, by Sir J. J. Thomson (1936).

The Life of Sir J. J. Thomson, by Lord Rayleigh (1942).

Application of Dynamics to Physics and Chemistry, by J. J. Thomson (1888).

Recent Researches in Electricity and Magnetism, by J. J. Thomson (1893).

Conduction of Electricity through Gases, by J. J. Thomson (1903).

Conduction of Electricity through Gases, by J. J. and G. P. Thomson: 2 vols. (1928–1933).

A History of the Cavendish Laboratory, 1871–1910 (1910).

'Cavendish Laboratory,' *Nature*, Supplement, **118**, 865 (1926).

'On the Masses of the Ions in Gases at Low Pressures,' J. J. Thomson, British Association, Dover, 1899.

'On the Electrical and Magnetic Effects produced by the Motion of Electrified Bodies,' J. J. Thomson, *Phil. Mag.*, April 1881.

The Progress of Physics, 1875–1908, by A. Schuster (1911).

'On the Illumination of Lines of Molecular Pressure, and the Trajectory of Molecules,' W. Crookes, Bakerian Lecture, *Phil. Trans. Roy. Soc.*, 170, 135 (1879).

'The Discharge of Electricity through Gases,' A. Schuster, Bakerian Lecture, *Proc. Roy. Soc.*, **47**, 526 (1890).

'On the Electric Discharge in Rarefied Gases,' E. Goldstein, *Phil. Mag.*, September 1880.

'Obituary Notices,' *Roy. Soc. Obit.*, **3**, 477 (1939–1941); *Proc. Phys. Soc.*, **53**, 305 (1941); *Nature*, **146**, 351 (1940).

ERNEST RUTHERFORD

1871–1937

I. THE NATURE OF HIS GENIUS

UNTIL Rutherford appeared, no experimental physicist had shown a genius comparable in magnitude with that of Faraday. It will be easier to say in the twenty-first century than it is today which was the greater.

Both accomplished the supreme achievement of introducing a new fundamental conception and establishing it by experiment: Faraday's electro-magnetic field and Rutherford's nuclear atom. In both cases they broke with classical ideas, Faraday with the idea of action at a distance employed by Newton in his conception of gravitation; Rutherford with the idea of mechanical stability in his nuclear atom.

Faraday and Rutherford had profound physical imaginations and made little use of mathematics. Their conceptions were purely physical. They were more sustained experimenters, but less philosophical or mathematical than Galileo, Newton, Clerk Maxwell and Hertz. Their genius is in complete contrast with that of pure theorists like Einstein and Planck, who used only mathematics, and never made experiments, to establish new ideas.

Rutherford's work was much more concentrated than Faraday's. It was almost entirely in nuclear physics (radioactivity, as he showed, is the property of the nuclei of atoms), whereas Faraday's extended over the multifarious phenomena of electricity, electro-magnetism, electrolysis, and regions of chemistry and metallurgy.

Yet, though restricted to the nucleus, it is possible that Rutherford's work will prove to be even more fundamental than Faraday's, for the nucleus may be the seat of actions which determine most of what happens outside it.

Rutherford accomplished three major achievements. The first was the conclusive proof, announced with Soddy in 1902, that radioactivity consists of the spontaneous transformation of atoms; the second was the utilization of radioactivity to reveal the structure of the atom;

and the third was to use this knowledge to disintegrate atoms, thus leading towards the discovery of their structures and the command of their internal energy.

Rutherford's ideas were always extremely simple. It is remarkable that such simple ideas could have been so potent in the twentieth century. One might have expected that after three centuries of intensive development of physics, ideas would necessarily have evolved a complicated subtlety and all simple ones would have been used up and exhausted.

Rutherford not only worked with the simplest ideas. Very many of his most brilliant experiments were made with simple apparatus. The most extraordinary feature of Rutherford's achievement is that, at this date, he was able to obtain such profoundly important results with such simple ideas and apparatus. It makes the analysis of his genius exceptionally difficult.

Personal acquaintance with Rutherford very quickly established the impression of a great man. He had an immense imaginative enthusiasm. He had an extraordinary zest for life. He had a strong interest in and memory for persons. Immediately after his death, men of very varied opinions and types showed how he had been understanding with them all.

Niels Bohr has referred to Rutherford's 'unique energy'. Langevin, a fellow-student with him under J. J. Thomson, when asked whether he had been friendly with him, replied that relations with Rutherford were not ordinary—'one could not be friendly with a force of nature', in the sense that one could not be familiar with the gods.

Imagination, energy, force were characteristics of Rutherford. Niels Bohr has said that Rutherford 'is not a clever man, he is a great man'. As Aldous Huxley said of Balzac and Tolstoy, he was larger than life. His qualities were normal and in no way peculiar, but present in supernormal degree. He was not at ease with clever people who put conundrums to him, and he used to bluster at them to conceal his discomfort.

He had no taste in art and music, and he wrote in a rugged style which nevertheless revealed greatness of mind in his first scientific paper, published in New Zealand before he had left for England. He read a great deal, especially novels, biographies, popular histories and detective stories, largely as a diversion and sedative after work. Four library subscriptions were scarcely sufficient to keep him supplied with reading material. With his strong memory, he was often able to recall surprising pieces of literary and general information.

Rutherford accepted the conventional aims of the society in which he found himself. From his infancy, he had a set of aims in life, which he never questioned. He attained many of them, one by one, from secondary school scholarships to supreme scientific achievements, and he enjoyed all his incidental honours.

He could not have achieved so much unless he had felt these successes worth while. Like some other very great men, such as Balzac, he had a capacity for taking the commonplace seriously and lending it some of his own genius. He had a similar combination of naïveté and insight.

His profoundest gifts were subconscious, while many of his conscious thoughts, expressions, and activities were ordinary. The conventional surface, deceptively intelligible, is the most misleading guide to the fire beneath. He adopted the title of Lord Rutherford of Nelson, the New Zealand town named after the great admiral. His genius, like Nelson's, was immediately obvious in action, but difficult to analyse.

It is necessary to try to explain, without depending on the instinctive impression of his personal acquaintance, how he accomplished so much, for this is not immediately clear from his scientific papers. At first sight these may appear disappointingly simple. There is scarcely an idea or an experiment in them that an intelligent non-scientific reader cannot understand. There is an apparent absence of surprising discovery: nothing like X-rays or the Wilson cloud-chamber.

These appearances are, of course, illusory. Rutherford's work appears simple because he could judge which aspects of fundamental phenomena were simple, and devise and carry out simple experiments to elucidate them. Other men, cleverer than he was in the ordinary sense, looked at the same phenomena and were baffled by their complexity. They were unable to judge where to start, and to guess what aspect, and more, what fundamental aspect, was susceptible of simple description and elucidation.

Further study of Rutherford's works shows that another factor in his deceptive simplicity was his instant readiness to throw over orthodox ideas without qualm. His simple picture is preserved by ignoring inconvenient postulates of classical physics. His sub-conscious scientific thought is revolutionary and in contrast with the conventionality of his scientific expression and of his external life.

Rutherford was subconsciously at home with explicitly original thinkers such as Bohr and Dirac, but consciously he was himself a conventional scientific thinker, which explains why we all imagine we can understand him.

Rutherford's works exhibit an extraordinary confidence and contain exceptionally few mistakes. These two qualities very rarely go together. But in Rutherford's case they were combined with apparently simple ideas. The triple combination produced the illusion that Rutherford had direct access to reality itself, about which he reported in familiar conversation.

As the characters of Shakespeare seem more real than living men, so Rutherford's atoms seem more real than actual matter. But Shakespeare's characters, in spite of their consonance with humanity, cannot come to life. In contrast, Rutherford proved by experiment that his atoms were actual, and has, as it were, brought his imaginings to life. And within ten years of his own death, these grew into the might and the terror, the promise and the threat of atomic explosions.

Physicists working in the same field as Rutherford know from experience the greatness of his work, because of the difficulty of equalling it. When their ideas are as simple, they generally prove to be unimportant, and when they have Rutherfordian confidence, their ideas generally prove to be wrong.

The transparent greatness of Rutherford's work becomes more and more inscrutable. Because Rutherford's genius was in an exceptional degree subconscious, much study will be necessary in order to reveal its complete nature and roots.

II. NEW ZEALAND

Rutherford was born on August 30th in 1871 at Brightwater, about thirteen miles south of Nelson, on the South Island of New Zealand. This was only twenty-nine years since the settlement of Nelson in 1842. New Zealand had been sighted by Tasman in 1642, two hundred years earlier, in the year of Newton's birth, but he had not dared to land, owing to fear of the natives. Captain Cook circled the islands in 1769, having with him Joseph Banks, the future famous President of the Royal Society. He made a masterly map of their coasts, but he also did not settle the islands.

The Maori population was originally large and war-like. It presently began to acquire occasional muskets from adventuring seamen, by trading one ton of dressed flax for one musket. The Maoris also contracted influenza from these adventurers, and their population was very rapidly reduced by the combination of firearms and disease. By the first half of the nineteenth century, the Maoris were willing to barter depopulated areas of their lands for trifles.

Nelson was the third of the British settlements, carried out by a private enterprise company which combined the vigour of the pioneer with an unusually progressive programme of development. This appealed to persons who were materially and morally discontented with the terrible conditions of early British capitalism, classically described by Engels in the *Condition of the Working Class in England* (1845).

Men and women of character and strength could either seek to change these conditions or escape from them. The settlers of New Zealand were mainly of the latter type. They were competent hard-working craftsmen, nonconformist in religion, who sought a better life through conventional self-help abroad rather than political action at home.

The prospectus of the company that established the Nelson settlement appealed to the interests of this class. It included an unusually large financial allocation for schools. Within thirteen years of the 1842 settlement Nelson College was founded. Special efforts were made to attract teachers, who were given high social status in the community.

The settlers sailed for New Zealand in parties of about 200, in ships only about 125 feet long. Among the early groups was George Rutherford, a wheelwright of Perth, who had been engaged to build a sawmill at Motueka, near Nelson. He sailed from Dundee in 1842 with his family, which included a son, James Rutherford, aged three, who became Rutherford's father.

The voyage, in a vessel of 300 tons, lasted more than six months. The Bay of Biscay was too rough to be crossed at the first attempt, and the ship had to return and wait in the Bristol Channel for calmer seas.

Then, life in the virgin forest after reaching Motueka was very hard. After some years the family moved to the neighbourhood of Brightwater, where they built a water-driven sawmill. The forest was laboriously cut, the ground cleared of stumps and set with wheat, every grain of which had been wiped by hand to remove rust disease. The corn was reaped with sickles and threshed with hand-flails, and carried on the back four miles to a mill to be ground into flour.

James Rutherford was given the rudiments of education by his family, and did not attend school until he was ten.

The voyage and the subsequent settling selected the strong and the sociable, for in those conditions not even the strongest could survive by himself. The members of the community had to work together, and the exercise and acceptance of leadership became a prominent communal habit. As a result of the preliminary depopulation, the colonization of New Zealand was attended by less than the usual amount of sustained warfare with the natives. The early struggles of the Rutherfords in New Zealand were with nature rather than with man.

The young Nelson agrarian democracy, relatively pacific and free from human exploitation, was stimulated in 1851, only nine years after its foundation, by demands from Australia for food and materials, caused by the Gold Rush. The New Zealand colonies were able to share in the profits of the Gold Rush, but were protected by thousands of miles of ocean from contact with the vicious society of speculators and gold-diggers.

Rutherford was notably free from mercenary motives throughout his life. His first professorship had a low salary, but exceptional opportunities for research; he resisted suggestions that he should seek a better-paid chair. He did not patent his radio detector, subsequently adopted by Marconi; he was violently opposed to patenting his atomic discoveries; on more than one occasion he almost threw well-known men out of his room at the Cavendish Laboratory for daring to suggest such a thing. When he died, he left only £7,402.

James Rutherford grew up in the pioneering life of saw-milling and farming. Meanwhile, in 1855, a twelve-year old girl named Martha

Thompson, daughter of a widow of Hornchurch in Essex, whose father had had scientific interests, was brought to New Zealand by her mother. The Thompsons landed in North Island after one of those long voyages, but presently moved to South Island owing to the hostility of the Maoris. Martha Thompson was one of the first women schoolteachers in New Zealand. She and James Rutherford met, and were married in 1866.

James and Martha Rutherford had seven sons and five daughters. James lived to the age of eighty-nine and Martha to ninety-two. Their fourth child, and second son, was Ernest, born in 1871.

The little wooden cottage was too small for the growing family. In 1875 they moved ten miles to Foxhill, where they built a larger bungalow.

James Rutherford practised his craft of wheelwright, and in addition to running his small farm, undertook some contracting for bridge building and railway construction.

Ernest found himself in a healthy, vigorous, inventive environment. His mother understood education. Both of his parents, in the midst of all their labours, were exceptionally carefree and happy.

Marsden has drawn attention to the great influence of the happy constructive atmosphere of Rutherford's early home life and youth, quoting the psychologists' observation that calm feeling in children leads to clear thinking.

Rutherford was free from complexes. He went to the village school when he was five. There was only one teacher, but he was good. Starting with his mother, he had a succession of exceptional teachers. At the age of ten he acquired a primer on science written by that excellent physicist Balfour Stewart, who had been J. J. Thomson's teacher at Manchester. There are traces of the influence of this book on Rutherford's later thought and experiments.

In 1882 James Rutherford moved to Havelock, a village on the eastern side of Nelson, where he erected a flax-mill for preparing the wild flax growing in neighbouring swamps and a water-driven sawmill for cutting railway sleepers. He secured a contract for 40,000 at 2*s.* 8*d.* each.

Ernest, now eleven, attended the Havelock Primary School, where he had an excellent teacher who taught extra Latin to good pupils for an hour before ordinary lessons began.

At the age of fifteen Rutherford competed for a county scholarship worth fifty guineas a year, to Nelson College. He was examined in his own school. The scholarship examinations were followed with keen public interest. Parents and neighbours used to peep into the room to see how the candidates were getting on. Rutherford won the scholarship, gaining 580 marks out of a possible 600.

He entered Nelson College, whose headmaster was W. J. Ford, the

3. Rutherford's parents in their garden in New Zealand

[face p. 48

famous cricketer and big hitter, who had been educated at Repton and Cambridge and had been a housemaster at Marlborough College. His chief master was W. S. Littlejohn, yet another excellent teacher, a classical scholar who was proficient in mathematics and science, and subsequently became head of the College, and then of Scotch College, Melbourne.

Nelson College was run on the lines of an English 'public' school. Owing to the nature of New Zealand society, its social class atmosphere was, however, very different from that of an English 'public' school. It was much more like an English country grammar school, in which some of the pupils are residential because they are farmers' sons, living too far away to go home at night.

In his first year Rutherford gained a prize for reading and a scholarship for history. He subsequently gained further prizes and scholarships for English literature, French, Classics, Latin, and mathematics.

He played forward in Rugby football, was good at singlestick and moderate at cricket. He became college librarian, carrying out his duties with 'genial despotism'. Finally, he became head boy.

His education was well balanced and science was not prominent in it. He gained much from personal instruction by Littlejohn.

In 1887 his father moved to Pungarehu, a remote and beautiful place near Mount Egmont in North Island, where he set up another flax farm. He experimented in flax preparation, devising a new method of soaking and a new kind of scraper for removing vegetable matter. He experimented, too, in the improvement of flax plants by selection.

Rutherford used to help on the farm during the vacations, and teach his numerous brothers and sisters. He spent happy hours shooting and fishing in the magnificent countryside.

He had the boy's usual taste for making mechanical models and guns and mending clocks. He made his own cameras and was a very keen photographer. But he was never a gadgeteer.

In his last year at Nelson College he gained a New Zealand University scholarship, being fourth in a list of ten. At the head of the list was W. S. Marris, two years Rutherford's junior. Marris was a brilliant classical scholar, who subsequently passed out easily first in the examination for the Indian Civil Service, and became the Governor of the United Provinces and Assam.

Rutherford and Marris entered Canterbury College, Christchurch, in 1890. Senior university scholarships for the best man of the year were in 1892 awarded to Marris for Latin, Rutherford for mathematics, and Chisholm of Otago for physical science. Rutherford was already twenty-one, and had not yet made a mark in physics. He was not at all precocious.

The professor of mathematics at Canterbury was a very capable teacher named C. H. H. Cook, a former fellow of St. John's College,

Cambridge, who gave Rutherford a good mathematical grounding. The professor of chemistry and physics was A. W. Bickerton, an original but eccentric scientist who published papers on anything from the equilibrium of gaseous cosmic spheres to chlorine as a cure for consumption.

During a meeting of the British Association in England in 1913, the Mayor of Leamington, who was seeing off a party of members at the railway station, went up to Bickerton, looked at the striking old gentleman, hesitated, lifted his hat and asked whether he happened to be Sir Oliver Lodge. 'Not so famous, but greater,' replied Bickerton, leaning out of the window as the train carried him away.

Rutherford may have acquired some of his boldness of imagination from Bickerton, and, as Marsden has said, Bickerton's speculations on the collision of novae may have influenced Rutherford in his formulation of the planetary theory of the atom.

Rutherford completed his university course by securing first-class honours in both mathematics and physics. Such a double first had been awarded only once before in the history of the university.

Rutherford does not seem to have begun the serious study of physics until 1892. Nor did he take a prominent part in the Canterbury Philosophical Institute, the official science society of the university. He was more interested in an unofficial College Science Society, which had a more radical outlook. The members were prepared to discuss anything, in the Bickertonian style. They opened a philosophical discussion on Evolution of the Elements. Evolution was still considered an irreligious subject, and the choice of such subjects caused some of the authorities to be critical of the society. When the members asked Rutherford to become their secretary in 1893, he did not accept the situation without persuasion.

Rutherford did not become seriously engaged in science until he had completed a very successful and varied education. This broad education made him a well-informed man, but though he won many prizes for literature and languages he had no talent for them. He showed no signs in later life of literary and linguistic distinction. His greatness of style was an uncultivated expression of a natural greatness of physical imagination and achievement.

Rutherford's wide but superficial acquaintance with history, literature, and language was like that of a talented and energetic foreigner who had learned the rudiments of an alien culture. He knew quite a lot about it, but had not entered into it.

He seems to have stepped straight into research from his last two years' studies of physics for his honours degree. He had to make practical class experiments on secondary circuits, and these experiments raised in his mind questions concerning the passage of transient currents in conductors. He looked up what Kelvin and Rayleigh had written on the subject, and noted Kelvin's paper on 'An accidental illustration of the shallowness of transient current in an iron bar'.

Rutherford recorded in this note that 'it occurred to me that Faraday's statement of the absolute equality of the time integral of the induced current on making and breaking the circuit . . . was only true . . . of very fine wires, and that the thickness of the wire in the secondary circuit would have the effect of making the current at make and break different. . . . It was not observed until later that the same idea had occurred to Sir W. Thomson . . . in . . . 1890.'

Rutherford showed in virtually his first research note his characteristic qualities of mind. He had queried Faraday and had found that Kelvin had had the same ideas as himself. He had unconsciously placed his scientific judgment in the same class as that of the greatest masters.

He was in fact wrong in querying Faraday and soon after noted that 'Faraday's result we now know to be absolutely true theoretically. . . .'

He made experiments on the behaviour of transient currents in iron, and proceeded from this to the study of the effect of oscillatory currents, which might be regarded as repeated transient currents, in iron. He extended his reading, and found that Lodge in his *Modern Views of Electricity* stated that the oscillatory discharge from a Leyden jar did not magnetize iron, no matter how thin it might be, for it is 'protected from magnetism by inverse currents induced in its outer skin, and accordingly does not get magnetized.'

J. A. Fleming agreed with Lodge, and quoted Hertz to the same effect. Rutherford noted that 'the subject of the magnetization of iron in these fields has been very little touched upon since the time that Henry experimented on the effect of Leyden-jar discharges on the magnetization of steel needles.'

He then proved by experiment that iron is strongly magnetic in rapidly varying fields, even when the frequency of oscillation exceeded 100,000,000 a second.

Rutherford used a Hertz oscillator for producing the high-frequency oscillatory currents. He gave demonstrations of radio waves with it to the College Science Society. His experiments were carried out in a cold, draughty cellar, used by the students as a cloak room.

He described his results in his first published paper, on the 'Magnetization of Iron by High-frequency Discharges', which appeared in the *Transactions of the New Zealand Institute* for 1894. He reported that 'a magnetized needle is a very convenient galvanometer for oscillatory circuits for frequencies up to 500,000,000 per second. If the molecules of iron can follow the changes of magnetic force, which is reversed 1,000,000,000 times a second, there can be little magnetic viscosity, and the molecules must move as freely as when under the influence of an alternating current of 100 per second.'

He concluded that 'the use of magnetized steel needles as "detectors" and "galvanometers" has been explained, and their possible use also for measurement of the intensity of electro-magnetic waves.'

The magnetized steel needle was placed inside a coil connected with the receiving aerial. In the final design it consisted of twenty pieces of steel wire seven thousandths of a centimetre thick and one centimetre long, insulated from each other but bound together in a bundle with shellac. When radio waves were picked up by the aerial, they produced oscillating currents in the coil, which in turn reduced the degree of magnetization in the needle. This could be observed by a magnetometer, and hence the reception of the radio waves detected.

Thus Rutherford's first paper contained a striking discovery. But in it there was an even more striking exhibition of his attitude of mind. He had been reading and quoting with enthusiasm J. J. Thomson's *Recent Researches in Electricity and Magnetism*, published in 1893. Thomson had shown that an iron cylinder absorbs considerably more energy than a copper cylinder.

Rutherford remarks that 'the experimental method pursued here is entirely different from Professor J. J. Thomson's, but the final results obtained are the same. The results are also quantitative, while Thomson's method only admitted of qualitative results.'

Already, in his first paper, Rutherford's voice was authentic, and he expressed himself in the magisterial style which was natural to him. One wonders whether this style would have survived if he had been educated in England. Such sentences in an English research student's first paper would almost certainly have been modified by his professor. The free atmosphere of Canterbury College, reflecting New Zealand society, and influenced by the absence of scientific authority and the unconventional mind of Bickerton, did not constrain Rutherford's style.

He applied his 'magnetic detector', consisting of bundles of fine steel needles inside a coil, to the detection of radio waves passing through brick walls over a distance of sixty feet. He had no difficulty in getting fellow-students and even young demonstrators to assist in the experiments.

In 1896 he published a second paper in the *Transactions of the New Zealand Institute*, in which he investigated magnetic viscosity. He invented a simple device of mechanical levers and circuits for measuring intervals of less than $\frac{1}{100000}$th of a second, and mentioned that it could be used for measuring the speed of projectiles. He used it to prove that the rearrangement of molecules when iron is magnetized takes less than $\frac{1}{10000}$th of a second. The decay of induction in iron is purely due to the reaction effect of induced currents, and not to molecular rearrangement. In iron and steel the decay was very rapid at first, and then very gradual. He showed that the magnetization in steel needles due to high-frequency currents was very shallow by dissolving the surface layer of metal with acid.

As described on page 15, Cambridge decided in 1895 to grant degrees to post-graduate research students from other universities. Rutherford became a candidate in 1894 for the research scholarship offered to New Zealand students every two or three years by the Commissioners of the Exhibition of 1851. The first place in his year was awarded to J. S. Maclaurin, an able chemist who was one year his senior, and had already published an important booklet on the chemical preparation of gold. He subsequently accomplished distinguished work on the cyanide process of gold refining.

Maclaurin did not accept the scholarship because he was married and had just been offered a post, so it was then offered to Rutherford. His mother said that when the news arrived, Rutherford was digging potatoes in the garden. She went out to him and told him of his success, at which he flung the spade aside and said, 'These are the last potatoes that I shall dig.'

Rutherford's youth in New Zealand was thoroughly fortunate. His home life was strenuous but healthy and happy. He learned how to help with the family chores. Both at home and at college he was helpful to others. Bickerton said that this endeared him to his fellow-students. His gifts for getting on with people and his boisterous repartee were developed, as Marsden has remarked, by his crowded family life, and the example of his parents.

Yet, as Marris, who shared rooms with him for a time in Canterbury, wrote, he was rather shy and vague, and nervy in examinations. The bluff and booming exterior was a shield for a sensitive, imaginative nature. Marris thought that at that time Rutherford had not found himself and was not yet conscious of his powers.

Rutherford presently took rooms in a widow's house and became engaged to her eldest daughter, Mary Newton, a fellow-student at the College.

Rutherford's mighty scientific gift did not make itself evident until he had completed his education and had become a young man. Faraday showed a similar absence of precocity.

Why did not this gift make itself evident earlier? Perhaps owing to the attachment and discipline of his family life. Rutherford had a profound affection for his exceptional parents, and a Calvinist training in duty and religion. If he had had a less strong affection for his parents and a weaker sense of duty to his brothers and sisters he might have withdrawn himself at an earlier stage into science, and would almost certainly not have been so good a scientist as he subsequently became.

His dutiful working through all of the stages of his extensive education in non-scientific subjects, for which he had no particular talent, not reaching science until near the end, is a striking exercise of the habits he had learned in his family. The potato-digging incident seems to show that he was performing this family task from duty and not from taste.

He reached the age of twenty-four a thoroughly strong, healthy, normal young man, in whom pioneer habits of work, initiative, and leadership were combined with an extensive education and even elements of the ideology of the English 'public' school system. In him the cultures of the non-capitalistic agrarian colonists and English imperialistic ruling classes met. He had the health and morals of the one, and something of the technique of the other.

Marx, in the thirty-third chapter of *Capital*, dealing with the Modern Theory of Colonization, commented on the attitude of settlers, owning their means of production, and forming an economic system diametrically opposed to the capitalist system. He illustrated the independent psychology of pioneers by the behaviour of the settlers in West Australia, who would not even make the bed or fetch a pail of water for Mr. Peel, the wealthy capitalist who had financed the expedition, and had believed that the master-servant relation that obtained in England would continue under the conditions of agrarian settlement.

Rutherford's social ideology was fundamentally that of the early non-capitalist pioneers who gained a living from the land by their own labour and implements, and organized themselves in an energetic, optimistic, agrarian democracy.

Rutherford was never entirely assimilated into European society, in which capitalism had long since destroyed the independence of the farmer and craftsman and their ideology, grinding society into a finely striated class system. He was non-European, not because he was born in the distant geographical location of New Zealand, but because he had grown up with the ideas and outlook of the early independent farmer and craftsman settlers. Thus one of the sources of his extraordinary energy arose from the particular social class from which he had sprung, with its simple unquestioning belief that survival was necessary, that life was worth living, and science worth doing. It was the duty of every man to strive to contribute his utmost to a community of equals, and it was due to his dignity to strive to contribute more than anyone else in his own line, by cutting more trees, clearing more ground, shooting more game, and exploring more new territory.

Rutherford carried this spirit into science. In 1902 he wrote to his mother: 'I have to keep going, as there are always people on my track. I have to publish my present work as rapidly as possible in order to keep in the race. The best sprinters in this road of investigation are Becquerel and the Curies in Paris, who have done a great deal of very important work in the subject of radioactive bodies during the last few years.'

Rutherford was a Homeric figure, a tribal chief, a warrior who led the hand-to-hand fighting in person, whose characteristics, like those of the Greek heroes, also sprang from an agrarian society having democratic features. He was passionately emulative, but hated meanness, and exploded with rage at the slightest suggestion of cheating or commercial explitation.

III. THE CAMBRIDGE RESEARCH STUDENT

Rutherford arrived in Cambridge in the autumn of 1895. He was the first of the new post-graduate research students, being followed in about an hour by J. S. Townsend from Dublin. He sought permission to work in the Cavendish Laboratory. He was very kindly received by J. J. Thomson, who gave him detailed advice on how to enter Cambridge research and academic life. Mrs. Thomson found lodgings for him, and superintended his introduction to the desiccated, donnish Cambridge family society. If the bucolic New Zealander had not received so much help, he might not have succeeded in accomplishing the difficult change from a pioneering agrarian society to an ancient English university, devoted for centuries to the training of the statesmen, officials, priests, and technicians of an imperial governing class.

Rutherford recorded some of his first impressions of English society in his first letters to his mother and fiancée. The Great Gate of Trinity College and the Great Court into which it leads appeared to him rather disappointing and dilapidated.

In his walks near Cambridge, he met the English agricultural labourers. 'You can't imagine how slow-moving, slow-thinking the English villager is. He is very different to anything one gets hold of in the colonies.' The class differences between the depressed, exploited English agricultural labourer and the New Zealand agrarian democrat were indeed striking.

Rutherford reported, too, that some of the dons he met at dinner struck him 'as very capable, especially in their conversation, and it is a pity so many of them fossilize as it were in Cambridge, and are not that use in the world they might be.'

Thomson's consideration for the student Rutherford was one of the most admirable acts of his own very distinguished life. He had about twenty-five research workers and could scarcely be expected to attend to their personal details. Apart from Rutherford's obvious talent, Thomson seems to have been touched by his coming all the way from New Zealand to work with him.

Rutherford immediately began research on his magnetic detector for radio waves, and sent graphic descriptions of the Cambridge scene to his fiancée and his family in New Zealand. Those that have survived present a remarkable picture of Cambridge life, and increase the regret of the loss in New Zealand of a life-long series of letters to his mother.

Rutherford soon had a magnetic detector set up and working. He was asked by J. J. to demonstrate it to the Physical Society, an unusual honour for a student, as lectures were usually given by well-known physicists. His lecture, with experiments, was a great success.

He worked on the extension of the range of the detector, increasing it to over half a mile, and securing for a time for Cambridge the long-distance radio record. Townsend spoke of Rutherford's work to Sir Robert Ball, a fellow Irishman and Lowndean Professor of astronomy, a famous popularizer of science. The possibilities of radio appealed to Ball's romantic imagination. He spoke enthusiastically of Rutherford's experiments in the Cambridge colleges and London clubs, pointing out the value of radio for signalling the positions of lighthouses through fog. He invited Rutherford to dine at King's College. Rutherford reported to his fiancée in New Zealand how he, a raw research student, had been received by the eminent fellows of King's, how he had been marshalled into dinner by Sir Robert before all the standing undergraduates, how he was placed on the Provost's right hand, and after dinner, how he took nuts and wine in the Combination Room, seated between Sir Robert and Sir George Humphry. 'Seated opposite to me was a Mr. Browning—a lecturer at King's—and he made himself very agreeable. He seemed an extremely well-informed man—in appearance he was a good deal like the typical John Bull one so often sees in *Punch*. He evidently knew such men as Goschen and Chamberlain personally very well. . . .'

Browning asked Rutherford to lunch on the next day. When Rutherford arrived, he found that the other guests were not expected until half an hour later. Browning plied him with questions about his work. The other guests were a Howard of Howard Castle and a young man named Goldsmid, who had just come into a fortune of £30,000 a year.

Rutherford sat dumbfounded through the meal, while Browning gave a beautiful exposition of his radio researches to the other guests. Rutherford reported that Browning was 'a bit snobby. From what I have heard, he professes to know all the people worth knowing in Europe. It is a common yarn about him that he said "The German Emperor was about the pleasantest emperor he had met." '

Rutherford was 'very glad to get away. The atmosphere in Browning's rooms at King's was very different from that at Pungarehu.' Thus within three months of his arrival at Cambridge Rutherford had had a resounding scientific and social success, to which there is no close parallel.

He had written to his fiancée that he was keenly interested in his magnetic detector because of its practical importance, and hence of the possibility of making money out of it. His Cambridge letters show much preoccupation with living expenses, jobs, and prospects. He wanted to secure satisfactory conditions for marrying, which he did not do until five years later, in 1900.

These economic preoccupations were, however, foreign to his temperament and outlook, for he dropped research on the magnetic detector and accepted an invitation from J.J. to collaborate with him in the elucidation of the conductivity in gases produced by the recently discovered X-rays.

Rutherford's dropping of the magnetic detector has profound significance. Here was a young man, a 'stranger in a strange land', as he referred to himself, who had arrived with his own brilliant discovery and invention, which had brought him an extraordinary scientific and social success and the possibility of a fortune. Instead of fostering this first child of his mind, with the possibility of wealth and an early marriage, he changed to a very different, purely academic field, not following his own ideas, but assisting in the development of those of another. The incident reveals Rutherford's fundamental humility and social discipline. Few men of proved original power choose to subordinate themselves to the aims and ideas of others, when they have already found a fine line of their own. It reveals, also, his profound admiration for J.J. and the ideal of pure, fundamental research.

One of Rutherford's favourite sayings was that it is very important for a scientist to know not only when to start, but also when to drop a research. There are many possible reasons for this. The subject and the technique may be exhausted, or not ripe, or not suited to the investigator's gifts.

Very able men may be misled by the bias of a school. For example, that of the German school in favour of wave-theories, which prevented Goldstein and Hertz from recognizing that cathode rays were particles.

There are notable examples of success in fundamental discovery by scientists who have turned their backs on financial gain, the most striking being that of Faraday, who discovered electromagnetic induction a few months after he had taken the very difficult decision to abandon a large income from commercial scientific consulting. Rutherford probably had a similar psychological release when he abandoned radio physics, with its promise of gain.

Rutherford's willingness to surrender his own original research and follow his professor sprang from the same roots as his will to pursue his long and varied educational studies in New Zealand before seriously starting on physics.

When Rutherford and the new post-graduate students began work in the Cavendish Laboratory they were at first coldly ignored by some of the junior members of staff, who regarded them as barbarian interlopers who had illegitimately gained access to Cambridge without passing through the usual years of undergraduate initiation. Rutherford invited them with ironical humility to assist him in his research, knowing that they knew nothing of it, for it was original. He thoroughly enjoyed their chagrin and laughed over their extreme friendliness to him after his swift success.

Rutherford's new subject of research arose out of the discovery of

X-rays, announced in November, 1895. He never tired of describing the profound impression that this made. The rays had very astonishing properties, whose nature presented a challenge. But psychologically and historically, they were much more than just another remarkable discovery. There had recently been several very striking discoveries, such as Hertz's demonstration in 1887 of the existence of radio waves and the discovery of argon by Rayleigh and Ramsay in 1894. Hertz's discovery was, however, only the experimental proof of what had been theoretically forecast by Clerk Maxwell; it did not involve any new physical idea. Argon and the inert gases were surprising, but chemically disappointing because they would not combine with anything. They appeared at first as curiosities, dead atoms which were apparently useless.

X-rays were very much more exciting, with their unknown physical nature and incredible power of throwing shadows of the skeletons of living people on a screen.

Classical physics had long felt the need for a radically new outlook. The old ideas had been ingeniously refined, stretched, and twisted to cover the increasing accumulation of new experimental knowledge. The leading investigators were uncomfortably aware that the old web of theory was becoming threadbare.

The investigation of the nature and extraordinary properties of X-rays acted as a release from the accumulated inhibitions of classical physics. It gave a tremendous stimulus to new and unconstrained scientific imagining.

Physicists everywhere, from J. J. Thomson down to the most modest amateur, immediately made experiments with the new rays. J.J., like others, soon discovered that the rays made gases electrically conducting, and conceived them at once as the most promising new tool in the difficult researches on the conduction of electricity in gases which he had already been carrying on, without very clear-cut results, for ten years.

J.J. suggested in February, 1896, as others elsewhere, that X-rays turn gases into conductors by splitting their molecules into electrified particles, like the ions which render electrolytes or salt solutions conducting. He invited Rutherford to collaborate with him in the investigation of the effect of X-rays on gases.

Progress with the magnetic detector had slowed down. Rutherford, with his habit of duty, his admiration for J.J., and his instant sense of the profound scientific significance of the new rays, threw over his own original work without leaving a sign that he ever regretted the decision or gave it another thought.

He started on his new subject after Easter, 1896. Almost at once he noted that the conductivity of the X-rayed column of gas increased with its length, the opposite of what occurs with a conducting wire.

Together with Thomson, he showed that X-rayed gases lost their conductivity when blown through porous plugs. This suggested that ions responsible for the conductivity were being filtered out.

Then they found that the conductivity was lost when a voltage difference was applied to the gas. Evidently, the electric field swept the positive and negative ions towards the electrodes, so removing the carriers of the current. They found that, for a given intensity of X-rays, the current which could be sent through the conducting gas could not be increased beyond a certain saturating value, even by the application of very high voltages.

They showed that these effects were easily explained on the theory that the X-rays produced quantities of positively and negatively charged ions in the gas.

They showed that the size of the saturation current could be calculated from the number of ions (which depends on the intensity of the irradiation by the X-rays), the size of their electric charges, the sum of the velocities of the positive and negative ions (which proceed at different velocities in opposite directions towards their respective electrodes), the distance between the electrodes in the gas, and the size of the voltage which is applied to the gas through them.

It immediately followed from their formula that the greater the distance between the electrodes the higher the saturation current, for the number of ions in a thick layer of gas would be greater than in a thin layer, and therefore capable of carrying more electricity. Thus Rutherford's early paradoxical observation was explained.

Thomson and Rutherford delivered this paper at the Liverpool Meeting of the British Association in 1896. Rutherford also demonstrated his magnetic detector, a long final account of which was published in the *Philosophical Transactions* of the Royal Society in 1897.

He continued by himself to elucidate the effects produced in gases by X-rays. He measured the actual velocities given to the positively and negatively charged ions by the voltages applied to them.

This early work of Rutherford's contributed much to the clarification of what happens in a conducting gas, and replaced hazy qualitative descriptions by precise quantitative formulae based on exact measurements.

The new precision and certainty that he gave to the subject, especially in the Cavendish Laboratory, was one of the essential preliminaries to J. J. Thomson's final proof in 1897 of the existence of the electron.

With his quantitative experimental methods and his inventive resource, he investigated the conductivity produced in gases by other available agents, such as ultra-violet rays.

The discovery of X-rays had prompted Henri Becquerel in January, 1896 to search for similar radiations, which, by analogy with the phosphorescence seen in X-ray tubes, might perhaps be produced in minerals which phosphoresce under the action of sunlight.

He decided to expose some fine crystals of potassium uranium sulphate to sunlight, and see whether they would emit a penetrating radiation, like X-rays, which would affect a photographic plate through a wrapping of black paper. He wrapped up the plate, placed the crystals near it, and exposed the whole to bright February sunshine. When he developed the plate he found that it had a strong photographic impression.

He duly announced to the Academy of Sciences that he had discovered that sunshine causes potassium uranium sulphate to produce rays like X-rays. There was little sunshine in the following weeks, so he put his wrapped plate and crystals into a drawer. He took them out in March, and decided to develop the plate without exposing the crystals to sunlight. He was amazed to find, nevertheless, that the plate had a very strong photographic impression. The crystals were able of themselves to emit rays like X-rays! He had discovered radioactivity! Yet, for nearly a year, apparently nobody followed him—everyone was absorbed in the X-ray craze.

Becquerel proved within a few days that the radiations were produced by other uranium compounds, and that they electrified gases in the same manner as X-rays, and consequently had the power of discharging electrified bodies by causing the electricity on them to leak away through the surrounding air.

Rutherford noted this effect, and included the new Becquerel rays for examination among the agents capable of producing electrical effects in gases. He began to apply the technique he had evolved for investigating the electrical effects of X-rays to those produced by Becquerel rays, early in 1897, and he was one of the first, after Becquerel himself, to start important work on the new rays. Marie Curie did not start research on Becquerel rays until the end of 1897.

In the summer of 1897 Rutherford applied for a renewal of his 1851 Exhibition scholarship for a third year, giving research on the electrical effects of uranium radiation as the main item in his programme. J. J. Thomson gave him the strongest possible recommendation, describing him as 'quite in the first rank of physicists.'

1897 was the year of J. J.'s own greatest achievement, the conclusive proof of the existence and determination of the mass of the electron.

It is significant that, in the midst of the excitement and glory of the discovery of the electron, Rutherford was already leaving for another field, instead of pursuing the brilliant vistas opened by the new fundamental particle, which had been revealed in the same Laboratory, in the same working rooms, before his eyes, and in the discovery of which, with others, he had assisted.

While Thomson was in the midst of his own work and direction of the brilliant laboratory, he never forgot to recommend Rutherford for every possible advantage, putting him in the way of examining, book-reviewing, and lecturing. Mrs. Thomson was equally sympathetic and helpful. Rutherford told her in 1896 of his engagement, and she invited his fiancée, Mary Newton, to stay with her for a week in 1897.

Rutherford's quantitative investigation of the electrical effects produced by uranium radiations enabled him to show that the ionization produced by them in gases was of exactly the same kind as that produced by X-rays, while the radiations themselves were different. He identified in them two components, one which had great ionizing power, but little penetration; and the other which had weak ionizing power but considerable penetration. He named them respectively the α and the β rays.

He made exact measurements of the penetrating powers of the α and β rays in various solids and gases, and found that their absorption in gases was proportional to the density. The α-rays were absorbed by a few millimetres of gas, so it was possible to measure the total ionization produced by them. Rutherford found that the absorption of a fixed quantity of the rays produced the same number of ions irrespective of the kind of gas. He showed that the rate of production of ions by the rays was directly proportional to the pressure of the gas. He measured the velocity of the positive and negative ions and the rate at which they recombined. He noted the anomalous behaviour of the radiation from thorium nitrate, which led subsequently to his discovery of the thorium emanation.

All of this was in his last paper written as an 1851 Exhibition scholar and published in January, 1899. This magnificent paper, fifty-four pages long, has the title 'Uranium Radiation and the Electrical Conduction produced by it'. It opens with Rutherford's characteristic directness, and shows that his whole attention was now fixed on the uranium radiations. Then he draws up a list of nineteen aspects which he has examined, the most important of which have just been recounted.

He mentioned that while the amount of α radiation depends mainly on the area of the uranium compound, that of β radiation depends also on the thickness of the layer of the compound, and he remarked that he did not have sufficient of the compound to make a thick layer, and hence could not investigate the β radiation fully. This shortage of material was, no doubt, one of the reasons why he concentrated on investigating the α-rays.

In the same paragraph, while speaking of the effects of thickness and state of subdivision of the uranium compounds, he comments that 'it is possible that the apparently very powerful radiation obtained from pitchblende by Curie may be partly due to the very fine state of division of the substance rather than to the presence of a new and powerful radiating substance.'

This was the Curies' discovery which immediately preceded their identification of polonium and radium. Rutherford was completely wrong. It is probably the biggest scientific misjudgment that he ever made, and is an illustration of the difference between the French and the British schools of radioactive research, the former excelling in the qualitative discoveries, and the latter in the quantitative. Besides discovering radioactivity and radium, the French school, in the persons of the Curies, also thought of the perfect terms for describing these revolutionary new things.

In 1898 the unusual research chair in physics at McGill University in Montreal fell vacant. This had been endowed with a modest salary of £500 per annum but with about the best equipped research laboratory of its kind in the world, by an eccentric tobacco millionaire named Macdonald. The previous professor had been H. L. Callendar, who had returned to England after a brief but brilliant occupancy. McGill searched for the very best successor they could find, and J.J. strongly advised them to take Rutherford.

McGill sent Professor John Cox, a charming man, learned in the history of mechanics and a disciple of Mach, to interview Rutherford. Cox, who was described by Rutherford to his fiancée as knowing 'very little about physics', had a reflective, philosophical view of science for which Rutherford had little inclination. Rutherford was an activist, one who advanced science rather than reflected on it. A man of Cox's vision and talents was needed to perceive the possibilities of Rutherford. Cox was the scientific statesman who gave Rutherford his chance.

After Rutherford arrived at Montreal, Cox relieved him as far as possible from administrative work, so that all of his energy would be free for research.

In the future, the contribution of such men as Cox will be appreciated. During the negotiations, Rutherford reported to his fiancée that he would be expected 'to form a research school in order to knock the shine out of the Yankees'. He told his mother that he would prefer the Montreal chair at £500 a year to one in New Zealand at £700 a year, because it offered better prospects for research and advancement.

Cox and the Principal of McGill, who was with him, offered Rutherford the chair. He accepted it just before his twenty-eighth birthday, and sailed for Canada on September 8th, 1898.

IV. EXCITED RADIOACTIVITY

Rutherford had not been in Montreal more than a few weeks before he was pressing for the delivery of uranium and thorium salts in order to resume research. He was considerably irritated by the uncritical praise of his predecessor, which was poured out to him nearly everywhere. Callendar was one of those supremely facile men who had had an academic career of almost unparalleled distinction. At Cambridge he had gained first-class honours in classics and mathematics, and began experimental research under J. J. Thomson in 1885. He immediately showed himself to be a superb manipulator, and J.J. suggested that he should investigate the possibility of improving methods of measuring temperature by means of the change that it produces in electrical resistance. Callendar adopted this suggestion and investigated the Siemens thermometer, which utilized the electrical resistance of platinum wire, but had not been a success. Callendar showed with extraordinary speed and skill how its defects could be overcome, and in a few months had produced a new instrument of precision, of vast importance for temperature measurements both in science and industry. For this work he was elected in 1894 a fellow of the Royal Society, after he had been appointed to the McGill chair in 1893, at the early age of thirty.

Rutherford wrote to his fiancée in 1899, at the end of his first twelve months in Montreal, that 'as a matter of fact, I don't quite class myself in the same order as Callendar, who was more an engineering type than a physicist, and who took more pride in making a piece of apparatus than in discovering a new scientific truth—but this between ourselves.'

Rutherford made this observation at the beginning of his independent career, and his judgment was correct. Callendar became the leading authority on steam measurements, of basic importance to engineers. He did not follow up his early observation that the curve of resistance of certain metals suggested that at very low temperatures their electrical resistance might disappear completely. If he had, he would have discovered a 'new scientific truth' in Rutherford's sense.

Rutherford was not interested in instrumental and manipulative elegance for its own sake. The kernel of his genius was imaginative, to see the possible explanation of phenomena, and in simple terms which could be tested. The device of apparatus was secondary. As long as it was good enough, with the expenditure of unlimited human energy and patience, to give the required results, he was satisfied. Often the apparatus was not good enough for others with less than his extraordinary energy and patience.

Before the end of 1899, he had four research workers. He did not have what he called a 'foreign' research worker, that is, one from Continental Europe, until 1904. The first was E. Godlewski of Lwow, and the most distinguished was Otto Hahn, who came to Montreal in 1905, and thirty-three years later discovered nuclear fission in uranium.

As already mentioned, Rutherford had made a few experiments in Cambridge to compare the recently discovered radiations from thorium with those from uranium, and he had noted that thorium nitrate appeared to be an unsteady source of radiation, as its electrical effect 'varied very capriciously, being sometimes five times as great as at others.' One of his first acts at Montreal was to set one of his research students, the electrical engineer R. B. Owens, on a systematic study of the thorium radiations. Owens soon found that they were very capricious indeed, and appeared to be affected by draughts of air! He left Montreal soon afterwards to work in Cambridge.

Rutherford at once seized on the phenomenon and investigated it intensively. Before the end of 1899, by a multitude of varied experiments, he had 'found that thorium compounds continuously emit radioactive particles of some kind, which retain their radioactive powers for several minutes. This "emanation" as it will be termed for shortness, has the power of ionizing the gas in its neighbourhood and of passing through thin layers of metals, and, with great ease, through considerable thicknesses of paper.'

He had found that 'the emanation possesses the power of producing radioactivity in all substances on which it falls.' This 'excited radioactivity' emitted radiations more penetrating than those from thorium. He preserved an affection for 'excited radioactivity' because it had his own initials, and seemed to express his own ardent nature. He had a secret hope, too, that he might have discovered artificial radioactivity. But this discovery eluded him. When the Joliot-Curies made it more than thirty years later, he sent his congratulations, writing that he had searched for the effect all his life, but they had found it, and it was well!

But to return to 1899: 'the emanation from thorium compounds thus has properties which the thorium itself does not possess.' The emanation particles were not electrically charged, because, unlike ions, they did not move in an electric field, and they could pass through cotton-wool and be bubbled through water. 'In this respect, it (emanation) acts like an ordinary gas.'

Rutherford had discovered that radioactive substances not only emitted radiations, they also threw out material particles, like atoms of an ordinary gas. Madame Curie had pointed out that radioactivity must be an atomic property. It seemed that radioactive atoms were capable of emitting other atoms, different from themselves. Hitherto, radioactive substances had been conceived as radiating. Their radiations might come from the interior of the atom or be condensed by the atom from external space. They had been regarded as if they were atomic transmitters, emitting radiations, but the atom itself had not been conceived as undergoing fundamental change.

Rutherford's discovery of emanation shifted the conception of radioactive change from that of radiative to explosive action, to atomic disintegration, with the emission not only of insubstantial rays but of substantial atomic fragments.

Rutherford had been in Montreal less than fifteen months. Besides settling in a new department without previous professorial experience, he had with prodigious industry started a research school and made a major and many secondary radioactive discoveries.

His private correspondence had evidently become less comprehensive, for at the end of 1899, in the midst of these herculean researches, he received a letter from his fiancée commenting that in his last letter he hadn't mentioned whether he was doing any research work. 'My dear girl,' he replied, 'I keep going steadily turning up at the Lab. five nights out of seven, till 11 or 12 o'clock, and generally make things buzz along. I sent off on Thursday another long paper for the press which is a very good one, even though I say so, and comprises 1,000 new facts which have been undreamt of. . . .'

This was the second of his papers on the emanation.

He left for New Zealand early in the summer of 1900, in order to marry Mary Newton, to whom he had been engaged for five long years. The delay was due to Rutherford's desire to pay off the debts he had incurred in order to go to Cambridge from New Zealand, for he had had to borrow the boat fare. He desired also to have established his position and saved enough to start married life in what he conceived to be proper comfort and style. The young couple arrived in Montreal in September 1900. Their daughter Eileen, their only child, was born in March 1901.

Meanwhile, a young Oxford chemist, F. W. Soddy, aged twenty-three, had joined the staff of the Chemistry Department at McGill. Rutherford needed a chemist to help him in the study of the emanation, for it appeared to be a new chemical atom, not a physical radiation. From the beginning of 1901 Soddy devoted the whole of his research time to collaboration with Rutherford. They applied chemical reagents of many kinds to the emanation, and found that they could not get any reaction with it whatever. They concluded in 1902 that it 'is a chemically inert gas analogous in nature to the members of the argon family.'

Rutherford and Soddy found that by chemical treatment they could separate from thorium a small concentrate which appeared to contain nearly the whole of the source of the emanation and the radiations. The radioactivity of the concentrate decayed, while that of the thorium recovered, and the respective rates of decay and recovery were exactly equal.

Now the Curies had independently discovered that radium was able to excite radioactivity in neighbouring objects, and Becquerel independently discovered that most of the radioactivity of uranium could be temporarily removed by chemical separation, but that the rates of recovery of the treated uranium and of the decay of the separated source were the same.

The accumulating results seemed to show that thorium was producing new radioactive substances, which in turn were producing others. This explained why the thorium temporarily lost a large part of its radioactivity, because its radioactive products had been removed. It recovered its full radioactivity as it produced from within itself its usual accompanying stock of radioactive products.

All of the different radioactive substances decayed according to the same law. The rate of decay was proportional to the amount of radioactive material present.

Within two years, Rutherford and Soddy elucidated and confirmed these facts and then gave a connected explanation of them in their famous Spontaneous Transformation Theory of Radioactivity. They published their experimental results and deductions in two great papers in 1902, on the Cause and Nature of Radioactivity. At the end of the first they say that since 'radioactivity is at once an atomic phenomenon and accompanied by chemical changes in which new types of matter are produced, these changes of matter must be occurring within the atom, and the radioactive elements must be undergoing spontaneous transformation.'

They remarked that 'Radioactivity can also be used to follow *chemical changes occurring in matter*. . . . It seems not unreasonable to hope . . . that radioactivity . . . affords the means of obtaining information of the processes occurring within the chemical atom.'

As soon as the nature of radioactivity had been discovered, Rutherford's imagination began to turn from the study of radioactivity itself to the use of it to discover the secrets of the interior of the atom.

Meanwhile he had succeeded in showing that the α-rays could be deflected by electric and magnetic fields. He found that they were positively charged and had a mass of the same order as the hydrogen atom, and he showed that their velocity was about one-twelfth of that of light. Thus it was evident that α-rays were atomic fragments being shot out of exploding atoms like projectiles. The β-rays, which were known to be electrons, 'were a secondary phenomenon, and . . . the α-rays play the most prominent part in the changes occurring in radioactive matter.'

Rutherford persuaded his millionaire patron, Macdonald, to provide the laboratory with a liquid-air machine. Though Macdonald made his money out of tobacco, he hated smoking, regarding it as a filthy habit. When he visited the laboratory, Rutherford used to rush round in advance, telling everyone to put their pipes out and open the windows.

With Soddy, Rutherford used the liquid-air machine to liquefy the emanations. He found that the thorium emanation liquefied at −120° C., and radium emanation at −150° C. Their proof that the emanations were radioactive inert gases, belonging to the argon group, was now complete.

They summarized their theory of Radioactive Change in a more general paper, published in 1903. In this they calculated from the known energies of the α-particles the amount of energy in atoms, that is, the magnitude of atomic energy. They concluded that 'the energy of radioactive change must . . . be at least twenty thousand times, and may be a million times, as great as the energy' in any chemical change, such as the burning of hydrogen in oxygen.

From the rate of output of energy they calculated that one gram of uranium would lose less than a milligram in weight in a million years, and that a gram of radium would lose such an amount in one year. Hence the 'life' of radium, the time it takes to lose half its weight, could not be more than a few thousand years.

'All these considerations point to the conclusion that the energy latent in the atom must be enormous' . . . and 'there is no reason to assume that this enormous store of energy is possessed by the radio-elements alone. It seems probable that atomic energy in general is of a similar, high order of magnitude, although the absence of change prevents its existence being manifested.... It must be taken into account in cosmical physics. The maintenance of solar energy, for example, no longer presents any fundamental difficulty, if the internal energy of the component elements is considered to be available, i.e. if processes of sub-atomic change are going on. It is interesting to note that Sir Norman Lockyer has interpreted the results of his spectroscopic researches on the latter view (*Inorganic Evolution*, 1900) although he regards the temperature as the cause rather than the effect of the process.'

These conclusions were expressed two years before Einstein published the Theory of Relativity in 1905, in which he showed that mass or matter, and energy, are equivalent.

Rutherford and Soddy estimated in this paper that the energy of an α-particle was so great that the electrical methods of detecting it were 'almost equal to the investigation of a single atom disintegrating.' By 1905 he had convinced himself that α-particles were helium atoms with two positive charges.

The great papers of 1902–3 produced a tremendous impression. The ideas and the experiments were easily understood, and their novel implications were thrilling.

Soddy had a notable share in this partnership. He was strong enough in talent to influence Rutherford and, besides chemical skill, he contributed a speculative quality, which lightened Rutherford's exceedingly powerful imagination, relentlessly bent to the demands of experimental procedure. When Soddy left Rutherford, his speculative gift became less disciplined.

Rutherford calculated the age of radioactive minerals from the proportions of their radioactive constituents. By this method he calculated that the age of one of his mineral specimens was 700,000,000 years. This provided an ample time-scale for the theories of biological evolution. He pointed out that the energy produced in the earth by radioactive minerals would tend to raise the temperature, and thus delay the earth's final cooling with the consequent extinction of life, which, according to Kelvin's calculations, would occur within twenty to forty million years if the energy of the earth's heat radiated into space was not replaced by new energy from some unknown process.

In addition to all his laboratory work, Rutherford wrote and corrected his frequent long papers with the utmost care, and he sent most of them to London, to be published in the *Philosophical Magazine*. This involved delays of months in transmission across the 3,000 miles of ocean, when every day counted in the battle for priority with the brilliant French school of Becquerel and the Curies.

Besides giving many special lectures in America, he visited Europe in the summer of 1903 and again in the spring of 1904. At the same time, he pressed forward with his first book, *Radioactivity*, which was published in 1904.

On New Year's Day, 1905, he wrote to his wife: 'I have been amazed when I reflect on the number of things I got through. There was my visit to St. Louis [where he gave a notable paper on Benjamin Franklin's contributions to the physics of electricity], followed by the lectures in the States, the publication of my book, the voyage to England and my lecture there, the publication of the Bakerian Lecture and my paper to the Electrical Congress, also the address at St. Louis, and finally the Rumford Medal, and the award of the Silliman Lectures. It would be difficult to do more in the course of one year.'

At Montreal Rutherford had few technical assistants and never more than a handful of research workers. He did a great deal of work with his own hands.

In 1905 he visited New Zealand again. He delayed his departure for a few weeks, because, as he explained in a letter to his wife, 'it is very important I should write it [his work] up, as they are all following my trail, and if I am to have a chance for a Nobel Prize in the next few years I must keep my work moving.'

Rutherford was now receiving many offers of chairs, especially in the United States. But he felt that 'America as well as Canada is on the periphery of the circle' of world science, and he rejected them all.

In 1906 Schuster enquired whether he would be prepared to succeed him at Manchester, as he was considering an early retirement. Rutherford replied favourably, and the chair was offered to him and duly accepted.

Before leaving for England, he visited California to give some of his many special lectures. He saw the devastation caused by the San Francisco earthquake, and went to stay with Jacques Loeb. He reported that 'I like him very much and I find we have a lot in common. He is very modest but a terrible enthusiast, worse than I am.' As Mrs. Loeb had no maid, Rutherford 'helped wash up the breakfast things while Robertson swept out the dining-room.'

Rutherford repeated in Loeb's laboratory some of the famous experiments on parthenogenesis. 'I first tried the normal method with the spermatozoon, and saw the egg develop in the course of a few hours, I then developed the beggar from the egg till he swam, by adding appropriate chemicals, without calling in the aid of the male.'

After most regretful farewells from his Montreal collaborators and friends, he sailed for Manchester in May 1907.

Marsden has remarked that it was probably very fortunate that Rutherford's first chair was in Canada. He might have been frustrated by the inhibited English social atmosphere if he had been appointed to a chair in England as a young man. When he came to Manchester, his fame was already great enough to break social barriers.

V. THE ATOM

Schuster is one of those scientists whose reputation increases with time. His work on the discharge of electricity through gases was underrated in the 1890s, especially by J.J., who was inclined to work out all ideas and experiments for himself, even at the cost of duplication.

Schuster had generally not quite sufficient rude strength to reap the harvest from his imaginative insight. But his retirement in favour of Rutherford completely fulfilled his foresight and hopes. Besides vacating the chair, at the age of fifty-six, he bequeathed to Rutherford an almost new laboratory (built in 1900), and he endowed a special readership in mathematical physics, which also had great consequences. Schuster had also seen that Rutherford was offered an exceptionally good salary by the university. A. S. Russell states that during his Manchester period Rutherford received £1,600 a year, a very high figure for those days. He had a staff of nine graduates, whose total salary was also £1,600 a year—an average of less than £200 a year, and the bottom four of these received £125 a year. The disparity of British scientists' salaries in 1911 compared with 1951 is striking.

There are some scientists who have earned a noble memory by creating opportunities for others, besides contributing discoveries themselves. Both Newton and Rutherford benefited from such generous vision, for as Schuster resigned his chair for Rutherford, so, two and a half centuries before, Isaac Barrow had resigned his for Newton. J. J. Thomson, too, had received his early opportunity through Rayleigh's retirement, but this had definitely not occurred in order to give a younger man a chance, but because Rayleigh had agreed to occupy the chair only under pressure and then only for a limited time.

Rutherford found that the laboratory at Manchester was very well equipped with apparatus and that there were fifteen research students, many more than he had had in one year at Montreal. He found, too, that Schuster had left his personal assistant, a young German named Hans Geiger, among the laboratory staff, and a 'young, energetic and exceptionally capable' lab. boy named William Kay.

Within three weeks he had got experiments going, and he wrote to his American friend Boltwood that Manchester 'is a pretty active place, and, but for its climate, has a number of advantages—a good set of colleagues, a hospitable and kindly people and no side anywhere. . . . I find the students here regard a full professor as little short of Lord God Almighty. It is quite refreshing after the critical attitude of Canadian students. It is always a good thing to feel you are appreciated.'

Rutherford drew up a list of thirty research topics which he and his research workers might investigate. Several of them dealt with the properties of his favourites, the α-particles. He and Soddy had noted in 1903 that these particles had so much energy that they were almost individually detectable by the electrical methods that had been devised for studying the ionization produced by X-rays.

In that very year, 1903, Crookes, and Elster and Geitel, had noticed that when screens of zinc sulphide were exposed to beams of α-particles, brilliant little scintillations could be seen where an α-particle hit the screen. Rutherford commented on this discovery, pointing out that though undoubtedly α-particles could produce scintillations, one did not know whether all the α-particles emitted by a radioactive source would do so. He thought this unlikely, and concluded that the method would be unreliable for quantitative investigations.

Instead of investigating the reliability of zinc sulphide screens for revealing every α-particle that hits them, Rutherford ignored the new technique. His mind was bent on his own idea of a method of electrical detection of single α-particles.

Soon after arriving at Manchester he and Geiger started on this problem. They devised a partially exhausted tube, with an electric wire down its centre, which could be kept at high voltage, just below sparking-point. α-particles could enter the tube through a thin window of mica. Each time one entered, it caused sufficient ionization, or electrical conductivity, in the residual air or gas to enable a sudden leak of electricity from the highly charged central wire to occur. This could be made to move a spot of light, so the entry of each α-particle was visibly recorded.

Rutherford and Geiger were at great pains in designing this counter to avoid pointed wires within the tube, in order to evade possible irregularities caused by electric discharges from the points.

With this counter, Rutherford and Geiger deduced that one gram of radium emitted 34,000,000,000 α-particles every second. This actual specification of matter, atom by atom, was the most concrete that had yet been given of the reality of the atomic theory of matter.

Rutherford then directed a counted number of α-particles into an insulated vessel, so that their collective charge of electricity could be measured. This done, he divided the result by the number of particles. When divided by two (because the α-particle has two charges) the size of the fundamental unit charge of electricity (equal to the size of the charge on the electron) was obtained.

Rutherford's result was $4{\cdot}65 \times 10^{-10}$ electrostatic units. This was some thirty per cent larger than earlier determinations made from experiments on electrons. But it was nearly equal to the figure calculated eight years before by Planck on the basis of his quantum theory.

Rutherford, up to this time, was strongly prejudiced against the quantum theory. He thought in mechanical images of the simplest type, and, indeed, years after he was complaining that the quantum theorists 'are quite content to explain everything on a certain assumption, and do not worry their heads about the real cause of the thing.'

He had similar early prejudices against Relativity. Wien, after being playfully provoked by him, said that Anglo-Saxons couldn't understand Relativity. 'No,' replied Rutherford, 'they've got too much sense.'

Of course, Rutherford was very much in the wrong here. But it was typical of Rutherford that he instantly flung away prejudices if these theories supported his own experiments. When he found his 4·65 was confirmed by Planck's result, he is reported to have said: 'There's more in old Planck than meets the eye.'

While Rutherford and Geiger developed the electrical counter, Regener in Berlin had taken up the neglected zinc sulphide scintillation method and shown how it could be used reliably. Rutherford heard the news with surprise, but he and Geiger immediately compared the improved method with their own electrical method, and found that both gave the same results.

Rutherford then threw over the electrical method, to which he had clung so tenaciously, and enthusiastically adopted the method that he had formerly spurned. He did not return to his own electrical method for some twenty years, and then only when others had introduced radio valves and counters which could work automatically at a wide range of speeds.

The scintillation method was much simpler than the electrical, but it was very tiring to the eyes. However, Geiger, who had 'worked like a slave' with the electrical counter, now proved 'a demon at the counting of scintillations and could count at intervals for a whole night without disturbing his equanimity.' Rutherford, on the contrary, 'damned vigorously and retired after two minutes.' But he grew to love his dark room in the basement, with hot-water pipes over which one tripped on the floor, and others under the ceiling, against which one bashed one's head.

The scintillation method was now applied to the investigation of the scattering of α-particles. Rutherford had noted in 1906 that when a beam of α-particles was directed on to a photographic plate, the image was much more diffused when the photograph was taken in air than when the apparatus was exhausted and the photograph made in a vacuum. Evidently the air had slightly spread out, or scattered, the α-particles. He found that some of the particles were being deflected about 2 degrees while traversing a distance of about three-thousandths of a centimetre. As the α-particle has enormous energy, a very great force acting over that distance would be necessary to produce even 2 degrees' deflection, in fact, he calculated, it would be about 100,000,000 volts per centimetre. Such were the electric fields of which Rutherford was speaking in 1906, while still at Montreal.

Geiger now examined the scattering quantitatively, exactly counting what percentage of particles were deflected and recording the exact size of the deflections. He found that the average deflection increased with the atomic weight of the material of the foil.

Meanwhile, in 1908, Rutherford was awarded the Nobel Prize for Chemistry. He was, of course, a physicist. In his speech at the banquet in Stockholm after the prize-giving, he said that he had dealt with many swift transformations in his time, but the swiftest of all was his own transformation from a physicist into a chemist.

Immediately after his return to Manchester with the Nobel Prize he attended as a student Professor Horace Lamb's class on the mathematical theory of probability. He wanted to learn the general ideas of the mathematical methods for drawing the most exact conclusions from the statistics of the counting of α-particles and their deflections.

In addition to this main interest, he made a very beautiful experiment, with the collaboration of Royds, which gave the direct proof that α-particles were helium atoms. They fired the α-particles through the walls of a very thin glass tube. Then they examined the spectrum of the gas inside the tube. The lines for helium gradually appeared, and grew stronger and stronger, as the bombardment with α-particles continued. Evidently, the α-particles, stopping inside the tube, formed themselves into helium gas.

Geiger was joined in his scattering experiments by a young research student, Marsden, aged eighteen. They made a very remarkable observation. They found that a number of α-particles shot at very thin gold leaf were turned through a sharp angle and sometimes almost directly reversed.

Now it had been supposed that occasional large deflections were built up from a series of small deviations, owing to several successive encounters with atoms in the gold leaf. It followed from the laws of probability that the chance of the occurrence of such a series of encounters was very small. But Geiger and Marsden found there was a relatively large number of these deflections. Hence most of them must be occurring, not through a succession of small deviations, but through a complete big deviation, occurring through one encounter with a single atom.

Now a single atom could cause the recoil of an α-particle only if it contained an intense concentration of force strong enough to fling back with comparable force the enormously energetic α-particle. Such an intense concentration must be due to a nucleus with a big electric charge and high mass.

This conception was contrary to the current trend of thought on atomic structure, which was dominated by J.J.'s picture of the atom as an internal group of electrons surrounded by a sphere of positive electrification. J.J., the first to prove conclusively the existence of the electron, had shown that electrons could be obtained from any kind of atom, and so was the first to prove by direct experiment that atoms have a structure. His school, by experiments on the scattering of X-rays and particles, had made general estimates of the number of electrons in various atoms, and had collected very important results in conformity with J.J.'s picture.

Rutherford had the difficult and delicate task of disagreeing with his old master on the fundamental picture of the atom, and showing that the experimental scattering data, apparently supporting J.J.'s views, could be equally well explained on his own theory.

In addition to this difference with authority and tradition, Rutherford also had to disagree with the laws of classical mechanics, for, according to them, his nuclear atom could not be a stable mechanism. Without being deflected from his search Rutherford pressed forward with his analysis. He recalled the Newtonian theory of gravitational attraction, which shows that, when one gravitating particle approaches another with a velocity greater than the 'velocity of escape' (e.g. the velocity with which a stone would have to be thrown to escape from the earth), then it sweeps before the nucleus in a hyperbolic curve.

He calculated what the orbits would be if the approaching particle were an α-particle and the approached nucleus had a charge 100 times that of an electron, i.e. about equal to that which was to be expected on the nucleus of an atom of gold.

Rutherford found that he could give a general explanation of Geiger and Marsden's observations on the actual tracks, or orbits, of α-particles scattered by gold leaf. His calculation showed that the α-particle must approach within a distance of about 3·4 million millionths of a centimetre from the centre of the nucleus. As the radius of the atom was supposed to be about one hundred millionths of a centimetre, the diameter of the atom must be about seven thousand times that of the nucleus.

Thus the atom was an extremely empty, spacious structure. The relative lack of influence of the negative charges in the atom suggested that they were distributed over the extensive volume around the very tiny, highly charged massive nucleus.

The number of electrons outside the atom was exactly equal to the number of positive charges on the nucleus. This explained why atoms in ordinary matter are uncharged. The negative and positive charges in the atom exactly balanced.

It was surprising that the nucleus, which contained many charges and therefore many particles, was smaller than the simple electron. As disintegrating radioactive atoms emitted α-particles and electrons, it seemed that some nuclei, at least, might consist of groups of these two kinds of particles somehow bound together in a very small volume.

Rutherford established his new model of the atom in 1911. It was his greatest achievement, and has been the basis of the immense science of Nuclear Physics, that is, the physics of Rutherford's nuclear atom.

Rutherford was not the first to propose a nuclear atom. Newton himself had already discussed such a conception, as S. I. Vavilov has pointed out, and he even deduced that the physical properties of matter might depend on the nucleus and the chemical on the outer part of the atom. Lenard showed by his studies of the passage of electrons through thin metal foils, begun in 1894, that atoms must be spacious structures, with impenetrable centres in them. He was not, however, able to give precise form to his ideas, because the data provided by the scattering of electrons is too ill-defined. Nagaoka also proposed a nuclear atom in 1904.

The idea of a nuclear atom is very simple. Little originality is needed in propounding it. The merit is in proving that it is true, and this belongs overwhelmingly to Rutherford.

Geiger has described how one day before Christmas in 1910, Rutherford came into his room, saying that now he knew what the atom looked like. C. G. Darwin, who was his Reader in Theoretical Physics, recalls how he was his guest for a Sunday supper at his house in Manchester, and happened to arrive half an hour after Rutherford had proved that his nuclear model was correct. Speaking of his proof of the nuclear theory Rutherford said that 'it is really very fine to see the things one has seen in imagination visibly demonstrated.'

Then, in April 1912, Niels Bohr, a young Danish physicist and son of the professor of physiology at Copenhagen, 'pulled out of Cambridge', as Rutherford put it. He had not been fortunate in his relations with J. J. Thomson. Rutherford put him on to experimental work, but he very soon changed over entirely to his application of quantum ideas to the mechanism of the Rutherford atom. After a few months, he went back to Copenhagen and worked his ideas out. He returned to Manchester later, and became reader in theoretical physics.

Rutherford at once perceived the quality of Bohr, and was sympathetic to him, though at first 'the mixture of Planck's ideas with the old mechanics makes it very difficult to form a physical idea of what is the basis of it all.' But the explanations of the hydrogen spectrum were 'very ingenious and seem to work out well.' Within a short time, Bohr found the key to the enormous complications of the spectra, and the general features of the periodic table of the chemical elements, by means of his Rutherford atom, surrounded by electrons moving in orbits according to quantum rules and jumps.

As soon as Rutherford saw that Bohr's quantum atomic theory would explain why his atom was stable and how it could work, he accepted it absolutely, and, in this particular connection, all classical Newtonian prejudices were abolished, as they always were when they were contrary to his own intellectual interests.

The collaboration between Rutherford and Bohr was one of the most brilliant, fertile and fortunate in the history of science, and particularly valuable because Rutherford's conscious classical prejudices might easily have betrayed his genius if there had not been a Bohr to show him the truth and fertility of quantum ideas.

Schuster saw his Readership in Theoretical Physics held in succession by Bateman, Darwin, and Bohr. Geiger returned to Germany in 1912, and then devised a new type of counter which would detect single α-particles or electrons. It depended on the use of a charged needle, and Rutherford wrote that it was 'rather surprising that the pointed conductor, which we tried to avoid so much, should prove after all the best thing.'

In addition to all these excitements, Rutherford had in 1910 engaged H. G. J. Moseley of Oxford, before he had graduated. Moseley was descended from fellows of the Royal Society on both sides of his family and was an old Etonian. He did not obtain a first-class degree, but Rutherford nevertheless retained him.

Moseley immediately made elegant experiments in radioactivity. Then, in 1912, Laue, Friedrich and Knipping announced their discovery that X-rays could be diffracted by crystals, and were therefore a form of wave-motion. W. L. Bragg, a young student at Cambridge, immediately showed how the wavelength of X-rays could be deduced from the diffraction patterns formed by their reflection from crystals.

Moseley seized on this method and applied it to measure exactly the wavelengths of the 'characteristic' radiations of the elements. Barkla had shown that when atoms are excited they emit radiations which are characteristic for the element to which they belong, and increase in penetrating power with the atomic weight; thus the characteristic radiation of gold is much more penetrating than that of aluminium. Barkla's discovery gave very important but imprecise indications about the electronic constitution of the atom. When Moseley applied the crystal-analysis method, he obtained a very simple clear pattern of lines, which showed that the system of characteristic radiations for all elements was the same, only with regular differences in wavelengths from element to element.

Moseley found that, in general, the wavelengths could be explained by Bohr's quantum theory of the Rutherford atom, on the assumption that the atoms in the periodic table were arranged in ascending order of nuclear charge: hydrogen with charge one, helium with charge two . . . up to uranium with charge ninety-two.

Thus the whole of the chemical properties of the elements was dependent on an integral number of electrical charges, not on a mass which varied irregularly from atom to atom, like the chemists' atomic weight. The number of the charges on Rutherford's nuclear atom was given an enormous extension of precision and implication. After his main results had been established, Moseley left Manchester in 1913 for Oxford, where he completed his brilliant investigation.

Rutherford had based his nuclear theory of the atom on the occurrence of violent atomic collisions, which Geiger and Marsden's discovery of the rebound of α-particles from metal foils had revealed. If an α-particle could rebound with enormous velocity, then the atom from which it rebounded must have been knocked forward with great velocity, like one billiard ball struck by another.

Marsden began a systematic experimental investigation in order to discover the rules of this atomic billiards. When a swift α-particle was fired into helium gas, consisting of relatively stationary α-particles surrounded by their pairs of electrons, the impinging α-particle exchanged its momentum with the struck α-particle, so that the latter shot forward with the velocity of the impinging particle, while the latter was reduced to rest.

When a hydrogen atom, one-quarter the mass of the α-particle, received a head-on collision, it was to be expected that it would shoot forward with a velocity 1·6 times that of the α-particle and with 0·64 of its energy. The struck hydrogen atom should, therefore, have enough energy to produce a scintillation on a zinc sulphide screen, and hence its range could be found.

Marsden fired α-particles into hydrogen, and found this to be true. Some of the knocked-out hydrogen atoms had a range of more than 1 metre, four times that of an α-particle that was knocked-on. Marsden noticed that some of these long-range hydrogen atoms seemed to come from the radioactive source providing the α-rays. This raised an extremely interesting question, for if it were true, there must be a fourth kind of radioactive radiation in addition to the α, β, and γ-rays. At this stage Marsden was appointed professor in New Zealand, and left Manchester early in 1915.

In the midst of this intense research activity, the First World War broke out. Rutherford and Moseley were at the time in Australia for a meeting of the British Association for the Advancement of Science.

The research workers of the Manchester laboratory were soon engaged in military activities. Niels Bohr returned to Denmark. Moseley enlisted as a signals officer, and was killed in August, 1915, in Gallipoli, while his brigade was trying to carry out an apparently foolhardy operation. Rutherford had already asked the authorities to transfer him to military scientific work, but official action was too late. The death of Moseley was one of the events which made the British authorities much more careful of their scientific talent in the Second World War.

There were many other deplorable scientific casualties. Among those on the German side was Reinganum, who had first shown the possibility of recording the tracks of α-particles in photographic films, a technique which, a generation later, has been developed by Zdanov, Powell, and others with remarkable results.

In 1915 Rutherford joined the Admiralty Board of Invention and Research, and started on the problem of submarine detection and destruction. He worked with great energy and enthusiasm, making experiments on the transmission of sound in water in the Manchester laboratory and at special experimental stations. But the results were not really successful. Instruments made in a laboratory workshop would work in the hands of scientists, but not when mass-produced and used by seamen.

The effective answer to the submarine was developed by his old Cavendish fellow-student, Langevin, who in France showed that beams of ultrasonic waves produced by piezo-quartz vibrators could be successfully used to locate submarines.

Rutherford returned to his empty laboratory in July, 1917. He had asked Marsden whether he would mind his pursuing that fascinating problem of whether atoms of hydrogen were shot out of some radioactive substances, and Marsden agreed, as he was engaged in war work.

Rutherford began a systematic study of the 'range of high-speed atoms in air and other gases' on September 8th, 1917. Sitting in the dark room, in the empty laboratory, with no one except his laboratory assistant, William Kay, he steadily counted the scintillations produced when atoms in air were knocked-out and hit the zinc sulphide screen. Within three days he noticed peculiar faint scintillations, which looked as if they were caused by hydrogen atoms. In fact, there seemed to be twice as many of these faint scintillations as was to be expected. He and Kay strained and rested their eyes in the dark, as they counted the faint flashes of light, and checked and confirmed each other's observations. Rutherford was forty-six years old, and his eyes had already shown the usual signs of ageing.

Where were those particles causing the extra faint flashes coming from? Which constituent of the air was their source? By September 28th he had found that the bombardment of nitrogen produced twice as many as oxygen. He then thoroughly examined all possible sources for the extra scintillations, proving conclusively that they could not have come from dirt or moisture.

Then, on November 9th, he tried the experiment with chemically prepared nitrogen, and found there were 25 per cent more of the faint scintillations than with air. There was no longer any doubt that the particles were coming from nitrogen, so he set out to discover exactly what they were.

On January 10th, 1918, he wondered whether the faint flashes were due to hydrogen atoms of mass 2 (Urey subsequently discovered such atoms of 'heavy hydrogen' in 1932).

Finally, early in 1918, he concluded from extensive comparisons of the range and scintillations of the particles coming from nitrogen with those produced by hydrogen atoms knocked-on by α-particles, that these particles coming from nitrogen were atoms of hydrogen of either mass 1 or mass 2.

He wrote a complete description of this war-time research in four papers, which were published in 1919. He said that it was difficult to avoid the conclusion that hydrogen atoms had been obtained from atoms of nitrogen, and that if this were true, then 'the nitrogen atom is disintegrated under the intense forces developed in a close collision with a swift α-particle, and (that) the hydrogen particle which is liberated formed a constituent part of the nitrogen nucleus.'

So the artificial disintegration of the atom had been achieved—while he was scientifically alone and single-handed, using experimental methods in his middle age which demanded the utmost physical attention to be expected only of a young man, when his laboratory was empty and the stimulus of a brilliant group of scientific collaborators was absent, and the depression of war hung over all spirits.

The discovery was one of the greatest in the history of science, and the circumstances of its achievement increased its glory.

No wonder that he strayed back from anti-submarine researches. Bumstead has recorded how Rutherford, after an occasion when he had failed to attend a meeting of officials concerned with war research, was asked why he had been unable to come. 'I have been engaged in experiments which suggest that the atom can be artificially disintegrated,' he replied. 'If it is true, it is of far greater importance than a war.'

In 1918, J. J. Thomson was appointed Master of Trinity College, Cambridge. Rutherford was offered the Cavendish professorship, which he accepted, and he moved to Cambridge in 1919.

VI. THE CAVENDISH LABORATORY

When J.J. became the Master of Trinity College, he resigned the Cavendish professorship, but he had not originally intended to resign also the directorship of the Cavendish Laboratory. Rutherford obviously could not become professor without becoming director of the

Laboratory, and he wrote in his straightforward way to J.J. saying that 'no advantages of the post could possibly compensate for any disturbance of our long-continued friendship or for any possible friction, whether open or latent, that might possibly arise if we do not have a clear mutual understanding with regard to the Laboratory and Research Students.'

His method of running the laboratory was different from J.J.'s. Rutherford definitely directed research, putting men on to different aspects of a general problem and taking them off if they were not succeeding. J.J. inspired others, suggesting problems to them, and then left them to find their own way of working them out.

J.J. was persuaded to resign the directorship also, and was appointed special professor, with research rooms and assistants of his own. The delicate situation of pupil succeeding master, complicated by the rivalries over the structure of the atom, was successfully resolved.

Rutherford swiftly reorganized the Laboratory for research on radioactivity. He brought Chadwick from Manchester and appointed him assistant director of research, to reduce the directive burden. Chadwick had spent four years as a prisoner of war at Ruhleben in Germany, but had been able to continue some scientific work. With him in Ruhleben had been a regular army officer, C. D. Ellis, who had scientific interests. When Ellis was released he came to work in Cambridge, and became a physicist. A naval officer, P. M. S. Blackett, entered Cambridge for a 'holiday course' after the war, and stayed to become a physicist. F. W. Aston, E. V. Appleton, and many others returned to Cambridge from war-service.

In his characteristic way, Rutherford at once got research going. He confirmed his disintegration of nitrogen, proving conclusively by magnetic deflection that the particles emitted from it were atoms of hydrogen. Then, in 1920, he reviewed his work on atomic disintegration and discussed its implications. It appeared that the nuclei of atoms contained electrons and particles of hydrogen. How could these be put together to form other nuclei? 'It seems very likely that one electron can also bind two hydrogen nuclei, and possibly also one hydrogen nucleus. In the one case, this entails the possible existence of an atom, of mass nearly 2, carrying one charge, which is to be regarded as an isotope of hydrogen. In the other case, it involves the idea of the possible existence of an atom of mass 1 which has zero nuclear charge. Such an atomic structure seems by no means impossible. . . . Such an atom would have very novel properties. Its external fields would be practically zero, except very close to the nucleus, and in consequence it would be able to move freely through matter. Its presence would be difficult to detect by the spectroscope, and it may be impossible to conceal it in a sealed vessel. On the other hand, it should enter readily the structure of atoms, and may either unite with the nucleus, or be disintegrated by its intense field, resulting in the escape of a charged hydrogen atom, or an electron or both.'

Here he forecast the possible existence of heavy hydrogen and the neutron, and in the same lecture he also discussed the possibility of the existence of hydrogen and helium atoms of mass 3. These prophecies were all fulfilled. But their originality should not be exaggerated. Harkins suggested slightly earlier the possibility of the existence of neutrons and J. J. Thomson held that he had discovered hydrogen of mass 3 by his parabola method.

Rutherford's forecast of the neutron proved to be of great importance, not because of its ingenuity, but because it familiarized the Laboratory with the idea of a neutron.

Rutherford named the hydrogen nucleus the proton in 1920. He chose this title partly out of regard for Prout's hypothesis, that all atoms are built out of one fundamental stuff, which was hydrogen; and also to avoid identifying it with the positive electron, which, he remarked in a lecture in 1922, might be discovered in the future, though there was as yet no evidence whatever of its existence.

Meanwhile Aston had got his mass-spectrograph to work separating isotopes, i.e., nuclei with the same charges but different masses. He showed that neon, chlorine, mercury and most of the ordinary elements consisted of mixtures of atoms of different integral masses. The atomic weight measured by the chemist was an average weight of the atoms in these mixtures. Aston's work introduced a brilliant simplification into the knowledge of the elements, for it showed that all atoms had masses very close to whole numbers, and the slight differences from whole numbers were a measure of the amount of energy needed to bind the particles in the nuclei together.

Rutherford embarked with Chadwick on a systematic study of the artificial disintegration of the lighter elements. They soon succeeded with aluminium, phosphorus and fluorine, and in general succeeded in disintegrating those with odd, but not with even, atomic numbers.

In 1921 searches were already being made for the neutron and artificial radioactivity. It was thought that the proton and electron might combine together, under the influence of strong electric discharges, and form neutrons, which might be accompanied by the emission of heat. But the experiments were unsuccessful.

Under Rutherford's inspiration, the Cavendish Laboratory was seething with ideas of what might be found. He inspired the development of technical methods for pursuing the search. His Japanese pupil Shimizu devised an automatic type of Wilson cloud chamber, which reduced the labour of photographing large numbers of cloud tracks of atomic particles. He arranged that the cloud tracks should be photographed simultaneously from two directions, so that their direction in space could be seen stereoscopically.

In 1921 a young Soviet physicist, P. L. Kapitza, arrived at the Cavendish Laboratory, and at first collaborated with Shimizu. Then he independently suggested that tracks of α-particles in Wilson chambers might be bent by a very powerful magnetic field obtained by short-circuiting accumulators. The experiment was tried and was a success. Kapitza produced fields of 80,000 gauss in 1923, and obtained photographs of heavily curved α-ray tracks.

He had contributed a new technique of intense magnetic fields. Blackett introduced a still more effective type of automatic cloud chamber, and began a systematic study of the appearance of the cloud tracks, drawing up rules for their interpretation so that the maximum amount of information could be inferred from them. This research was based on the study of 440,000 tracks, and was a fundamental contribution to the use of the Wilson chamber and the technique of atomic physics.

A spectacular outcome of this work was his exquisite photograph, in 1925, of the interaction of an α-particle with a nitrogen atom, leading to the emission of a proton and the synthesis of an oxygen nucleus. Striking as this was, however, Blackett's laborious study of cloud-tracks was probably more important, for it helped to turn the Wilson chamber from a qualitative instrument into one capable of giving reasonably precise quantitative information.

In 1924, Chadwick was still thinking about the neutron, and wrote to Rutherford that 'we shall have to make a real search' for it.

The rapid increase in detailed knowledge of atoms and nuclei provided the basis for theoretical advances. These all started outside Britain, and showed the weakness of the British theoretical school. They were coordinated in Copenhagen, in the theoretical school led by Bohr, with great contributions from Paris, where Louis de Broglie in 1924 proposed the wave-theory of matter; from Göttingen in 1925, where Heisenberg proposed the new quantum mechanics, based on the mathematical description of observable quantities only, and the rigid exclusion of unobservable hypothetical entities; and from Zurich in 1926, when Schrödinger proposed the new wave-mechanics.

There was no major Cambridge theoretical contribution until Dirac showed in 1928 how to combine the new quantum mechanics with the theory of relativity, and discussed the meanings of the negative roots of his relativistic form of the wave-equation. What could an electron in a state of 'negative energy' mean? He interpreted it as a 'hole' in the universe, consisting of the absence of a negative electron. He showed that such a 'hole' would behave like a positive particle, and as the positive electron was unknown, he tried to identify the 'hole' as a proton. It was shown, however, that the 'hole' must behave as if it had the same mass as an electron.

When the positive electron was subsequently discovered, Dirac immediately calculated from his equation what some of its properties would be, explaining observations and guiding the experimentalists to new experiments.

Apart from Dirac, no major theorist appeared in Cambridge in Rutherford's time. Dirac himself was an extension of the Continental school of theorists to Cambridge, rather than a product of a native British school of theorists. It seems that British physicists during the last hundred years have been too indisciplined to produce a school of great theorists.

Rutherford concentrated entirely on the experimental sphere. He said in 1925 that the theorists 'play games with their symbols, but we, in the Cavendish, turn out the real solid facts of Nature.' Rutherford behaved with deep consideration towards great theorists such as Bohr and Dirac, but he tended to regard lesser theorists as intellectual servants of the experimentalists. This was a wholesome corrective to the mathematical playing-about, or *spielerei*, with formulae, the vice of the theorist, which sometimes degenerates into enervating speculation and mental arrogance. But it did not stimulate the growth of a school of theoretical physics comparable with that of the Cavendish experimentalists. Perhaps it was not possible to have two such great schools simultaneously in the same place, and the greatness of the Cavendish school could be secured only by a concentration on experimental physics, while theoretical physics was in the main developed abroad, in Continental Europe.

Rutherford applied his technique of atomic disintegration to acquire information about the structure of the nucleus. He was deeply impressed by the fact that α-particles of 10,000,000 electron volts energy shot at uranium nuclei are incapable of getting inside, while α-particles of 5,000,000 electron volts energy are able to escape. Evidently there must be some kind of barrier, an electrical potential barrier, inside the uranium nucleus, of at least 10,000,000 electron volts, through which, nevertheless, 5,000,000 electron-volt α-particles could escape.

Gurney and Condon, and Gamow, in 1927–28, showed that this paradox could be explained by the new wave-theory of matter. According to this theory, the position of a particle at any moment is defined by a statistical law, and can be stated only as a mathematical probability. There is therefore a chance, the size of which can be calculated, that the α-particle may at a particular moment be outside instead of inside the potential barrier in the nucleus. In that case, it escapes. Thus the mechanics of the nucleus is in this respect quite non-Newtonian.

Rutherford welcomed the new theory, and tried, with his colleagues, to picture the nucleus as a vibrating system of particles, emitting from time to time α- and β-particles, and in the course of intra-nuclear rearrangements changing from one excited state to another, emitting energy in the form of γ-rays or wave-radiations.

By 1929 the nucleus had lost its Rutherfordian simplicity, and the

Cavendish work in this direction exhibited an uncertain intellectual groping. Rutherford warmly welcomed in this year the news from Paris that Rosenblum had succeeded in resolving the α-rays into many different groups of uniform velocity, that is, he had discovered their fine structure. He received this as an important step towards the clarification of the confusion in the picture of the nucleus. Rosenblum's success had been dependent on the use of the great electromagnet of the French Academy of Sciences, designed by Aimé Cotton.

In Cambridge Kapitza had concentrated on producing bigger and bigger fields. He designed a dynamo which gave a momentary current of 13,500 amperes at 1,000 volts, producing a field of 270,000 gauss. Kapitza was assisted in this work by J. D. Cockcroft, an electrical engineer from Manchester, who had come to Cambridge to graduate in applied mathematics after the First World War. He had served his apprenticeship in the works where Kapitza's machine was made, and was able to help in many ways.

Kapitza went on to design new forms of liquefiers for hydrogen and helium, in order to make experiments with his very strong magnetic fields at very low temperatures.

After assisting Kapitza, Cockcroft helped Ellis to design a special magnet for repeating and extending Rosenblum's resolution of the beams of α-particles. Wilson-chamber studies had been concerned with the properties of straight cloud-tracks, of the amount of ionization along them, and their lengths, and the measurement of the energy-exchanges in the collisions of atoms, as deduced from the ionization along their tracks.

The application of magnetic fields for bending the tracks in cloud chambers was first introduced systematically by Skobeltzyn in Paris and Leningrad.

Blackett began to transfer his interest and technique from atomic physics to cosmic rays. He improved upon Skobeltzyn's method by designing a cloud chamber which enabled the cosmic ray to take its own snapshot, as it were, while being deflected by a magnetic field. The chamber operated only when the cosmic ray went through it. This introduced a big economy in cloud-chamber cosmic ray research. Blackett had emancipated himself from the purely Rutherford tradition, and was beginning to strike out on his own characteristic line of work and thought.

About 1929–30, there were signs that the pure line of Rutherfordian research was ending. But one more glorious burst of discovery was impending. It was not the result of his immediate work, the intellectual return from current research, but rather the result of investments of ideas made long ago, and suddenly coming to fruition, while he himself was engaged in other things.

Ever since he had first realized that the rebounding of α-particles, discovered by Geiger and Marsden, was due to collisions between atomic nuclei, the violence of atomic encounters was vividly in his mind. He at once began to think of the possibilities of producing artificial atomic projectiles even more energetic than α-particles. As early as 1913 he was saying that 'it is a matter of pressing importance at the present time to devise electrical machines of the highest possible voltages.' At Cambridge he had this on his programme. T. E. Allibone and E. T. S. Walton started on the design and construction of high-voltage machines. Cockcroft also began to construct a machine for producing very high-voltage direct current, while he was still primarily assisting Kapitza.

Such was the situation in the Cavendish Laboratory about 1930. There was intense activity, and never were so many gifted experimental physicists confined in such a small space. At times all of Rutherford's binding force was needed in order to hold this highly excited and almost spontaneously explosive nucleus of discoverers together.

Very able men were competing for a portion of the modest resources of the Laboratory, and some seemed, in the opinion of their colleagues, to get far more than their share. Nor did it appear that those who had discovered most always received the largest grants for apparatus. Tempers were not improved when Rutherford replied, in answer to expostulations, that expensive apparatus was not necessary in order to make discoveries. He had a tendency to give excessive consideration to some research workers. Perhaps this was due partly to psychological reasons. He had had no son and his only daughter died in 1930. It may be that in these instances Rutherford was transferring some of his very strong parental feelings.

Rutherford was prejudiced in favour of men from the Dominions, which could easily be understood. He was rather prejudiced against Continentals, and even when he was President of the Academic Assistance Council, he did not specially encourage the many brilliant refugees to settle in the Cavendish Laboratory. Consequently, most of these went on to America. If they could have been settled in Britain, it is possible that theoretical physics in Britain would have flourished much more strongly and the British position in the development of atomic energy might have been much stronger than it is today.

Rutherford's boyish feelings of rivalry towards other branches of physics had some unfortunate consequences. His prejudices against the quantum theory were circumvented by the interposition of Bohr. The Theory of Relativity did not impinge closely on his field. He was not enthusiastically in favour of inviting Einstein to accept a position in Cambridge after 1933.

His prejudice against X-ray crystal analysis had more serious results. He fiercely opposed a very constructive suggestion that a new chair and department of X-ray crystal analysis should be created in Cambridge. History has not confirmed his judgment in this matter. At dinner in Trinity one night, he found a physicist from abroad sitting beside him. 'What's your subject?' he asked. 'X-ray analysis,' replied one of its founders. 'That's not fundamental physics,' Rutherford remarked.

Besides directing the Laboratory he had an increasing burden of public duties. He was President of the Royal Society from 1925–30. In the latter year he was awarded the Order of Merit, and in 1931 a barony was conferred on him. In 1930 he became Chairman of the Advisory Council of the Department of Scientific and Industrial Research. He performed the duties associated with these positions thoroughly and conscientiously.

In addition to all these, he gave innumerable special lectures and presidential addresses to societies. His stamina was very remarkable.

In the midst of all this comprehensive life, news came from France that Irène and Frederic Joliot-Curie had observed a remarkable phenomenon which they interpreted as due to a new interaction between waves and matter. They had found that the radiations discovered by Bothe and Becker, which are produced when beryllium is bombarded with α-particles, were able to knock large numbers of atoms of hydrogen out of substances such as paraffin wax, which contains a great deal of hydrogen.

Bothe and Becker's rays were of great penetrating power, and were supposed, on analogy with γ-rays, to be wave-radiations. The Joliot-Curies accepted this view, but they found that the energy of the atoms of hydrogen knocked forward was far larger than could be explained by the orthodox theory of the interaction between waves and particles. They therefore concluded that the interaction between Bothe-Becker rays and particles was governed by a new law.

With the long tradition of the search for the neutron behind him, Chadwick at once suspected that the hydrogen atoms were being knocked forward by neutrons, not waves. He swiftly made a number of experimental measurements which showed that all the observations could be reconciled on the orthodox law of interaction if the particles were neutrons of very nearly mass 1.

Rutherford found his forecasts of the properties of the neutron in 1920 substantially confirmed. There was glorious enthusiasm in the Cavendish Laboratory. Feather immediately showed that neutrons had, as Rutherford had expected, unusual powers of penetrating and disintegrating nuclei. At this time Rutherford daily telephoned the Laboratory after midnight for the latest results.

Very soon Heisenberg recast the theory of the constitution of the nuclei of atoms in terms of combinations of protons and neutrons.

While the excitement was at its height, Cockcroft and Walton announced their successful disintegration of atoms of lithium by protons accelerated in a steady electric field of 600,000 volts.

Rutherford organized scientific field-days for his triumphant captains at meetings of the Royal Society. Two of the most important experimental discoveries of modern times had been made within six weeks.

In September 1932, C. D. Anderson published a single photograph of a Wilson-chamber picture of the track of a particle that had lost speed in passing through a lead plate. The track had been curved by a magnetic field in a degree and a direction which showed that it was probably due to a particle with the mass of an electron, but bearing a positive charge. Early in 1933, Blackett and Occhialini published a paper containing numerous tracks of positive electrons, and of pairs of positive and negative electrons, and of whole groups of tracks, which Blackett named 'showers'.

Blackett immediately identified Dirac's 'holes' with positive electrons. He explained the electron pairs in showers as due to photons or waves which were converted into material particles in the neighbourhood of the nucleus of an atom.

Chadwick, Blackett and Occhialini confirmed this by experiments with γ-rays from radioactive sources. Thus the phenomenon of the 'materialization' of wave-radiation was established. Similar results were obtained almost simultaneously by Anderson and by the Joliot-Curies. Dee and others made splendid cloud-chamber photographs of the atomic disintegrations by accelerated particles.

In 1936 Chadwick and Goldhaber demonstrated for the first time the disintegration of an atom by wave-radiation, when they split a nucleus of heavy hydrogen into a proton and a neutron by bombardment with γ-rays.

The height of achievement in 1932 necessarily soon passed, for many reasons. Rutherford was sixty-one years old in that year: there was no parallel in the history of science for this degree of scientific achievement at this age. The gifted men that assembled in Cambridge after the First World War were maturing, and looking for their own laboratories and fields of work. Blackett left in 1934; Chadwick in 1935; Oliphant in 1937.

The rise of Hitler in 1933 destroyed the old international unity of the scientific world. Thousands of persecuted scientists and scholars left Germany. Rutherford accepted the presidency of the Academic Assistance Council, in order to help those who had come to Britain. He took the chair at a huge meeting in the Albert Hall, to start a campaign for support and funds, at which Einstein made the first public speech and appearance in his life.

Rutherford had secured considerable funds from the Royal Society and the Department of Scientific and Industrial Research for the building of a magnetic and low-temperature laboratory for Kapitza. He had supported, too, the award of a Royal Society professorship to Kapitza. During 1934, Kapitza made one of his periodical visits to his native land. The Soviet authorities, in view of the threat caused by the rise of Nazi power, could no longer approve his absence from the Soviet Union, which had already lasted thirteen years. They felt that it was now absolutely essential that those who were needed by their country should stay and work at home. They were unable to grant him permission to return to Cambridge.

There were some hard feelings at first, but everyone concerned behaved with dignity and judgment. Rutherford did not attack the Soviet Union, but negotiated a very fair settlement of the situation. The Soviet Government undertook to purchase Kapitza's apparatus for £30,000.

The Royal Society Mond Laboratory, which had been built for Kapitza, was empty. Cambridge, too, was the poorer for the loss of Kapitza's personality, for his Russian cultural tradition was a stimulating contrast to the strong and sometimes stereotyped English tradition. But Kapitza's return to the Soviet Union was morally correct.

The growth of Nazi power caused the British Government to seek Rutherford's advice on the utilization of science for defence. So while his men were dispersing, his preoccupations with scientific advice on industrial matters, on military problems, on academic freedom, were increasing.

Then, in 1936, the Cavendish Laboratory received a donation of £250,000 from Sir Herbert Austin, the motor-car manufacturer, who was subsequently raised to the peerage. Rutherford was immediately deep in plans for a new big laboratory.

He accepted an invitation to deliver the Presidential Address to the Indian Science Congress in January, 1938. But on October 14th, 1937, he became ill, owing to a slight hernia. A small but dangerous operation became necessary. It was performed, but the internal organs did not recover their functions, and he died on October 19th.

Thus Rutherford departed suddenly, in the midst of a healthy, happy, triumphant and glorious life.

VII. THE SCIENTIST AS HERO

Rutherford departed almost like a Greek hero, taken into heaven at the summit of his glory to join the immortal gods, who had decided that their valiant son should not live to see his victories bartered away.

At the end of 1938, little more than twelve months after his death, his old pupil, Otto Hahn, discovered the phenomenon of the fission of the nucleus of the uranium atom, following a pointer of the Joliot-Curies, much as Chadwick had done in the discovery of the neutron.

Early in 1939 Joliot-Curie and his colleagues, and others somewhat later, showed that at least two neutrons were released in each uranium fission caused by one neutron, thus indicating the practical possibility of establishing a nuclear chain reaction which might release atomic energy on a vast scale.

Rutherford had always been very cautious in his public statements on the possibility of the release of atomic energy, and this caution was communicated to many who were close to him, appearing in some as doubt and convinced disbelief. Thus, in addition to the dispersal of his school, which had begun in his lifetime, atomic physicists in Britain were, on the question of the release of atomic energy, depressed by a pessimistic tradition.

In the preparations for defence against the threatened Nazi attacks, the possibilities of the utilization of atomic energy do not seem to have been rated very high. This is apparently confirmed by the fact that many of his greatest colleagues, such as Cockcroft and Oliphant, dropped atomic physics and engaged in the development of radar.

It would be easy to understand that this policy was quite in keeping with Rutherford's explicit views. But it was entirely contrary to his genius. Rutherford always threw over his most strongly expressed prejudices and views as soon as he recognized the slightest hint of the possibility of scientific advance in spite of them.

But unfortunately, in 1939, Rutherford was no longer alive to deny, in his characteristic way, his publicly expressed views by his scientific actions, and his pupils were unable to keep his tradition of revolutionary action by departing from his conservative attitude on the release of atomic energy.

When the Nazis started the Second World War in 1939, the dispersion of the Rutherfordian school of atomic physics was complete. But far worse, the Rutherfordian outlook had vanished. The courageous optimism to pursue the release of atomic energy was stronger outside than inside Britain, and in particular in Paris.

When the occupation of France by the Nazis became inevitable, Joliot-Curie sent Halban and Kowarski to England to continue his experiments, while he and his wife remained behind to face the Nazis and stand by their own oppressed people. Halban and Kowarski were provided with facilities in the Cavendish Laboratory, and during the autumn of 1940 secured experimental results which had an important influence in removing the atmosphere of pessimism concerning the release of atomic energy. By 1941 the physicists in Britain led the world in preliminary research and development.

But the recovery and advance were too late. A large part of British scientific talent had become involved in other urgent developments that promised a quicker return in military strength, and British industrial resources were already fully mobilized for other tasks. More than 80 per cent of British industrial capacity was engaged in armaments production, whereas at no time during the war was more than 50 per cent of the much greater American industrial capacity engaged on armaments. Consequently, the reserve of industrial capacity in the U.S. which could be devoted to novel projects was very much greater than in Britain. In addition, Britain could be bombed and might also be invaded, or at least atomic laboratories might have been temporarily occupied by parachutists and their secrets captured.

The heads of the British Government decided in 1942 to restrict the work in Britain broadly to the experimental measurement of nuclear data, and in 1943 most of the leading atomic physicists left Britain to join in the American research laboratories and projects for atomic energy, and the making of the first atomic bombs.

Thus by the end of the war there were virtually no atomic physicists in Britain, there was no atomic energy establishment; the remains of the Rutherford school were at work in the United States. The knowledge of atomic energy was presently treated as American private property, and after the end of the war a number of new advances were even concealed from the pupils of Rutherford.

It may well be asked whether such a situation need have arisen. Were the heads of the British Government too precipitate in handing over British atomic science without adequate guarantees of reciprocity? Were the economic and military reasons sufficient for this surrender, or was there a desire to get atomic energy into American hands for other reasons?

It is true that British industry was very fully mobilized in 1942. But had the correct industrial strategy been chosen? Would it not have been possible to reduce the heavy bomber programme, which absorbed so much effort and gave such inadequate military returns? A fuller analysis of the British situation in 1942 might show that if there had been the vision and the will, the British could have developed atomic energy and atomic weapons, if not in Britain then within the Commonwealth.

Rutherford would not have been daunted by lack of facilities. As he was fond of saying, 'I, myself, could do research work at the North Pole.' It seems doubtful, however, that he would have succeeded in dealing with the political problems involved in the development of atomic energy.

What happened was contrary to Rutherford's principles. He had refused all offers of posts in the United States. He had never accepted American money for research endowment, though in his latter years the merest hint from him would have been met by a very large sum.

He believed in British independence and he considered that the British could very well afford their own research. He believed the British Empire could pay its way, and that begging abroad was beneath it. Certainly, it was beneath him. He was in the habit of saying, 'The British Empire is O.K.' and did what he could to make it so.

The contrast between his spirit and the events in the ten years after his death indicates that, in spite of his conscientious public labours, his influence on British national scientific policy did not penetrate very deeply. His great prestige and attractive personality commanded attention, but his heroic simplicity was not in tune with the decadent politicians and men-of-affairs of his day. They enjoyed the simple and noble spirit which was so refreshing compared with their own, they were very polite to him and loaded him with honours. But as soon as he was gone, they relapsed to their customary levels.

The difference between Rutherford and his political contemporaries was vividly exemplified at a public meeting during the Faraday–British Association celebrations in 1931. Ramsay MacDonald had formed his National Government and happened to make his first public appearance at this meeting as its head. His tortured expression of conscious hypocrisy had an intensity that caused almost physical pain to many in the audience. Rutherford's frank honesty contrasted startlingly with MacDonald's crookedness. It was obvious that his qualities could not function in MacDonald's world, and consequently he could not have a major influence in it. A healthier, simpler form of society than that which existed in England in the 1930s would have been needed in order to give scope to his fundamental human qualities.

Rutherford lacked the kind of political ability needed to deal with such men as MacDonald and Baldwin, or with the sordid problems of the exploitation of atomic energy. He had not been very successful with military scientific matters in the First World War, and he would probably have been still less successful with those of the Second World War.

In general, he took a generous and kindly view of people, seeing their best side and minimizing their weaknesses. This disqualified him for politics. He was not so shrewd a judge as J.J. of a person's qualities.

Rutherford's public activities cannot be regarded as having been of major importance. If he had had the kind of political insight and ability which would have enabled him to deal with the social problems of his day, he would not have been able to enjoy innocent satisfaction in conventional honours, which were already hollow.

Rutherford was obstinately opposed while he was still living by many who contended that his concentration of ability into atomic physics was harming the nation. Ought not the nation's talent to be spread more evenly among the various branches of physics? It was characteristic of many of those who supported this view that they were opposed on principle to the planning of scientific research. And it was characteristic of Rutherford that, in this as in so many other questions, he was on the right side. His posthumous presidential address to the Indian Science Congress explicitly called for the planning of science in a modern nation.

He sympathized with the strikers in the great British industrial battles of 1911. He was interested in the movement for the organization of scientific workers. He was the President of the Academic Assistance Council.

He was violently opposed to personal participation in the commercial exploitation of scientific discovery. He had no use for position for its own sake, and expected only the opportunity to do scientific research and be paid enough to be able to work efficiently. Throughout his life his income was moderate, and his arrangements for retirement in his small country cottage were very modest.

Rutherford's qualities were those of an agrarian pioneer and democrat from the colonies, who had been brought by a declining imperialism to its centre. He was in it, but not of it. He accepted its outer forms, and was never absorbed into its decaying content. His real life was in the science that he pursued in the spirit of the society from which he had sprung. He was not an inhibited English professional man, and had a power of energetic endeavour that this class had lost.

His nuclear theory of the atom is probably more important for mankind even than Faraday's conception of the electromagnetic field. The idea is simple and unoriginal, and not profound and difficult like Faraday's. But the simplicity of the experiments by which it was established, and his personal and directive contributions in carrying them out show perhaps the greatest genius in the whole history of experimental physics. His only competitor is Faraday, and he had the advantage over him of being able to inspire and direct a host of the ablest collaborators. Rutherford was fortunate in finding a supremely important field of research which was peculiarly suited to his gifts. He was born to play with particles: every α-particle was one of his personal friends, as Hardy had said of Ramanujan's acquaintance with positive integers. Would he have been as successful in a period when particulate ideas were not a key to fundamental discovery? His exceptional early successes with radio research suggest that he would have had the versatility to make discoveries of comparable power in other fields, if those had been equally ready for cultivation. His dropping of radio research for atomic research may be interpreted as a sign of genius in scientific judgment in successfully choosing the most promising field. His complete concentration on a single field may not be a sign of narrowness of range; it may also be interpreted as an efficient economy, so that the greatest possible result should be obtained from a given effort.

As a citizen, Rutherford was the descendant of a simpler and healthier society than that which he entered and illuminated, but did not change. He did not live long enough to see exactly how corruptly contemporary society used the discoveries to which he had contributed so much.

Rutherford was still young in spirit when he died, and those whom the gods love die young.

BIBLIOGRAPHY

Rutherford, by A. S. Eve (1939).

Lord Rutherford, by N. Feather (1940).

Man of Power, by Ivor B. N. Evans (1939).

'Rutherford: Life and Work after the Year 1919,' by J. D. Cockcroft, *Proc. Phys. Soc.*, **58**, 625 (1946).

'Lord Rutherford 1907–19: a Partial Portrait,' by A. S. Russell, *Proc. Phys. Soc.*, A **64**, 217 (1951).

'Rutherford,' by M. L. E. Oliphant, *Proc. Phys. Soc.*, **59**, 144 (1947).

'Rutherford,' by H. R. Robinson, *Proc. Phys. Soc.*, **55**, 161 (1943).

'Rutherford,' by E. Marsden, *Proc. Phys. Soc.*, A **63**, 305 (1950).

Radioactive Substances, and their Radiations, by E. Rutherford (1913).

Radiations from Radioactive Substances, by E. Rutherford, J. Chadwick, C. D. Ellis. 2 vols. (1930).

Hommage à Rutherford, by Irene Joliot-Curie and others (1947).

Obituary Notices: *Proc. Phys. Soc.*, **50**, 441 (1938); *Nature*, **146**, 746, 1937; *Roy. Soc. Obit.*, **2**, 395 (1938).

'Magnetization of Iron by High-frequency Discharges,' E. Rutherford, *Trans. New Zealand Institute*, **27** (1895).

'Magnetic Viscosity,' *Trans. New Zealand Institute*, **28** (1896).

'A Magnetic Detector of Electrical Waves,' E. Rutherford, *Phil. Trans. Royal Soc.*, **189**, 1 (1897).

'On a Diffuse Reflection of the α-Particles,' H. Geiger and E. Marsden, *Proc. Roy. Soc.*, A **82**, 495 (1909).

Bakerian Lecture: 'Nuclear Constitution of Atoms,' E. Rutherford, *Proc. Roy. Soc.*, A **97**, 374 (1920).

'The Cause and Nature of Radioactivity,' E. Rutherford and F. Soddy, Part I, *Phil. Mag.*, Sept. 1902; Part II, *Phil. Mag.*, Nov. 1902.

'The Scattering of α- and β-Particles by Matter and the Structure of the Atom,' E. Rutherford, *Phil. Mag.*, May 1911.

'The Passage of α-Particles through Hydrogen,' E. Marsden, Part I, *Phil. Mag.*, May 1914; Part II, E. Marsden and W. C. Lantsberry, *Phil. Mag.*, Aug. 1915.

'Collision of α-Particles with Light Atoms,' E. Rutherford, Parts I, II, III, IV, *Phil. Mag.*, June 1919.

JAMES HOPWOOD JEANS

1877–1946

I. THE EXPOSITOR

JEANS was read more widely than any other British writer on physical science in the decade preceding the Second World War. The British public acquired its ideas on the modern physical theories of relativity, quanta, and their application to the explanation of the operation and history of the universe more through him than any other writer.

What were the qualities which gave his writings their vast circulation? What was the character of his style? What was the outlook which he communicated to his vast circle of readers? Why were his works so popular?

His style was transparent, trenchant, and dignified. His scientific narrative flowed like a grand river under complete control. The reader was confidently navigated through a flood of ideas and facts. Jeans had positive answers to all problems: yes or no. His shades of meaning were always sharply distinguished, there was no shading of one idea into another. As Milne has said, he was graphic and fluent.

His favourite mode of illustration was by proportion. He arranged the magnitudes of the universe in ascending and descending orders like the notes in a fugue. His style was orchestral, and no doubt arose from the same source as his passion for the contrapuntal music of Bach. He often played the music of this master on the organ for three or four hours a day.

In addition to his clear, bold, rushing style, he had another major qualification: scientific authority. This arose from his original contributions first to mathematical physics and then to the mathematical theory of cosmogony: the evolution of the universe, the stars, and the solar system.

In his first field of mathematical physics, he gave a more rigorous proof of Clerk Maxwell's law of the distribution of the velocities of molecules in a gas which is the foundation of the kinetic theory of gases: that they consist of millions of flying colliding minute particles. He worked out more exactly the law of equipartition, or distribution of energy, between the possible modes of vibration in a system. He corrected Rayleigh's formula for calculating, according to the classical electromagnetic theory, the distribution of energy among the wave-lengths of the radiation inside an enclosure at uniform temperature.

In 1904, the year before Einstein's discovery of the theory of relativity, Jeans suggested that energy might be released in vast amounts in the sun and stars by the annihilation of matter, through the combination, or rushing together, of negatively charged electrons with positively charged bodies to form bursts of radioactive energy.

In this period he wrote two classical textbooks: the *Dynamical Theory of Gases*, published in 1904, and the *Mathematical Theory of Electricity and Magnetism*, published in 1908. These were standard textbooks of a generation of British students of mathematical physics.

He completed his work in this field with his *Report on Radiation and the Quantum Theory*, written for the Physical Society of London and published in 1914. This brilliant report persuaded all unprejudiced British physicists to accept at last Planck's quantum theory, which had been proposed fourteen years earlier, in 1900.

Jeans had, during those years, tried every conceivable device, in a long series of papers, to evade Planck's theory or prove it wrong. His surrender after a most determined struggle, combined with his dazzling expository powers, made his report all the more convincing.

Though his conversion was apparently complete, he thenceforth left this field of scientific battle and did not return to it. He transferred all his effort into the theory of cosmogony. He applied the technique that he had developed for analysing the motions of particles in the dynamical theory of gases to the behaviour of matter in nebulae. He elucidated errors in Laplace's nebular hypothesis of the origin of the stars.

He analysed the behaviour of rotating masses of fluid, and in particular the stability of pear-shaped rotating masses. His teacher, George Howard Darwin, the mathematical astronomer and one of Charles Darwin's talented sons, had concluded that they were stable, whereas the Russian mathematician Liapounoff had demonstrated that they were unstable. Liapounoff suggested that the mistake was due to the use of divergent series, and Jeans found the actual place where Darwin had gone wrong.

All of the previous investigators had assumed that the rotating masses consisted of incompressible fluids, or of combinations of very rigid and very tenuous substances. Jeans solved the problem of the behaviour of rotating masses of *compressible* fluids, probably his greatest achievement.

The rotation of the star formed by gravitational condensation of nebular matter was started by the approach of a second body. As the now-rotating star aged, it shrank, but its angular momentum remained constant. Consequently, it rotated faster as it became smaller.

Jeans showed how it might whirl so fast that it changed from a flattened sphere into a lenticular, or lens-shaped, body with a sharp edge around its equator. Presently this became unstable, and jets of matter would spurt out of the rim.

Jeans suggested that the planets had been formed by condensations of the matter ejected from the sun in such a crisis thousands of millions of years ago.

He discussed the behaviour of the universe of stars as if they formed a gas whose molecules never collided. He calculated that the chances of two stars approaching each other very closely were extremely small, so he concluded that planetary systems must be very rare, and that indeed our solar system, with the earth, life, and man, may be unique in the universe.

He discussed the origin and behaviour of binary stars produced by the fission of rotating masses. He attacked the obscure mechanics of spiral nebulae, and he investigated the evolution and radiation of gaseous stars. In 1918 he deduced the possible length of the life of the sun and stars. He expounded these researches in his *Problems of Cosmogony and Stellar Dynamics*, which was published in 1919, and is the most original and greatest of his books.

He published a revised and extended version under the title of *Astronomy and Cosmogony* in 1928. This was less well digested and less sure than the earlier work.

Then Jeans suddenly abandoned fundamental research and commenced to write his expositions of astronomy and physics for the public. The first of these was *The Universe Around Us*, published in 1929, which had an immense success.

It was followed in 1930 by *The Mysterious Universe*, which became the best known of all his books, one which, as Tallulah Bankhead said, every girl should read. He wrote another half-a-dozen volumes, including *Science and Music*, in which he combined his two mental passions; and *The Growth of Physical Science*, which was published posthumously in 1947.

This was his performance. We have to trace its growth and analyse its nature.

II. THE CITY SCHOOLBOY

Jeans was born at Ormskirk in Lancashire on September 11th, 1877. His father was William Tullock Jeans, a parliamentary journalist, and his mother, whose maiden name was Hopwood, came from Stockport in Lancashire.

William Tullock Jeans was a Scotsman from Elgin, whose father and grandfather had been owners of local newspapers. Thus Jeans came of two consecutive generations of journalists. It was not entirely surprising that he should have been a fluent writer, with a sense of how to attract the general reader, and should have become one of the great journalists, at his worst one of the great 'yellow' journalists, of science.

William Tullock Jeans was not only a journalist. He was interested in science, and conceived it a duty to explain the importance and effects of science to the public. He wrote two books with this object: *The Creators of the Age of Steel* (1884) and the *Lives of the Electricians* (1887), the second of which was intended to be the first of a series of such books.

He inscribed Cobden's saying 'Our wealth, commerce, and manufactures grow out of the skilled labour of men working in metals' on the title-page of his first book. But, unfortunately, the public were not interested in direct accounts of metallurgy, so they were unaware of how much this science had contributed to their welfare. He therefore conceived that some of the important processes of metallurgy might be presented in a more attractive way through the medium of biographical accounts of the greatest metallurgical innovators, such as Bessemer, Siemens, and Whitworth.

W. T. Jeans was evidently passionately interested in science and had read considerably. He quotes J. D. Forbes' interesting essay on the *Progress of Mathematical and Physical Science* and Bacon's remark that the ancient Greeks awarded divine honours to inventors, 'but only heroic honours to those who did good service in the state.'

In both books he quotes Goethe's view that the main object of biography is to exhibit man in relation to the features of his time, and he says that this has been the guiding principle of his writing. The *Creators of the Age of Steel* contains information about the founders of modern metallurgy, their ancestry, and the nature and economic and social significance of their inventions. But the book, though intended to be popular, is heavy in style, and infinitely less sprightly than his son's writings on difficult investigations in applied mathematics.

The *Lives of the Electricians* is lighter. William Tullock Jeans evidently loved John Tyndall; 'by his teaching and writings he has probably done more than any other man in England to kindle a love of science among the masses; and by his life he has set an example to students of science which cannot be too widely known or appreciated.'

He published this when his son was ten years old. Jeans grew up in an atmosphere of admiration for science, and of John Tyndall, the great expositor, in particular. William Tullock Jeans' other subjects were Wheatstone and Morse, whose electric telegraph inventions had had such profound economic and social implications.

Two other points in William Tullock Jeans' books are also worth notice. In the second he quotes with approval Dumas' condemnation of biographers who 'penetrate the meanest details' of their subjects' existence for 'the gratification of a paltry curiosity.' He says that the biographer should deal only with the public facts of his subject's life, and should not reveal those that are properly private.

William Tullock Jeans had strong Victorian inhibitions, which prevented him from entering into the psychology of his subjects and also, perhaps, that of his relatives. He repressed the sensitive feelings underlying psychological insight. Perhaps the most revealing of his remarks is another quotation from Goethe, that 'it is a pious wish of all fathers to see what they have themselves failed to attain realized in their sons.'

It would not be surprising that an important influence on James Hopwood Jeans' life had been his father's pious wish to attain through him scientific ambitions in which he had himself failed.

Jeans' mother was descended from a family of Lancashire clock and instrument makers, so he may have received mechanical aptitude from that side of his family.

His childhood seems not to have been very happy. He was oppressed by formal religious observances. This is rather striking, as his father had such a strong admiration for the free-thinking John Tyndall.

He was precocious, being able to tell the time at the age of three and to read at four. He was trained to read the first leader of the *Times* each morning to his parents. There are cases of the balance of infants' minds having been disturbed by this practice.

He had a remarkable memory for numbers. At the age of seven he happened to find his father's table of seven-figure logarithms. Without knowing what they meant, he memorized the first score of entries, and retained them nearly to the end of his life.

The main interest of his childhood was in clocks. He took to pieces old clocks given to him by his father, and, among other operations, used to boil them in order to make them go. At the age of nine he wrote a little MS. book on clocks, containing nine pages bound in blue paper, with an explanation of the escapement principle, and descriptions, with diagrams, of how to make a clock out of bits of tin and other oddments.

Jeans learned to play the piano at an early age and the organ at the age of twelve.

He was sent to the Merchant Taylors' School in 1890, when he was thirteen years old. The School at that time was in Charterhouse Square in the City of London. It was one of the leading day-schools, whose boys were the sons of middle-class City and professional men. Many of Jeans' schoolfellows became notable administrators in the Civil Service, academic life, and the City. The school produced a hard-working, competent middle-class type, capable and hard-headed, and without the aristocratic tradition inculcated in the leading residential schools. There are deep differences in social ideas between the day and residential schools, and Jeans was definitely moulded by the former sort.

He had an exceptionally good mathematics teacher, S. O. Roberts, who had taken a high degree at Cambridge, and inspired him with a passion for mathematics.

Jeans was awarded a mathematical scholarship at Trinity College, Cambridge, and entered the university in 1896. He gained the place of Second Wrangler, or second place, in the mathematical examination, after only two years' study, instead of the usual three years', and in 1900 shared a Smith's Prize with G. H. Hardy, who subsequently became the most distinguished British mathematician of his generation.

Jeans took the course of practical physics in the Cavendish Laboratory in 1899–1900, during the exciting period after J. J. Thomson's discovery of the electron. His first scientific paper was on the striated electrical discharge in gases at low pressure, and it remained his only paper on experimental physics.

In spite of his early interest in clocks and mechanisms, Jeans was lacking in feeling for experimental physics. Thenceforth he concentrated on the exploration and development of the mathematical description of mechanical and physical processes.

III. THE APPLIED MATHEMATICIAN

Jeans was awarded the Isaac Newton Studentship in 1900 and was elected a fellow of Trinity College in 1901. His work was to become a very characteristic and distinguished example of Cambridge and Trinity mathematical physics at the beginning of the twentieth century. It consisted in the application of a brilliantly clear and accomplished mathematical technique, which had been elaborated by the successors of Isaac Newton, and latterly by Stokes, Kelvin, Rayleigh, and Routh, for the description of the behaviour of the subjects of Newtonian natural philosophy. These were the infinitely hard billiard-ball atoms, elastic solids, continuous fluids, and other idealized conceptions of the bodies met with in everyday life.

Jeans began by applying this technique to the partially and wholly unsolved problems bequeathed by Clerk Maxwell in his theory of gases as flying and colliding particles, and in his wave theory of the radiation of energy, of light, and electricity.

Jeans' first important paper was on the *Distribution of Molecular Energy*. It was, appropriately to the subject, communicated by J. J. Thomson to the *Philosophical Transactions* of the Royal Society and published in 1901. It dealt with the problem of the distribution of energy among the molecules of a gas. Maxwell had been the first to publish the theorem that if a gas consists of two kinds of molecules, then the mean or average energy of each kind of molecule is the same; the molecule of lower mass will travel at higher speed, so that its energy will be the same as that of the molecule of higher mass travelling at lower speed. Maxwell's theorem was extended and generalized by Boltzmann.

In his memoir, Jeans discussed some problems raised by the Maxwell-Boltzmann theorem. This is how he introduces the subject:

'It is found by experiment that the ratio of the two specific heats of certain monatomic gases (e.g. mercury, argon) is $1\frac{2}{3}$. If we admit that the energy of these gases is distributed in the manner indicated by Boltzmann's Theorem, then this theorem leaves no escape from the conclusion that the molecules of these gases must be rigid and geometrically perfect spheres. A similar difficulty arises in connection with other gases: the number of degrees of freedom which a consideration of the ratio in question leads us to expect a molecule of gas to possess is always less than the number which the spectrum of the glowing gas shows to actually exist. Further, Boltzmann's Theorem excludes the possibility of the ratio of the two specific heats having any values except one of a certain series of values, whereas experiment shows that the ratio is not always equal to one of this series, although it is generally very near to such a value. Finally Boltzmann's Theorem leaves no room for a variation of this ratio with the temperature although such a variation is known to exist.

'In the present paper I have tried to suggest a way by which it is possible to escape from this dilemma. As there is not sufficient known about the constitution of a molecule to enable it to be completely specified as a dynamical system, the paper is limited to the consideration of two imaginary types of molecules.

'The conclusions arrived at are the same in each case. In the first place the distribution of energy which is given by Boltzmann's Theorem is the only distribution which is permanent *under the conditions postulated by this theorem.* And in the second place, this law of distribution may break down entirely as soon as we admit an interaction, no matter how small, between the molecules and the surrounding ether.'

Jeans deduced that 'the second law of thermodynamics will only hold in special cases for a gas which is emitting radiation of any kind.'

He tried to explain the dilemma that the specific heats of certain gases were not in agreement with the law of equipartition of energy, by supposing that there was a departure from the conservation of energy, some of the energy being dissipated into the ether.

'In fact, Boltzmann's Theorem rests on the assumption that the molecules of a gas form a conservative dynamical system, and it will appear that the introduction of a small dissipation function may entirely invalidate the conclusions of the theorem. Thus we may regard the Boltzmann distribution as unstable, in the sense that a slight deviation from perfect conservation of energy may result in a complete redistribution of the total energy, and it will appear that this new distribution of energy will lead to values for the ratios of the two specific heats which are not open to the objections mentioned above.'

The dissipation function that he proposed might be analogous to the frictional drag of the surrounding medium on the rotation of the molecules. 'It is not suggested that a reaction of this type exists in nature, but this reaction may be taken as being the simplest reaction possible.'

In this paper Jeans introduced the idea that a gas may have a principal temperature and subsidiary temperatures: 'the state of a gas may be regarded as depending upon a principal temperature *T*, and also on a number of subsidiary temperatures . . . each of these corresponding to one (or possibly more) of the degrees of freedom of the molecule.'

Milne remarks that this idea has frequently been utilized in subsequent researches in the theory of the properties of the atmospheres of hot gases on the sun and stars.

This memoir, communicated to the Royal Society before Jeans was twenty-three years old, reveals his clarity, mathematical resource, power, and elegance. There is a ruthless, hard, diamond-like quality so characteristically expressed in his opening phrase, 'This theorem leaves no escape from the conclusion that the molecules of these gases must be rigid and geometrically perfect spheres.' In positive, unhesitating language he proposes to amend the law of conservation of energy and allow deviations from the second law of thermodynamics.

A true physicist would not dare to play with fundamental physical concepts in this way. Jeans, in fact, is playing with mathematical concepts as if they were physical principles. As a virtuoso in applied mathematics he commands admiration; as a physicist he already had a confidence that dizzied the physicist in direct experimental contact with material nature.

This notable paper foreshadowed one line of Jeans' life-work: researches on the mathematical theory of gases. Another magnificent paper sent to the Royal Society in the following year, before he was twenty-four years old, and published in 1902, foreshadowed the other main line of his researches. This was on the problem of the *Stability of a Spherical Nebula.*

It was communicated to the Royal Society, again appropriately to the subject, by his old teacher, Professor G. H. Darwin, whose paper of 1888, *On the Mechanical Conditions of a Swarm of Meteorites, and on Theories of Cosmogony*, he quotes in his opening paragraphs. Darwin remarked that: 'The principal question involved in the nebular hypothesis seems to be the stability of a rotating mass of gas, but, unfortunately, this has remained up to now an untouched field of research. We can only judge of probable results from the investigations which have been made concerning the stability of a rotating mass of liquid.'

Jeans comments that 'In so far as the two cases are parallel, the argument by analogy will, of course, be valid enough, but the compressibility of a gas makes possible in the gaseous nebula a whole series of vibrations which have no counterpart in a liquid, and no inference as to the stability of these motions can be drawn from an examination of the behaviour of a liquid. Thus, although there will be unstable vibrations in a rotating mass of gas similar to those which are known to exist in a rotating liquid, it does not at all follow that a rotating gas will become unstable, in the first place, through vibrations which have a counterpart in a rotating liquid: it is at any rate conceivable that the vibrations through which the gas first becomes unstable are vibrations in which the compressibility of the gas plays so prominent a part, that no vibration of the kind can occur in a liquid. If this is so, the conditions of the formation of planetary systems will be widely different in the two cases.'

He then says that in order to answer this question he has examined 'the stability of a mass of gravitating gas', and has found that, 'on the whole, the results are not such as could have been predicted by analogy from the results in the case of a gravitating liquid. . . . In the case of an incompressible liquid we may say that gravitation tends to stability, and rotation to instability; the liquid becomes unstable as soon as the second factor preponderates over the first. The gravitational tendency to stability arises in this case from the surface inequalities. . . . As soon as we pass to the consideration of a compressible gas the case is entirely different . . . when the gas is compressible, the tendency of gravitation may be towards instability. The gravitation of the surface inequalities will as before tend towards stability, but when we are dealing with a gaseous nebula, it is impossible to suppose that a discontinuity of density can occur such as would be necessary if this tendency were to come into operation. Rotation as before will tend to instability, and the factor which makes for stability will be the elasticity of the gas.'

There was nothing improbable in the supposition that a gaseous symmetrical nebula may become unstable even in the absence of rotation.

Jeans concluded that a spherical nebula at rest and in absolute equilibrium would be stable. If, however, the lag in the cooling of the inside of the nebula be taken into account, owing to the time taken for the central heat to travel to the surface, as radiative heat is lost from the surface into space, then the nebula may become unstable.

If the nebula rotates slowly it will acquire the shape of the Maclaurin spheroids, or flattened spheres. At a much higher speed of rotation these turn into Jacobian ellipsoids, which have three unequal axes, rather like an almond.

Jeans concluded that a uniform distribution of matter in infinite space would be unstable, independently of the mean temperature or density of the distribution.

In a footnote he remarks that: 'An interesting field of speculation is opened by regarding the stars themselves as molecules of a quasi-gas. If space were Euclidean and unbounded, there would be no objection to this procedure, and we should be led to the conclusion that the matter of the universe must become more and more concentrated in the course of time. If space is non-Euclidean, this concentration might reach a limit as soon as the coarsegrainedness of the structure attained a value so great that the distance between individual units became comparable with the radii of curvation of space.'

So in his first important paper on cosmogony, Jeans had considered the possibility that the universe might be non-Euclidean, and space might be curved.

It appeared that an infinite mass of matter in the gaseous or meteoritic state necessarily assumes the form of a number of nebulae, being clusters round points of maximum density. The mean density of the primitive distribution of matter appeared to be about 10^{-23} that of water.

The rotating gaseous nebula, when it becomes unstable, does not split into two nearly equal parts. One is much larger than the other. He suggested that the splitting of a nebula in a critical state of instability might be precipitated by the penetration of a wandering meteorite or by a quasi-tide caused by a distant mass, such as another nebula or a star.

The plane of revolution of the satellite formed by the fission would, if the rotation of the nebula was sufficiently small, be largely determined by the path of the meteorite or by the position or path of the attracting mass. It would not depend much upon the axis of rotation of the nebula.

Thus, in his first memoir on cosmogony, Jeans proceeded beyond Poincaré and Darwin, his predecessors, who had considered the rotation only of liquid bodies. He started to take into account the properties of compressibility and gaseousness, found in real nebulae.

Then he introduced the theoretical conception of *gravitational instability*. The distribution of matter can never be perfectly uniform, the centres of maximum density tending to gather more and more matter to themselves, and thus ultimately upsetting equilibrium.

Poincaré had first shown in a famous paper published in 1885 that a liquid sphere rotating with increasing speed, after passing through Maclaurin's flattened spheroidal shape into Jacobi's ellipsoidal shape, presently developed a lop-sided bulge, with slight concavities, which G. H. Darwin described as pear-shaped. The mathematics of the rotating fluid body, capable of changing its shape in three directions in space, was very formidable. Jeans therefore introduced a simplification by considering how a rotating liquid *cylinder* might behave, which could change its shape in only two directions, that is, only in the plane of rotation at right angles to the axis. He found that at slow speeds of rotation the cylinder expands in diameter, but its cross-section remains circular. At a certain critical speed, its cross-section becomes elliptical, and at a still higher critical speed, it begins to bulge on one side, so that it becomes pear-shaped. Thus the cylinder went through changes of shape analogous to those of Poincaré's rotating sphere.

But Jeans was able to take the calculations in his simpler problem several stages farther, so that he could prove that the slight concavity in the pear-shape deepened into a furrow, and then into a neck, which suggested very strongly that the pear-shaped figure would ultimately split into two parts. Jeans not only proved the surprising qualitative result that one part would be much larger than the other; he calculated the quantitative difference. The diameter of the larger part, which he called the 'primary', was about four times that of the smaller, which he called the 'satellite'.

The behaviour of Jeans' cylinders, though different in many ways, gave much insight into that of Poincaré's pear-shaped masses. This second notable contribution on fluid mechanics in relation to cosmogony was published in 1903.

Poincaré's genius, which tended always to work with methods of the greatest mathematical generality, was not inclined to tedious arithmetical calculations, and he did not determine whether his pear-shaped rotating figures could be stable. Darwin published a long paper in 1902 in which he claimed that he had almost certainly proved that they could be stable. Jeans had not succeeded in determining whether his rotating cylinders could be stable when they had acquired a pear-shaped cross-section, so his analogue did not convey any decisive suggestion on the behaviour of the pear-shaped masses.

The phenomenon of fission, which has such a fundamental role in the evolution of nebulae and stars, has recently been found to be not less fundamental in the splitting of the atom. Bohr and Wheeler have given the first qualitative account of the fission of the uranium atom by treating it as a liquid sphere undergoing violent distortions owing to excitation through entry of a neutron.

By 1903 Jeans had already demonstrated his great powers in his two main lines of research, the mathematical theory of gases and the mathematical theory of cosmogony. He was elected a fellow of the Royal Society in 1906, at the early age of twenty-eight.

In 1902 he suffered a serious illness. He was attacked by tuberculosis of the joints, and had to spend about a year in three sanatoria. Fortunately, he recovered completely. He occupied his vacant hours in the sanatoria by writing a work on the mathematical theory of gases, incorporating many of his recent researches, and giving a masterly and elegant review of the subject. This was published as his *Dynamical Theory of Gases*, which at once became a standard treatise.

He was appointed a university lecturer at Cambridge in 1904, but in

1905 he became Professor of Applied Mathematics at the University of Princeton, New Jersey. In 1907 he married Charlotte Tiffany Mitchell, the heiress and daughter of Alfred Mitchell, a traveller and explorer, related to the wealthy Tiffany family of New York jewellers.

Jeans remained in the United States until 1910, when he returned to Cambridge to become Stokes Lecturer. He resigned from this appointment in 1912. He did not need to earn a salary.

In 1913 Sir George Darwin died, and his chair, the Plumian Professorship of Astronomy, fell vacant. Jeans was deeply disappointed when he was not appointed to succeed the teacher in whose field he had worked so brilliantly. He was still only thirty-six years of age, but he was passed over for a man yet five years his junior: Arthur Stanley Eddington.

Gifted though Jeans was, it cannot be said that the electors made a mistake. As he was well-to-do, he was not in hardship through lack of an appointment, and though he was an effective lecturer, he was inclined to be impatient and had no taste for leading a team of research students.

For the rest of his life, he followed his personal intellectual interests as he wished. He continued his fundamental researches.

In 1919 he became one of the joint Secretaries of the Royal Society. He held this post for the usual period of ten years. Not having any other appointment, he was able to devote much more time to the conduct of the society than was customary in a secretary. He had a large part in carrying through the extensive reorganization that had become necessary. The Society had gradually acquired an endowment, which had grown to the considerable sum of about £1,000,000. The management of this large property had become a serious business matter, and Jeans, who was himself wealthy and apt in financial matters, helped to reorganize the society's investments on a more modern basis.

He undertook also to improve the standard of the *Proceedings* of the Royal Society. During his secretaryship and editorship of the physical series, the *Proceedings* displaced the *Philosophical Magazine* as the leading British learned journal for the physical sciences. On occasion he arranged for the publication of important papers at great speed: Dirac's famous paper on the negative roots of the wave-equation of the negative electron, which contained the forecast of the existence, through his 'theory of holes', of the positive electron, was published within two weeks of the receipt of the manuscript.

Unfortunately, during the Second World War, the results of this excellent reform fell away and the old delays in publication returned. These are vexatious in the modern world-wide and intense conditions of research, where a man's contribution may be undervalued through the loss of priority in publication.

Jeans became well-to-do through his first marriage. When he died in 1946, his fortune was valued at £256,054, one of the very largest ever

4. Sir James Jeans

[face p. 104

left by a British scientist. As his brusque manner suggested, he was a very capable businessman. The stranger meeting him in the train coming up to London from the country might easily have taken his neat, round, bustling figure, clean-shaven face and clear, alert head for that of a wealthy stockbroker with intellectual hobbies.

Jeans' passion for music was profound. After he had settled in a house in Dorking in Surrey in 1918, he installed an organ on which he often played for many hours of the day. He knew all of Bach's organ compositions by heart. But he would never play in front of other people, even his closest friends.

After his first wife died in 1934, he married in the following year the distinguished young Viennese organist, more than thirty years his junior, Suzanne Hock. She bore him two sons and a daughter, in whom he had great joy and pride.

After his second marriage, he built a second organ for his wife, and had the organ rooms acoustically insulated, so that he and his wife could play simultaneously without disturbing each other.

He became a director of the Royal College of Music in 1937 and an Honorary Member of the Students' Union of the Royal College of Music. He was widely read in the scientific aspects of music, on which he wrote the most restrained, but best, of his popular books.

So, as a wealthy independent gentleman, without institutional or academic responsibilities and able to follow his inclinations, he lived out his life. His best period of fundamental research ended in about 1918. After a further decade of less certain achievement, he abandoned fundamental research and began to write popular books. He was knighted in 1928. In 1935 he was appointed Professor of Astronomy in the Royal Institution, which involved only the delivery of occasional semi-popular lectures. He retired from this position in February 1946, after heart-attacks, the first signs of which had troubled him in 1917; and on September 16th, 1946, he died of coronary thrombosis.

IV. PHYSICAL ENIGMAS

The nineteenth century bequeathed to the twentieth two physical enigmas, both arising out of the profound discoveries of Clerk Maxwell. The first was the result of the Michelson–Morley experiment (which had been suggested by Clerk Maxwell), that the speed of light in empty space is independent of the speed of motion of its source. The second concerned the distribution of energy among the different wavelengths of radiation emitted by a hot body. (In technical language, the radiation from a black body, or from an enclosure in which radiation is in equilibrium.)

If the light from a hot body, such as a star, is examined, it is found that the amounts of energy transmitted by the different wavelengths or colours of light, those corresponding to red, blue, ultra-violet, etc., vary according to the temperature of the star. As the temperature rises, more and more energy is found to come from the shorter wavelengths. Thus the light from the very hot stars is much bluer than that from the cooler stars, which is red.

The observed distribution of energy among the various wavelengths emitted from the surface of a star or hot body did not agree with that deduced from the electromagnetic theory. The best theorists and mathematicians, among whom were Lorentz, Fitzgerald, Kelvin, Rayleigh, Larmor, and Poincaré, wrestled with this intractable discrepancy.

Jeans strode into the fray in the height of his youthful mathematical strength and power. He concentrated on the second of these enigmas, and its related problems. He improved the mathematical methods for describing the properties of matter in statistical terms. With their aid, he discovered the correct formula for deducing the distribution of energy among the various wavelengths, on the assumption that a radiating body obeys the laws of Newtonian, or classical, dynamics.

While Jeans' paper was in the press, Rayleigh published a similar formula, but containing a numerical factor 64. Rayleigh immediately agreed that Jeans' value of 8 for this factor was the correct one. The two papers establishing the Rayleigh-Jeans formula appeared in 1905.

The formula indicated that in any radiating system which operated according to the laws of Newtonian, or classical, mechanics, the energy of the system tended to concentrate into the shorter wavelengths, ultimately producing what was called the 'ultra-violet catastrophe', when all the radiation had become concentrated in the shortest waves.

Meanwhile, the German master of thermodynamics, Max Planck, an old pupil of Kirchhoff, had since 1879 devoted himself to the study of energy and radiation, and showed in 1900 that the observed distribution of energy among the radiations from hot bodies could be exactly explained on the assumption that energy is emitted not in continuously varying amounts, but in multiples of finite packets.

When an atom radiated energy, it did not send out waves of every wavelength. On the contrary, it behaved as if it contained what Planck called 'Hertzian oscillators', or little radio stations, each operating on a fixed wavelength. When the radiation from a hot body warmed up a cold body, the atoms of the hot body sent out waves of particular wavelengths, which were picked up by the atoms in the cold body, which were able to absorb their energy in the same way that a radio-receiver absorbs the energy of the waves it picks up.

Planck found that the energy in any wave-radiation was emitted in packets equal to the length of the wave multiplied by a constant which he called *h*. For this particular wavelength, energy could be emitted only in multiples of this size of packet.

Planck's assumption gave the right answer, but it was contrary to the principles of Newtonian mechanics. According to these, the energy carried by a particular wavelength could be continuously varied, and fixed at any specified amount, by increasing or decreasing the strength of the wave.

On the Newtonian theory, the atom could be conceived as a solar system. The radiations which it emitted were regarded as due to loss of energy of the revolving planetary electrons, owing to loss of speed. The electrons were supposed to lose their speed of revolution continuously, like the planets which are continuously slowing down in their revolutions round the sun. As the electrons slowed down, they would send out waves of continuously changing wavelengths.

But Planck's assumption indicated that the revolving electrons did not change their speed of revolution continuously. They must change suddenly from one fixed speed of revolution to another, in order to emit radiation in energy-packets of standard sizes.

Planck propounded his quantum theory to account for the observed properties of radiation. It had the appearance of a mathematical trick which enabled the theorist to obtain the right answer for this particular problem. There were no other important problems to which the quantum idea seemed immediately applicable.

Physicists, both theoretical and experimental, were exceedingly loth to admit, on the evidence of one rather esoteric phenomenon, that a fundamental departure from the Newtonian conception of nature was possible. They believed that the Newtonian conception of continuity in natural processes, with its two centuries of tremendous successes in the explanation of material phenomena, could not contain a limitation in its foundations. They tried every conceivable trick, with Jeans among their leaders, to find some way of reconciling the facts of radiation with the Newtonian conception of a continuous change. Even Planck himself, though confident from the first of the correctness of his quantum theory, hoped to the end of his long life that some kind of higher synthesis would be discovered, by which quantum theory and Newtonian theory would be reconciled.

But as the years passed, more and more anomalies were found to have their explanation in the quantum theory. In 1905 Einstein showed that the peculiarities of the photo-electric effect and of specific heats could be explained by the quantum theory. It was for this work, not for the Theory of Relativity, that he was awarded a Nobel Prize in 1921.

After fighting the quantum theory at every point, Jeans was finally convinced that it must be accepted.

The Physical Society of London invited him to write a *Report on Radiation and the Quantum Theory*. This was completed and published in 1914. It did more than any other work published in the English language to secure acceptance for the quantum theory, owing to its brilliant expository qualities and even more to its being rooted in the prejudices and habitual modes of thought of the English school of mathematical physics descending from Newton.

Jeans begins his *Report* with an illustration of the paradoxical nature of the properties of radiation. Suppose a lump of iron is inside an enclosure whose inner walls are perfectly reflecting surfaces. It is found that an equilibrium is established, in which the iron steadily absorbs as much energy as it emits. The amount of wave-energy coming into the iron from the surrounding space or ether is equal to the amount of wave-radiation emitted into the space by the iron. Yet it is easily shown that the amount of energy in the iron is a million million times greater than the amount in the surrounding space.

Jeans illustrates this paradox by a typical Newtonian illustration: 'A very little consideration will show that this state of things is different from what might be expected by analogy from other systems which are known to obey the ordinary dynamical laws. Consider, for instance, a tank of water (to represent the ether) in which is floated a system of corks (to represent atoms of matter) connected by light springs or elastics so that they can oscillate relatively to one another. Suppose that initially the surface of the water is at rest. Let the system of corks be set into violent oscillation and placed on the surface of the water. The motion of the corks will set up waves in the water, and these waves will spread all over the surface of the water, undergoing reflection when they meet the walls of the tank. We know that ultimately the corks will be reduced to rest; the energy of their motion will be transformed first into the energy of waves and ripples on the surface of the water, and then, owing to the viscosity of the water, into heat energy in the water. A final state in which the corks continue to oscillate with extreme vigour while the water has almost no energy is unthinkable; we expect a final state in which practically all the energy has found its way into the water.'

This tendency for the energy of any system moving in a medium to be transferred to the medium, and ultimately to be found in the shortest vibrations of which the medium is capable, is a direct consequence of Newton's laws and can be rigorously proved. 'This tendency is not observed in the crucial phenomenon of radiation; the inference is that the radiation phenomenon is determined by laws other than the Newtonian laws.'

Planck searched for a new kind of law, from which it would follow that the amount of energy gained by the surrounding space or ether is equal to the amount absorbed. He found such a phenomenon could occur if change occurred in nature, not in continuous amounts, as with Newton, but in discontinuous amounts, and he succeeded in formulating the law mathematically. In 1910, Jeans showed that there were strong reasons for concluding that Planck's theory of the partition or distribution of energy in radiation indicated that motion must be fundamentally discontinuous.

Finally, Poincaré, in his last paper, and one of his finest, which was published in 1912, conclusively proved that if the final partition of energy is that given by Planck's law, then no system of laws of motion can possibly lead to Planck's law, except one based on the quantum theory of discontinuous motion or change.

After outlining the mathematical arguments, Jeans extended his review: 'black body radiation has yielded what information there is to be extracted from it; to complete our knowledge we must survey other phenomena of Nature.' The quantum laws so far obtained merely indicated that exchanges of energy must occur in multiples of finite packets, they did not indicate 'when or why these exchanges may be expected to take place. Finally, they give no indication at all as to the seat of these quanta of energy, whether they are to be looked for in the ether or in matter.'

'In searching for other phenomena on which the Newtonian laws break down it is natural to examine closely those in which the temperature is very low, or the frequency of a vibration very high, or the 'action' of the process very small.'

The phenomena of specific heats at low temperatures, of the line spectra of the elements, and of the photo-electric effect were of these respective types. He discussed line spectra first. 'So long as physicists thought in terms of the Newtonian mechanics, almost every characteristic of the line spectrum was a source of difficulty.'

The line spectra of even the simplest atoms, such as hydrogen and helium, are very complicated. Yet, on Newtonian mechanics, the number of lines ought not to exceed three times the number of electrons in the atom. There was experimental evidence that, in general, an atom does not contain more optically active electrons than the number of its atomic weight. Summarizing a paper of Rayleigh, published in 1906, as the atom radiated energy, the orbits of its electrons would change. Consequently, on Newtonian principles, the frequencies of the oscillations about these orbits, which were supposed to produce the line spectra, would change. Hence 'each atom in a gas would have its own spectrum, so that the spectroscope ought, on this view, to show continuous radiation and not sharp lines.'

Since Rayleigh's paper of 1906, 'our knowledge of atomic structure has progressed from speculation to something approaching fairly near to certainty.' He then summarizes Rutherford's nuclear theory of the atom, and the first attempts, made by J. W. Nicholson, to give a quantum theory of its mode of operation. Very soon after this, in July, 1913, Niels Bohr began to publish his famous theory. Jeans describes it as 'very remarkable and intensely interesting.' It 'explains so many facts, and its numerical agreement with observation is so complete, that little doubt will be felt that it is at least based on a very substantial substratum of truth.' He swiftly outlines Bohr's theory, and its explanation of the hydrogen spectrum, and that certain lines previously believed to belong to hydrogen really belonged to helium.

A remarkably complete description could be given on the assumption that the electrons in the atom circulated, in general, in fixed orbits. They occasionally jumped from one orbit to another, and when this happened a quantum of energy in the form of radiation was emitted or absorbed. The wavelength of the radiation was precisely defined by the size of the quantum and Planck's constant h.

Enough has been said, says Jeans, 'to indicate how rich a field is opened by the application of the quantum theory to problems of atomic structure.'

Jeans next discussed the photo-electric effect. It is found that when light falls on certain metals, such as zinc, electrons jump out. Their velocity, or energy, never exceeds a certain maximum value, and is quite unaffected by the intensity of the light. If the light is very weak, the electrons jump with just the same velocity, but there are fewer of them.

Einstein pointed out in 1905 that this could be explained by supposing that the emitted electrons require a definite quantum of energy to free themselves from the metal. He suggested that light, and radiation itself, travelled in such quantum packets of energy, which he called 'photons'. When a 'photon' of appropriate energy fell on the metal, an electron of corresponding energy jumped out.

Jeans remarks that Einstein's theory 'falls into place very naturally as a logical extension of Bohr's theory.'

Then he discusses the law of Dulong and Petit, that the product of the atomic weight and the specific heat of an element is, at ordinary temperatures, approximately equal to 6. This has been explained by the classical kinetic theory of matter. But Nernst, Lindemann and others showed that at very low temperatures there are wide departures from the law.

Einstein gave the first adequate explanation by supposing that the energy in the solid did not reside solely in the kinetic energy of its atoms, but that these were absolutely at rest. Every atom had, however, three modes of vibration, all of the same frequency.

On the assumption that the energy of these vibrations was governed by Planck's formula, he obtained an expression which gave a much closer approximation to the observed results at low temperatures.

In 1912 Debye gave a perfected quantum theory of specific heats which 'both from its complete naturalness and from its agreement with experiment, seems destined to be final.'

Jeans concluded his report with a discussion on the physical basis for the quantum theory. The quantum theory was 'believed to have disclosed in nature a certain atomicity of a kind unsuspected by the older mechanics . . . the real atomicity, if it exists, is that of the entity measured by h, . . . which is physically of the dimensions of energy multiplied by time, or action. But an attempt to imagine a universe in which action is atomic leads the mind into a state of hopeless confusion.'

He discusses the atomicity of the electric charge e, and remarks that it 'has never received a physical explanation. It is in no way implied in the classical explanations of Maxwell. Any attempt to refer back the atomicity of e to the structure of the ether simply discloses the fact that the fundamental equations of the ether are not yet fully known.'

Bohr's theory showed that the emission of radiation by the atom must be catastrophic in the sense that a whole quantum is emitted or none at all. The photo-electric effect showed that the complementary action of absorption of radiation by the atom also occurred in whole quanta or not at all. Poincaré's analysis seemed to show that Planck's radiation formula could be true only if radiation *cannot* vary continuously.

Jeans then compares British and Continental physical outlooks. 'The modern school of British physicists have been inclined to regard the ether as a completely real substance—indeed to some of them the ether is the primary real substance of the universe. At the other extreme stands the relativity school, who almost deny that the ether has any reality at all.'

If the ether is ignored as 'too devoid of substantiality', the means through which the changes of energy occur must be the mechanism by which the radiation is produced. 'Almost all the Continental physicists have regarded the quantum theory from this point of view, which has the disadvantage that it is necessary to specify exactly the mechanism of radiation . . . so far no success has attended any explanation of why the energy of any type of mechanism should be incapable of changing except by quanta.'

Jeans ended his report by a quotation from Poincaré containing the verdict that: 'It is impossible at present to predict the final issue.'

It is evident that Jeans' conscious mathematical intelligence accepted the incontrovertibility of the quantum explanations and especially the finality of Poincaré's great mathematics. But the way he speaks of the ether, and his remark that British physicists regarded it as a 'completely real substance' and indeed the primary substance of the universe, shows that, for his physical intuition, the ether was still the most real thing.

The facts of nature required the 'quantum theory, and nothing but the quantum theory'. This theory made the ether superfluous, and yet he could not disbelieve in the ether.

After writing his Report, Jeans virtually ended his researches in mathematical physics and devoted the whole of his powers to cosmogony, where continuity and Newtonian mechanics still reigned.

He had made the inapplicability of classical mechanics to atomic physics incontrovertibly clear in his sharp, flashing, antithetical phrases. But he could not surrender Newtonian modes of thought. He preferred to retreat from the empire of the atom.

Jeans illustrated in himself the difference between the applied mathematician and the theoretical physicist. He was primarily interested in the operations which could be performed on the mathematical descriptions of physical ideas. He was not primarily interested in the physical ideas themselves, and was therefore not primarily a theoretical physicist. He failed as a theoretical physicist of the very first rank. He was shackled by the British tradition of applied mathematics, in contrast with the Continental tradition of theoretical physics, and unlike a Newton or a Clerk Maxwell he had not the genius of physical insight and intuition which enabled them to sense when the shackles were false and could be consciously or unconsciously ignored.

He made his most imaginative physical suggestion in 1904, in an attempt to explain radioactivity as due to the mutual annihilation of 'positive and negative' ions in violent collisions. Following Osborne Reynolds, he conceived these 'ions' as 'grains' in the ether. Such 'grains' should obey Maxwell's law of distribution of velocities, so that a few 'grains' of very high velocity should always be present. The coalescence of the positive and negative ions would 'result in the disappearance of a certain amount of mass. There would, therefore, be conservation neither of mass nor of material energy: the process of radioactivity would consist in an increase of material energy at the expense of the destruction of a certain amount of matter.'

Jeans did not develop this physical idea, though more than a decade later he used it effectively in the theory of the evolution of the universe and stars.

In the following year, 1905, Einstein showed that the equivalence of mass and energy followed at once from his first theory of relativity.

Jeans' abandonment of mathematical physics after 1914 was a silent acknowledgement of unfitness, a sign of his own lack of physical insight, and the supersession of the old British school of applied mathematics. Jeans had in a high degree the strengths and the weaknesses of this school, and his inability to adapt himself to the revolutionary quantum ideas, and develop them creatively, helps to explain why British theoretical physics during the last half-century has been so disappointing.

The spirit of theoretical physics could not be learned from Jeans. The influence of his widely used treatises and textbooks may not have been entirely good from that point of view. Owing to his gifts and limitations, he was one of the most significant characters in the unsatisfactory story of British theoretical physics in the first half of the twentieth century.

V. THE EVOLVING UNIVERSE

With his *Report*, Jeans took leave of the Atom and henceforth devoted his creative abilities to the investigation of the mechanics of the Universe.

He had followed his papers of 1902–3 on the mechanics of rotating masses of liquid, and gravitating masses of gas or particles in nebulae with an analysis of the behaviour of a gravitating sphere. He concluded that the origin of the planets must be due rather to gravitational instability than to rotational instability.

Then, in 1905, the Russian mathematician Liapounoff published a paper: 'Sur un Problème de Tchebycheff' in the *Memoirs* of the St. Petersburg Academy, in which he concluded that Sir George Darwin's confident opinion that the pear-shaped rotating masses 'by something short of an absolute algebraic argument', were stable, was incorrect.

Liapounoff described how, at the beginning of the 1880s, 'on the invitation of the illustrious Tchebycheff' to himself and to other Russian mathematicians, he had investigated the conditions under which the equilibrium of rotating ellipsoidal masses of liquid ceases to exist. He worked on the problem in 1882–83 and published a paper on it in 1884. He was unable, however, to overcome at that time some of the difficulties presented by the further approximations.

In 1885 Poincaré published his famous memoir. Liapounoff's paper of 1884 was overlooked and was not widely known until 1904, when a French translation was published in the *Memoirs* of the Academy of Toulouse.

Since 1882–83, Liapounoff 'did not have the occasion to occupy himself with the question' again until 1903. Finding that his result was contrary to that of George Darwin, he verified his calculations very carefully several times, but could not find any significant error, so he concluded that his result was correct. He believed that Darwin had been misled by the use of an infinite series containing an infinity of elliptical integrals, while he had himself used an algebraic function. He concluded that the pear-shaped figures, related to rotating ellipsoids of viscous liquids, were unstable. 'I do not insist, however, on this conclusion, because the considerations on which it is based may not be regarded as well-founded.'

Liapounoff's paper unsettled the subject again, and hence all deductions from it about the origin of heavenly bodies were again in doubt. Jeans returned to the problem now with undivided vigour. The transfer of his attention from physics to astronomy was marked by his joining the Royal Astronomical Society in 1909. He published his first paper in the *Monthly Notices* of that Society in 1913.

During the period of the First World War, he re-investigated the stability of rotating liquid bodies, and published a series of masterly papers in which he proved conclusively that the rotating pear-shaped bodies must be unstable. He devised a more manageable method of handling the problem, based on the notion of the gravitational potential of a distorted ellipsoid. He showed that no conclusions could be drawn from calculations which included only terms of the first and second order of size, and that it was necessary to include terms of the third order. He undertook this extension of the laborious calculations, obtaining the conclusive result, and exactly identifying George Darwin's error.

Jeans then extended his researches on the stability of rotating masses of incompressible liquid, to that of rotating masses of compressible and heterogeneous fluids, a field in which little had been accomplished previously. He expounded these classical results in his Adams Prize Essay, which he published in 1919 under the title *Problems of Cosmogony and Stellar Dynamics.*

This book, published when Jeans was forty-two years old, exhibits his powers at their finest. He starts by explaining that he has attempted to follow the path opened by Maclaurin, Laplace, Jacobi, Roche, Kelvin, Poincaré, and George Darwin. It will be noted that all of these were applied mathematicians, in whom, with the exception of Kelvin, with his strong bias towards physics and engineering, mathematical modes of thought were uppermost.

He quotes Darwin's remark in 1911 that further progress in cosmogony was blocked by 'ignorance of the figures of equilibrium of rotating gaseous masses'. Darwin had said that 'human life is too short to permit us to watch the leisurely procedure of cosmical evolution, but the celestial museum contains so many exhibits that it may become possible, by the aid of theory, to piece together bit by bit the processes through which stars pass in the course of their evolution.'

Jeans' object was to contribute to the theory by which the history of the celestial exhibits might be interpreted by building for it 'a framework of absolute mathematical truth'.

He quoted Herschel in support of disciplined conjecture, and he said he would propose a definite scheme of cosmogony, 'not in the belief that it will prove to be true, but in the hope that it may in some degree help others ultimately to find the truth.'

He attributes the origin of scientific, as contrasted with speculative, cosmogony to Galileo's discovery of Jupiter's moons, which revealed in one comprehensive glance the actual existence of a planetary system. This in itself virtually confirmed the Copernican theory of the solar system, and in fact indicated that at least two planetary systems existed. Hence there were probably others. It was natural to enquire how such planetary systems could arise.

The telescopic observations begun by Galileo and the gravitational conception founded by Newton provided the data and theory for the first suggestions.

In 1755 Kant published his nebular hypothesis of the origin of the whole stellar universe. He supposed that the universe had begun as a cold nebula at rest. It fell in through the mutual gravitation of its particles, and in the process of contraction began to rotate. Kant did not realize that if the nebula was not originally rotating, no subsequent rotation could arise through internal gravitation: rotation could not arise out of no rotation, owing to the law of the conservation of angular momentum. Kant's hypothesis followed that of Thomas Wright of Durham.

He supposed that as it condensed and rotated, it formed rings, like Saturn's rings. These became unstable and agglomerated into planets. The planets, in turn, continued to contract and form more rings, out of which the moons and satellites agglomerated.

Laplace, who was apparently unacquainted with Kant's theory, suggested in 1796 that the solar system had evolved out of a hot, lens-shaped nebula. With his better command of theoretical mechanics, he avoided Kant's mistake and endowed his nebula with rotation from the beginning. He supposed that the nebula would cool by radiation from the surface, and would contract by internal gravitation. In order to obey the conservation of momentum, the speed of rotation would increase with contraction. Under these conditions, he supposed that revolving rings of matter split off from the outer layers of the main mass, and out of these the planets condensed.

Jeans remarks that up to the discovery of spiral structure in nebulae (first seen with the Earl of Rosse's six-foot reflecting telescope, which was built in 1842), there had been a tendency 'to try to prove that Saturn's rings (the most sensational astronomical objects then known) formed an intermediary stage in the evolution of planetary systems'.

He suggested that it would be more scientific to investigate the abstract problem of how masses of matter of the kind found in the spaces of the universe behaved under the effect of various kinds of forces. When the different possible forms of behaviour had been worked out, 'possibly not much speculation will be required to identify' examples of them in the actual universe. Here Jeans' method is very clearly expressed.

The kind of forces whose effects are to be abstractly investigated are those of mutual gravitation between the constituent particles of the astronomical matter; the gravitational effects due to other bodies, which may be described as tidal forces; the forces due to rotation, collisions, etc. Special attention was to be given to the effects of these forces in time, so that their evolutionary significance could be elucidated.

The founders of mathematical cosmogony, who made the first exact analyses of the effects of rotation on celestial bodies, supposed, in order to simplify their problem, that they were made of liquids or incompressible fluids. Jeans proposed, in the first place, to expound their results and then to extend them. After this, he would work out the behaviour of rotating masses of compressible matter, that is, of gaseous matter, which is the kind generally found in nebulae and stars.

He discussed first the effects of internal gravitation on a mass of freely rotating liquid. This had a bearing on the theory of planetary evolution as due to rotation. Then he discussed the case of a non-rotating mass acted upon tidally by another mass, and of two masses rotating around each other and acting tidally on each other, as in double-stars.

He expounded Roche's demonstration that a satellite, revolving around a planet of the same density, will break up if it comes within a distance of about $2\frac{1}{2}$ times the radius of a sphere made by combining the material of both satellite and planet.

He described Darwin's 'paper of very great importance' in which he elucidated the mutual tidal distortion produced in two close liquid bodies. If a planet raises a tide on a satellite, which is resisted by friction, the effect is to transfer energy of rotation of the planet into energy of movement in the satellite. Consequently, the satellite moves more quickly and expands its orbit, drifting farther and farther away.

Jeans showed that the tidal distortion in a liquid sphere due to the passage of a neighbouring body caused the sphere to be transformed into Maclaurin and Jacobian ellipsoids, and then into the pear-shaped body, with the probability that each of the two parts into which the pear-shaped body divided, itself split into two parts, making four in all.

After showing that rotation and tidal forces may cause fission in a liquid, incompressible body, Jeans proceeded to work out a general theory by which the effects of these forces on compressible bodies, such as the gaseous nebulae and stars, could be deduced. He showed that a compressible spherical body might be changed by rotational and tidal forces into a lens-shaped body. This might break up, under these forces, by the ejection of streams of matter from points on its equator. Under certain conditions, which he analysed precisely, the centrifugal and tidal forces acting at these points could be such that a jet of matter would flow out of the body. He pointed out that this result was of the utmost importance for cosmogony.

Jeans then considered the effects of radiation of heat from a spherical mass of gas. He showed that as long as it is of low density, radiation will cause it to increase in density and temperature, and thus gradually condense into stars. Hence it was reasonable to conclude that the earliest matter is that of lowest density: which confirmed the hypothesis that the stars originate out of nebulae.

Matter once in existence would either disperse into space, or contract continually. 'The problem of tracing out the history of the astronomical universe is seen to be closely related to the abstract problem of following out the sequence of events in a rotating mass of gas. . . . Our typical nebulous mass of gas may now be briefly recapitulated. It has been supposed to come into existence in an entirely unknown way, probably forming at first an irregular mass of comparatively cold gas at a very low density. This will contract under its own gravitation and would in time assume a spherical form except that it is repeatedly being disturbed by tidal forces from passing masses. The effect of these is to set up a slow rotation which continually increases as the mass contracts. The mass assumes at first a spheroidal form, then a pseudo-spheroidal form, until . . . a sharp edge is formed round the equator. The figure of the mass is now lenticular in shape, and any further contraction results in matter being thrown off from the periphery or equator of the lens.'

Here Jeans' style is seen in its most characteristic quality: grand, almost cavalier, and slightly school-masterly. He writes with confidence on the evolution of nebulae without mentioning the theories of relativity and the quantum, and takes absolute space and time for granted.

He illustrated his theory of lenticular nebulae ejecting streams of matter from their equators by striking photographs of actual nebulae, whose appearance corresponded closely with the various mathematical shapes he had predicted for them. If the rotating nebula forms a ring by the splitting-off of outer layers, it could be shown that the mass of the ring must be of the same order as that of the central mass of the nebula. 'This result seems to dispose of Laplace's theory of the formation of the solar system', for according to this theory, the planets were supposed to have been formed out of a ring whose mass was small compared with that of the central body.

Here, again, Jeans is inclined to be sweeping. He dispatches a theory on one mathematical point. He is inclined to attribute too much to mathematical considerations. The business of a scientist is to be right, not logical.

He conjectured that spiral nebulae are masses of gas or dust rotating too swiftly to be in statical equilibrium, and he offered an explanation of the origin of their arms, which are such a striking feature. He supposed that the nebulae had passed through the spheroidal into the sharp-edged lenticular shape, and that subsequently matter was ejected along two arms originating from the sharp edge. The points of origin of the arms were determined by the minute tidal forces set up in the nebula by the rest of the matter in the universe.

The matter in the spiral nebulae appeared to be condensing into star-clusters similar to those observed in our own local Galaxy. We might regard the Andromeda nebula as 'a picture of the primeval nebula' out of which our own Galaxy has condensed.

The changes in the velocities of the stars may be regarded as due to the forces exerted by the main body of matter in the universe. The pull of near stars can be entirely neglected.

The motion of the stars can be compared with that of the molecules in gas, but with the collisions left out. It can be described by the formulae of the kinetic theory of gases, modified to omit collisions. The chance of a collision between stars in the present state of the universe is less than one in a million million years.

However, though stars continually contract in size, they steadily disperse, so that the spiral nebula that contains them continually grows in size. Thus, at an earlier period, when the parent nebula was much smaller, the constituent stars were much closer together, and collisions were much more frequent. 'Single stars must have behaved like monatomic molecules and binary stars like diatomic molecules. There is however the welcome difference that we understand the dynamics of a binary star, whereas we do not yet understand the dynamics of a diatomic molecule.'

Jeans now considered the origin and evolution of the solar system. He had described how the spiral nebulae, twin-stars, ring nebulae, and moving star-clusters might arise from rotating masses of matter. The solar system could not have originated through rotation. As Babinet pointed out in 1861, if the planets were originally produced out of the sun by rotation, the original sun would have been negligibly heavier owing to its containing the matter of the planets, which are very light compared with the mass of the sun. But the angular momentum of the original sun would have to have had sixty times its present value to equal that of the present solar system, which contains several large planets revolving at a great distance from the sun, and thus giving the solar system as a whole a large angular momentum.

Accordingly, the original sun must have rotated on its axis sixty times as fast, and had a period of revolution of 10 hours. This, however, is not fast enough to cause fission by rotation, as can be seen in the case of Jupiter, which revolves on its axis at about this rate, and appears to be very stable.

Nor can the ejection of matter from the equator of the sun through rotation provide an explanation of the origin of the planets. Jeans showed that the matter forming the earth would have had to have been ejected within a period of three years, or more probably a few months, in order to allow for condensation in gaseous form. This period being inconceivably short, the supposition would be that the material was ejected in the solid or liquid state. But then, disintegration by rotation of a solid or liquid body produces pairs of bodies of comparable size, such as binary stars, quite unlike the sun and planets.

Jeans therefore abandoned the rotational theory of the origin of the solar system and preferred the tidal theory, in which a mass is supposed to have passed near the sun, drawing a filament of matter out of it, which has subsequently condensed into the numerous comparatively small planets. The plane of the sun's rotation and that of the planets differs by six degrees. This cannot be explained by the rotational theory, according to which both planes should be the same.

The tidal theory offers the explanation that the plane of the approaching body was slightly inclined to that of the sun's rotation, and so the material was dragged out of the sun at an angle six degrees from the perpendicular to the sun's axis.

Jeans described the encounters between stars which might lead to the formation of planets through tidal action as transitory or slow, according to the time in which the two masses remained in close interaction.

He showed that an approaching star of double the sun's mass and travelling with a relative speed of forty kilometres a second would have to come into grazing contact with the sun, at that date occupying a volume as great as the present orbit of Neptune, in order to break it up by tidal action. The chance of such an encounter in the universe as it is at present was only once in 30,000 million years, which is much longer than the age of the universe.

Thus, if the universe always had its present general features, tidal encounters would be excessively improbable, and the origin of the solar system as due to them altogether unlikely. But in a much earlier epoch, nearer the beginning of the universe, the stars were much closer together, and moving with much slower relative motion. Under those conditions, slow encounters capable of producing planets out of stars by tidal action might occur once in 3 million years, a reasonable frequency.

The sun would be broken up when it was young, its size large, and density low. The gas in the filament drawn out by the tidal forces would condense into planets of about the size of Jupiter and Saturn.

The satellites of these major planets could not, however, be due to a condensation of gas. They were probably born in the solid or liquid state, by matter from the outer layers of those planets, which cooled very rapidly as soon as it left their surfaces.

The satellites of Mars and of Uranus and Neptune were to be explained by a second tidal cataclysm arising from a close approach of the planets to the sun, in which gas was ejected from the as yet uncondensed atmospheres of these planets. This secondary filament would condense into bodies of about the size of the observed satellites.

The 'genesis of the earth-moon system' remained indefinite, and the high inclination of the plane of revolution of the satellites of Uranus and the retrograde motion of the satellite of Neptune seemed to preclude a tidal explanation of their origin.

Jeans described his account of the tidal theory of the origin of the solar system as a 'vague sketch'. The theory was beset with difficulties and required an enormous amount of further mathematical investigation. But it seemed more satisfactory to him than the rotational theory.

Jeans then considered the time required for the solar system to have evolved to its present stage, and the amount and sources of energy necessary to propel this evolution. Kelvin had calculated long ago that the sun could shine at its present rate only for twenty million years if its source of energy was solely due to the work done by the falling-in of the sun's matter under its own gravitation.

Geology showed, however, that the evolution of the earth's crust required at least 250 million years, at a fairly steady temperature.

A single rotation of the Milky Way takes at the present rate about 320 million years. As the stars move with a remarkable uniformity of speed, the Milky Way has probably revolved at least ten times, for otherwise the stars would not have had time to settle down to a relatively steady state of motion.

Thus the Milky Way would be 3,200 million years old if its speed of rotation were constant. But it was probably much faster at an early date, taking then only 160,000 years, and gradually getting slower by geometrical progression down to its present speed. Allowing for this progressive decrease in speed of rotation, one may estimate that the Milky Way has been revolving for about 560 million years. This, then, is the order of age of the Milky Way and of the sun and earth.

The sun cannot, however, have been shining through energy provided by gravitational contraction for more than twenty million years. Nor could chemical change or radioactivity provide an adequate source. Where did the sun's energy come from?

'It is possible that other sub-atomic changes, unknown to us, may provide more energy . . .' Mere rearrangement of electrons in atoms would not provide enough, but 'there remains apparently as a last resource the possibility of energy being created by the destruction of matter, as for instance by positive and negative charges rushing together and annihilating one another.'

Taking the intrinsic energy of an electron to be mc^2, where c is the velocity of light, the reduction of the sun's mass by only 1 per cent 'would set free sufficient energy to furnish radiation at the present rate for 150,000 million years.' The annihilation of matter could therefore provide more than ample supplies of energy for the observed processes of the universe.

Jeans ended his book with the observation that 'the time for arriving at conclusions in cosmogony has not yet come.' His object had been rather to consider various hypotheses, leaving to future investigators, with more observational and mathematical knowledge, to decide which was the most probable.

He provisionally concluded that some hundreds of millions of years ago, the material of our Galaxy, the Milky Way, consisted of a single mass of very tenuous cold gas in slow rotation.

Owing to loss of energy by radiation, it contracted, and its angular velocity increased until it assumed a lenticular shape. Beyond a certain critical speed, matter was ejected from the sharp edge, providing the nebula with spiral arms. The gas in the arms gradually condensed into individual masses. These continued to shrink, accompanied by a rise in temperature, and thus became luminous stars. Their velocity of rotation increased, so that many became binary stars by the process of fission.

Occasionally two stars passed relatively near to one another. It was conjectured that 300 million years ago the sun experienced such an encounter, a larger star passing within a distance of about the sun's diameter from its surface.

At this time, the sun was dark and cold and its density was very low, its diameter being comparable with that of the present orbit of Neptune. The approaching star pulled out filaments of the sun's gas, which became colder and condensed into liquid masses at its ends, and perhaps partially in the middle, forming planets.

Owing to the orbital velocity given to these planets by the passing star, many of them would have very elliptic orbits, and thus would pass near to the sun. When they passed very near, the sun would draw out filaments from them by its tidal action, thus providing material which would condense into satellites.

The earth–moon system is exceptional, for the moon is relatively much bigger than the usual satellite. And the solar system itself 'appears to be exceptional, and for aught we know may be unique, in the system of the stars.'

So Jeans concludes his greatest book.

VI. THE SCIENTIFIC WRITER

Jeans did not mention the theories of relativity and the quantum in his *Problems of Cosmogony and Stellar Dynamics*. During the ten years following the publication of this book, he made some attempts to apply the newer physical theories to astrophysics and cosmogony. He summarized this work in his *Astronomy and Cosmogony*, published in 1928, which he described as a sequel of the earlier book. It was a bigger and more ambitious volume, but it was less satisfactory. He showed himself not at ease with the new physics, and his confidence drifted into mistakes.

His theory of liquid stars, based on his own views of atomic properties and of how atoms would behave inside stars, was held to be unconvincing, and seemed to show a lack of insight into modern atomic physics. Throughout the book he generally gave too much prominence to his personal views.

Jeans himself evidently felt that he had concluded his period of creative discovery, and he did not publish any more contributions to pure research.

In 1929 he published the first of his books for the general reader. This was *The Universe Around Us*. He described the book as 'A brief account, written in simple language, of the methods and results of modern astronomical research.' His ideal had been to make 'the entire book intelligible to readers with no scientific knowledge.'

He strove to put the substance of his two learned works on cosmogony into simple language. As before, he ascribed the beginning of modern cosmogony to Galileo's telescopic discoveries: 'the revolution of thought initiated by Galileo's observations of January 7th, 1610, proved to be the most catastrophic in the history of the race. The cataclysm was not confined to the realms of abstract thought.'

Galileo had shown that Aristotle, Ptolemy, and the majority of those who had thought about the structure of the universe during the last two thousand years had been 'utterly and hopelessly wrong. In estimating his position in the universe, man had up to now been guided mainly by his own desires, and his self-esteem; long fed on boundless hopes, he had spurned the simpler fare offered by patient scientific thought. Inexorable facts now dethroned him from his self-arrogated station at the centre of the universe; henceforth he must reconcile himself to the humble position of the inhabitant of a speck of dust, and adjust his views on the meaning of human life accordingly.

'The adjustment was not made at once. Human vanity, reinforced by the authority of the Church, contrived to make a rough road for those who dared draw attention to the earth's insignificant position in the universe. Galileo was forced to abjure his beliefs.'

So Jeans majestically describes the impact of a great astronomical discovery on human thought and social behaviour. Why does man study astronomy? Because he 'wishes to probe further into the past and future than his brief span of life permits.'

'The wish does not originate solely in mere intellectual curiosity . . . it has deeper roots and a more personal interest. Before he can understand himself, man must first understand the universe from which all his sense perceptions are drawn.'

In addition to this, astronomy offers at least three other interests, the utilitarian, scientific, and aesthetic. It started mainly as a utilitarian science; 'it provided measures of time, and enabled mankind to keep a tally on the flight of the seasons, it taught him to find his way across the trackless desert, and later, across the trackless ocean.'

This was not intrinsically absurd, for astronomy is still largely concerned with foretelling the future. Astrology went wrong in supposing that the empires, kings, and individuals of the earth were so important that the scheme of the universe could be bound up with their fates. 'As soon as man began to realise, even faintly, his own insignificance in the universe, astrology died a natural and inevitable death.'

Astronomy had recently acquired a new interest through becoming 'an integral part of the general body of science', the advance of which might depend on that in any of its parts. For instance, astronomy might throw light on physics and chemistry, by revealing physical and chemical phenomena that cannot be produced in laboratories on the earth. A star is now regarded as 'an experiment on a heroic scale, a high-temperature crucible in which nature herself operates with ranges of temperature and pressure far beyond those available' to us.

Jeans now begins his comparison of magnitudes, or orchestra of numbers, which is his chief technique of illustration, and the most characteristic feature of his expository style. It no doubt sprang from the same source as his passion for music, especially for that of Bach.

Science advances along an unbroken front in its investigation of matter: 'from electrons of a fraction of a millionth of a millionth of an inch in diameter, to nebulae whose diameters are measured in hundreds of thousands of millions of millions of miles.'

Matter in nebulae exists in a state at least a million times less dense, and in some stars about a million times more dense, than any that can be produced on earth.

Astronomy presents us with a similar extension of the range of time. Together with geology and biology it gives us the following time-scale of development:

Age of earth	about	2,000,000,000	years
Age of life on earth	,,	300,000,000	,,
Age of man on earth	,,	300,000	,,
Age of astronomical science	,,	3,000	,,
Age of telescopic astronomy	,,	300	,,

Astronomy is a very recent phenomenon. Its 'total age is only a hundredth part of the age of man.' And 'during 99,999 parts out of the 100,000 out of its existence, life on earth was hardly concerned about anything beyond the earth.'

Here Jeans seems to be definitely wrong. There is evidence from burials that sub-human ancestors believed in a future life. It would have been more correct to say that until the last one-hundred-thousandth of its existence, life had not gained any considerable knowledge of what happens outside the earth.

Jeans unfolds his story by describing how the sky, the atom, and time are explored. Then he outlines the evolution of the universe and the stars. He explains how William Herschel had begun to elucidate the structure of the Galaxy, showing that the stars were distributed in it in a biscuit-like shape, which resembled that of the nebulae. This led Herschel to regard the latter as 'island universes'.

Their size is colossal. The photograph of the Andromeda nebula, reproduced in his book, would, says Jeans, 'have to be enlarged to the size of the whole of Europe before a body of the size of the earth became visible in it, even under a powerful microscope'.

As Newton had suggested, the distance of the stars might be estimated from their brightness. The star Altair appears about as bright as the planet Saturn, yet, as Kepler had maintained, Altair probably emitted as much light as the sun. Saturn shines only by reflected light, reflecting not more than one part in 10,000 million of the light emitted by the sun. Hence Altair must be about 10,000 million times as intrinsically bright as Saturn. As apparent brightness falls off with the square of the distance, the distance of Altair must be about 100,000 times that of Saturn; 100,000 being the square root of 10,000,000,000.

The speeds of the planets and stars are on the average about 1000 times that of an express train. The earth travels round the sun at 18½ miles a second, or 1200 times the speed of an express train. The sun moves through the stars at about 800 times this speed.

In spite of the similarity of speed of the planets and stars, the movement of the planets is evident day by day, while that of the stars is inappreciable even in centuries. This shows how distant are the stars. In fact, the nearest one is 25 million million miles away, or 4·27 light years—the time taken by light to cover the distance at 186,000 miles a second.

The distances of the island-universes can be calculated from the apparent brightness of Cepheid stars detected in them. These stars vary regularly in brightness, 'as though someone were throwing armfuls of fuel on to a bonfire at perfectly regular intervals'.

Miss Leavitt made the very remarkable discovery that their absolute brightness or candle-power depends on their period. Thus once a Cepheid star is detected, and its period measured, its distance can be simply derived. A Cepheid whose life fluctuates in a period of ten days is about 1600 times as luminous as the sun. If it has the apparent brightness of a star of magnitude 16, i.e. 'that we receive as much light from it as from a single candle at a distance of 10,000 miles,' then it is distant 3,600,000 light years.

Then the numbers of stars are estimated. About 3000 are visible to the perfect unaided eye and about 1,500 million in a 100-inch telescope, 'about equal to the number of men, women and children in the world.'

Shapley has estimated that the Galaxy or Milky Way contains 100,000 million stars.

About two million island-universes are visible in the 100-inch telescope, each containing thousands of millions of stars. Their average distance apart is about 2 million light years, and the most distant of them is about 140 million light years away.

Jeans then explains that relativity theory indicates that space itself may be limited in extent, and that we can already see into a large fraction of it. 'Our astronomer-explorers may not as yet have circumnavigated the globe, but they are perhaps discovering America, and we can well imagine that even the next generation will have completed the circumnavigation of space, and will think of a finite but unbounded space in the same way, and with the same ease, as we think of the finite but unbounded surface of the earth.'

Jeans now constructs a model of the distances of the universe. He takes the circumference of a pin-head as the equivalent of the orbit of the earth. The sun is 'a minute speck of dust 1/3400 of an inch in diameter', while the earth is so small that it cannot 'be seen at all even in the most powerful of microscopes'. On this scale, the nearest star will be 225 yards away, and the model must be a mile long and a mile high if it is to contain the hundred nearest stars.

The Milky Way would stretch more than 7000 miles on its longest diameter, and the whole galaxy would be about the size of the American continent.

After we have completed the model of the Milky Way, we would have to travel another 30,000 miles before we came to the nearest island-universe.

'So we go on building our model—a family of thousands of millions of stars every 30,000 miles or so—until we have two million such families. The model now stretches for about four million miles in every direction.'

The total number of stars must be about equal to the total number of specks of dust in London. 'Empty Waterloo Station of everything except six specks of dust, and it is still far more crowded with dust than space is with stars.'

Throughout the model, the mean distance of a speck of dust, that is, a star, from its nearest neighbour would be about 80 miles. 'The universe consists in the main not of stars but of desolate emptiness.'

So much for the size of the universe: now we pass to the atom. 'A pint of water contains $1{\cdot}89 \times 10^{25}$ molecules, each weighing $1{\cdot}06 \times 10^{-24}$ ounces. If these molecules were placed end to end, they would form a chain capable of encircling the earth over 200 million times. . . . If we think of the molecules as tiny seeds, the total amount of seed needed to sow the whole earth at the rate of 100 million molecules to the square inch could be put into a pint pot.'

The molecules of ordinary air travel at about 500 yards a second, the speed of a rifle bullet. Each molecule collides with some other about 3000 million times a second, and travels a distance of about 1/160,000 inch between successive collisions.

As for the atom, small though it is, its volume is about 1,000 million million times that of an electron, and still greater than that of the nucleus which contains the vast atomic energy.

These are the atomic and sub-atomic sizes. What of those of light-waves and other radiations? The wavelength of the shortest visible violet light is about 0·000015 inch. X-rays are about a hundredth to a thousandth of this, while the γ-rays from radioactive substances are of the order of one ten thousand millionth of an inch.

Radiation, being a form of energy, has mass. When it strikes an object it exerts a pressure on it, though this may be very small. 'All the radiation emitted from a 50 horsepower searchlight working continuously for a century weighs only about a twentieth of an ounce.'

Jeans even expresses the 'ultimate fate of an ounce of uranium' by the equation:

$$1 \text{ ounce of uranium} = \begin{cases} 0{\cdot}8653 \text{ ounce lead} \\ 0{\cdot}1345 \quad \text{,,} \quad \text{helium} \\ 0{\cdot}0002 \quad \text{,,} \quad \text{radiation} \end{cases}$$

Having described the size of the atom, he gives an idea of how it works according to the quantum theory. The atom changes its state only by finite amounts of energy. 'The problem of shifting an atomic system is like that of extracting a box of matches from a penny-in-the-slot machine; it can only be done by a special implement, to wit a penny, which must be of precisely the right size and weight—a coin which is either too small or *too large*, too light or *too heavy*, is doomed to fail. If we pour radiation of the wrong wave-length on to an atom, we may reproduce the comedy of the millionaire whose total wealth will not procure him a box of matches because he has not a loose penny, or we may reproduce the tragedy of the child who cannot obtain a slab of chocolate because its hoarded wealth consists of farthings and half-pence.'

After exploring the atom, he explores time. His eloquence reaches its height when he expounds his favourite subject and his first love in research—the equipartition of energy.

The conditions which are necessary for it to be true 'happen to be amazingly simple, indeed it is difficult to believe that such wide consequences can follow from such simple conditions. They amount to practically nothing beyond a law of continuity and a law of causation.'

It follows from the law that in a mixed system of bodies interacting on each other, they ultimately reach a condition in which, on the average, heavy bodies are moving slowly and light ones faster, so that the energy of each one is, on the average, the same.

It is possible to calculate the time which a system will take in order to reach such a condition. Conversely, when equipartition of energy prevails, it is possible to calculate the age of the system.

Now observation shows that the stars are in such a state: the heavy ones moving slowly and the light ones fast. The question of how long the stars must have interacted gravitationally to reach such a condition 'now becomes one of absolutely fundamental importance, for *the answer tells us the ages of the stars.*' Reverting to the Waterloo Station model: 'how long must six particles of dust, floating blindly about in Waterloo Station', at a speed of an inch in five years, take to reach a state of equal energy, through close meetings with other specks of dust?

The mathematician can make a similar calculation for the stars, and his answer is that the time taken will be from five to ten millions of millions of years. 'This, then, must be the length of life of the stars.'

How can the stars keep shining for such enormous periods of time? Only by the annihilation of matter. But the amount of energy in matter is so large that the supply is ample. 'The sun's present store of atoms would, at the present rate of breakage, last for fifteen million million years.' This means that in the sun only one atom in every 15 million million is used to provide energy.

'In the ordinary combustion of coal we are merely skimming off the topmost cream of the energy contained in coal, with the consequence that 99·999999994 per cent of the total weight remains behind in the form of smoke, cinders or ash. Annihilation leaves nothing behind.' If we could annihilate a pound of coal it would provide sufficient energy to keep the British nation going for a fortnight.

Jeans now proceeds to explain how the universe is built up. Somehow the molecules become collected together in stars. 'Why should the molecules in space herd together in this way, when the molecules in the rooms in which I am writing and you are reading do not?'

The latter are prevented by the earth's gravitational pull on them. In distant space this local force is absent, and the mutual gravitation between the molecules is not interfered with. They gradually attract each other and collect into nebulae, and then condense into stars.

The closeness of packing of the atoms may be very fine; van Maanen's star is about the size of the earth, and yet its weight is comparable with that of the sun. Its density is probably about 300,000 times that of water.

This is due to the stripping of the outer parts of the atom owing to the extremely high temperature inside stars of this kind. The quanta of radiation which fly about at such temperatures are able to strip all, or nearly all, of the electrons from the atoms, 'just as surely as, on a hot day, a lump of ice breaks up into its constituent molecules'.

The structure of a carbon atom can be compared to six wasps (or electrons) buzzing about in Waterloo Station. If the atom is broken up so that the wasps are made to fly closer together, its total volume will be much reduced. 'Six wasps which can roam throughout the whole of Waterloo Station can nevertheless be packed inside a very small box.'

The temperature of the centre of the sun is probably about 50 million degrees. A pin-head, a cubic millimetre in size, would radiate so fast at this temperature that an engine of 3,000 million million horse-power would be required to keep it going. In empty space, the pin-head 'would emit enough heat to kill anyone who ventured within a thousand miles of it.'

Then Jeans expounds his own theories of the existence of super-radioactive atoms which supply energy inside the stars, and of liquid stars consisting of stripped atoms so close together that they jostle each other like molecules in an ordinary liquid. Neither of these theories has been generally accepted.

The sun contains enough matter to continue to shine at its present rate for 15 million million years. 'By contrast, van Maanen's star can, and probably will, go on radiating at its present rate for 15 million million years without losing more than about a thousandth part of its present weight.'

When a star falls into the white dwarf state, like van Maanen's star, with its atoms very closely packed, and scarcely able to radiate, it seems to gain a new lease of life for a period of thousands of millions of years—'we can only wonder to what purpose.'

The sun weighed 360,000 million tons more yesterday than today, and its mass is such that it can keep up this loss of weight by pouring out light and radiation for 15 million million years. By contrast, the earth's daily loss of weight owing to the radiations emitted from its radio-active materials is only about 90 lb.

How can the stupendous outpouring of energy in all the stars of the universe be kept up? Are there cyclic processes at work, by which the outpouring is in some way perpetually renewed?

Jeans holds that 'the second law of thermodynamics rules out any such possibility.' Experience shows that we must think of energy 'not only in terms of quantity, but also in terms of quality.' Its quantity always remains the same, but its quality always changes in the same direction, from a few quanta of high energy to many quanta of low energy, and it is as impossible to reconstitute the original high energy quanta 'as it was to put Humpty-Dumpty back on his wall'.

'Change can occur only in the one direction, which leads to the heat-death', where all the universe is at a uniform temperature and nothing happens. 'With universes as with mortals, the only possible life is progress to the grave.' Many, giving rein to their fancy, have imagined that somehow the ultimate uniform heat energy may be reformed into new electrons and protons. But science cannot support such fancies. 'Perhaps it is as well; it is hard to see what advantage could accrue from an eternal reiteration of the same theme, or even from endless variations of it.'

What would happen if all the matter of the universe were annihilated into radiation? Space is so exceedingly empty that the enormous amount of radiation produced would raise its temperature by only one six thousandth part of a degree.

For all we know, the radiation from thousands of annihilated material universes may now be wandering through space without our noticing it. The present matter of the universe cannot have existed for ever. An upper limit of its age is, perhaps, 200 million million years. We are forced to imagine a time when matter was created. 'In some way matter which had not previously existed, came, or was brought, into being.'

We may imagine radiation of very high energy being poured into space and materializing into particles. 'If we want a concrete picture of such a creation, we may think of the finger of God agitating the ether.'

Modern theories of relativity suggest, however, that it is meaningless to speak of space and time existing at all before matter existed. Travelling back in the relativistic picture brings us not to its creation but to its edge: 'the creation of the picture lies as much outside the picture as the artist is outside his canvas.'

'On this view, discussing the creation of the universe in terms of time and space is like trying to discover the artist and the action of painting, by going to the edge of the picture.'

This brings us very near to those philosophical systems which regard the universe as a thought in the mind of its creator. This reduces all discussion of material creation to futility. The plain man, who recognizes that the human mind cannot comprehend the full plan of the universe, decides that his own efforts shall stop this side of the creation of matter. This point of view is perhaps the most justifiable philosophically.

Jeans aligns himself with Poincaré, who definitely stopped trying to explain phenomena, and resigned himself to describing them in the simplest possible way.

The astronomer must faithfully record the message he receives from the universe. It is of obvious interest to philosophy, religion, and mankind in general, but 'it is not the business of the astronomer to decode it.' It is for others to try to understand and explain the ultimate meaning of the knowledge he learns and writes down.

Jeans concludes that life must be limited to a tiny fraction of the universe. Three centuries ago, Giordano Bruno suffered martyrdom for believing in the plurality of worlds. We can still only guess as to the meaning of this life which is so rare. 'Is it the final climax towards which the whole creation moves', for which all the transformation of matter through millions of millions of years is 'only an incredibly extravagant preparation'? Or is it an unimportant by-product of natural processes? Or a disease, 'which affects matter in its old age'? Or is it the 'only reality, which creates, instead of being created by, the colossal masses of the stars and nebulae and the almost inconceivably long vistas of astronomical time'?

'It is not for the astronomer to select between these alternative guesses.'

The immediate outlook for the human race is not unhopeful. The luminosity of the sun will probably not decrease by more than 3 per cent during the next 150,000 million year.

Present conditions on the earth may remain fairly constant for the next million million years. This is five hundred times the past age of the earth and three million times the period through which humanity has so far existed. Looked at in terms of time, the outlook for humanity is 'one of almost endless possibility and hope'. The material universe may astronomically be near the end of complete dissolution into radiation. 'But as the inhabitants of the earth, we are living at the very beginning of time. We have come into being in the fresh glory of the dawn, and a day of almost unthinkable length stretches before us with unimaginable opportunities for accomplishment. Our descendants of far-off ages looking down this long vista of time from the other end, will see our present age as the misty morning of the world's history; our contemporaries of today will appear as dim heroic figures who fought their way through jungles of ignorance, error, and superstition to discover truth, to learn how to harness the forces of nature, and to make a world worthy for mankind to live in.'

'The main message of astronomy is one of hope to the race and responsibility to the individual—of responsibility because we are drawing plans and laying foundations for a longer future than we can well imagine.'

So Jeans ended his first book designed for the general reader. It had a great success, and it immediately established his fame as an expositor of science to those without technical scientific knowledge.

VII. THE ENTERTAINER

The success of *The Universe Around Us,* in which Jeans made a sustained effort at explanation for the non-technical reader, increased the demands on him for lectures to general audiences. He was asked to give the Rede Lecture at Cambridge in 1930, in which he attempted to describe in the simplest language the implications of recent astronomy and physics for 'our outlook on the universe as a whole, and on our views as to the significance of human life'. He decided to expand the lecture into a book, which might be read as a sequel to *The Universe Around Us*, and he named it *The Mysterious Universe*. He described his interpretations of modern scientific theory as those of 'a stranger in the realms of philosophical thought', and remarked that many would disagree with them—they were 'written to this end'.

This book, the sales of which ran into six figures, the kind of numbers of which Jeans loved to speak, was written in a not altogether serious style, a little like that of an after-dinner speech to a scientific society. The scientific explanations became thinner and the philosophical judgments more confident and even reckless. Caution seemed to be sacrificed to effect.

He outlined how the triumphs of Galileo and Newton had started a movement 'to interpret the whole material universe as a machine'; Newton had shown how comets, which had hitherto been regarded as portents of the deaths of kings, obeyed the universal law of gravitation, and had written: 'would that the rest of the phenomena of nature could be deduced by a like kind of reasoning from mechanical principles.' By the latter half of the nineteenth century, Helmholtz declared that 'the final aim of all natural science is to resolve itself into mechanics.' The characteristic of a machine, such as a clock, is that its action is completely and absolutely determined by its structure: it operates by a continuous causation, which leaves no possibility of variation.

Each success of the mechanical interpretation of nature made belief in free-will more difficult, and this produced a natural reaction towards idealist philosophies.

While philosophers were still debating whether it might be possible to construct a machine which could think like Newton, compose like Bach or paint like Michelangelo, Planck showed that even the comparatively simple phenomenon of radiation of heat and light could not be explained on principles of continuous operation, like an ordinary machine. He introduced his quantum theory of discontinuous action in order to give a logical account of the observed properties of radiation

Jeans said that the quantum theory 'marked the end of the mechanical age in science.' This view of Jeans' does not seem to be correct. The quantum theory showed only that small machines like atoms do not work in the same way as large machines like steam-engines. The existence of the science of Quantum Mechanics shows that there is a mechanics of atomic operations, which applies to atomic reality just as well as ordinary mechanics applies to the reality of steam-engines.

Jeans then describes the spontaneous disintegration of radium, in which atoms explode by chance. It can be predicted that one out of every 2000 atoms in a piece of radium will explode within a year, but it is impossible to predict when any particular atom will explode. It seems to happen 'because in some way fate knocks at the door'.

It had been suggested that radioactive atoms exploded because they were struck by cosmic rays. This had been disproved by taking radium down a mine: it disintegrated at the same rate when protected by miles of earth as on the surface. It had been suggested, also, that cosmic radiation may cause changes in the germ-plasm of living things and thus be responsible for evolution, and turning 'monkeys into men'.

We had long thought of the workings of nature as representing the acme of precision, and we had cherished a belief that though our own man-made machines were inaccurate, 'the innermost workings of the atom would exemplify absolute accuracy and precision.' But Heisenberg had now shown that 'nature abhors accuracy and precision above all things.' His famous Uncertainty Principle indicated that the position and speed of an electron cannot be simultaneously determined with complete precision.

Jeans describes nature as permitting herself a certain margin of error. 'It is as though the position and motion of the electron had been marked on the two different faces of a lantern slide. If we put the slide in a bad lantern, we can focus half-way between the two faces, and shall see both the position and motion of the electron tolerably clearly. With a perfect lantern, we could not do this; the more we focused on one, the more blurred the other would become.

'The imperfect lantern is the old science . . . the perfect lantern in the new science . . . merely shows us that the specifications of position and motion lie in two different planes of reality.'

Jeans' ingenious illustration misses the point. He assumes that the old ideas of complete mechanical determinism are naturally right and precision is to be measured by coincidence with the old ideas. It is a mistake to suppose that Heisenberg's Uncertainty Principle is not precise: on the contrary, it gives an absolutely precise description of atomic reality. It is expressed in a mathematical formula, containing Planck's constant h.

Jeans describes h as 'mysterious'. But it is impossible to see why one constant of nature should be any more mysterious than another.

Jeans remarks that if we toss a million tons of half-pence, 500,000 tons will come down heads and 500,000 tons tails, and we shall always get that result. We might regard this as evidence of the uniformity of nature but 'in actual fact it is an instance only of the operation of the purely mathematical laws of chance.'

Here he seems to be putting the cart before the horse: are not the laws of probability to be deduced from how real halfpennies fall, rather than the way they fall be deduced from mathematical laws?

He remarks that the number of atoms involved in any ordinary piece of matter is enormously greater than one million: hence its properties can be calculated with all the more precision. 'It is easy to see how the illusion of determinacy—if it is an illusion—crept into science.'

Jeans contends that because atomic phenomena are better described by probability rather than continuous mathematical functions, 'the new physics contains more room than did the old mechanical picture of life and consciousness to exist within the picture itself'; for aught we know, 'the gods which play the part of fate to the atoms of our brains may be our own minds.'

The discovery that material particles have wave-properties suggests that 'we live in a universe of waves, and nothing but waves'. The whole story of creation 'can be told with perfect accuracy and completeness in the six words: God said, "Let there be light." '

He makes a persistent and almost dogged attempt to outline relativity theory and its implications for physics and astronomy. He describes how many years' arduous experimenting had seemed to show that the forces of nature were 'parties to a perfectly organized conspiracy to conceal the earth's motion through the ether'. This indicated that a natural law was involved. The unscientific inventor may exclaim in despair that the forces of nature are conspiring against him when he fails to make his perpetual motion machine work. 'The zealous but unenlightened social reformer and the ignorant politician are alike apt to see conspiracies of the deepest dye behind the operation of those economic laws which make it impossible to extract a quart out of a pint pot.'

Somehow or other, though, Jeans extracted £256,000 out of society.

The Michelson-Morley experiment showed that a mechanical ether, the trembling jelly imagined by the nineteenth-century physicists, could not exist. The signal for its despatch was a short paper published by Einstein in 1905. 'With its publication, the study of the inner workings of nature passed from the engineer-scientist to the mathematician.'

Did it? Einstein was then an electrical engineer working in the Swiss Patent Office. It is more reasonable to say that his engineering training enabled him to perceive the physical reality, and thus know what kind of mathematics to look for, than to suppose that his mere mathematical training, rather a modest one, enabled him, without physical insight, to make such a profound discovery.

Jeans makes a brave attempt to give a notion of the space-time continuum. He tells us that 'we shall not measure time in ordinary seconds, but in terms of a mysterious unit equal to a second multiplied by $\sqrt{-1}$ (the square root of -1). Mathematicians speak of $\sqrt{-1}$ as an "imaginary" number, because it has no existence outside their imaginations.'

But $\sqrt{-1}$ can be given a geometrical meaning as part of the definition of what mathematicians call a complex number. Just as ordinary numbers may be regarded as displacements along a straight line, so "imaginary" numbers may be regarded as displacements at right angles to this line—they are perfectly concrete conceptions, and they are not more mysterious than any other numbers.

The theory of relativity abolishes the force of gravitation. Matter itself becomes merely a crumple in space, a large or a small crumple. Bodies move towards each other because they bend space in their vicinity; they are constrained by their self-produced kinks in space to follow their observed paths.

Electrical phenomena will probably be explained in a similar way, though a satisfactory theory has not yet been found.

Jeans concludes that a 'soap-bubble with irregularities on its surface is perhaps the best representation, in terms of simple and familiar materials, of the new universe revealed to us by the theory of relativity.' The substance out of which this bubble is blown, is 'empty space welded on to empty time'.

It may be that time, from the 'beginning to the end of eternity is spread before us in the picture, but we are in contact with only one instant' like a bicycle wheel which touches only one point of the road. Or, as Weyl said, events do not happen, we merely come across them. But Plato had written twenty-three centuries earlier, 'We say "was, is, will be", but the truth is that "is" can alone properly be used.'

We must 'interpret consciousness as something residing entirely outside the picture.'

'The essential fact is simply that *all* the pictures which science now draws of nature, and which alone seem capable of according with observational fact, are *mathematical* pictures.'

Here we must enquire whether biology and geology, in which there is little mathematics, are sciences? The statement is untrue for all objects above the sub-microscopic, for they can be seen in operation and working drawings can be made of them, which are not entirely mathematical. Even sub-atomic phenomena can be given a certain non-mathematical visibility with such apparatuses as the Wilson cloud chamber.

Jeans argues that relativity and quantum theory show that 'we are not yet in contact with ultimate reality.' He is therefore fascinated by Plato's famous theory of shadows. 'We are still imprisoned in our cave', and all we can do is to describe the movements of the shadows thrown on the prison walls by the real objects existing in the world outside. We can do this more fully and clearly by mathematics than in any other way.

There is no more reason why we should understand the shadows which reality throws on the wall, than a dog that strays into a biology lecture room should understand a cinematograph film of tissue-culture growth, which happened to be being shown at the moment.

'Nature seems very conversant with the rules of pure mathematics, as our mathematicians have formulated them . . . out of their own inner consciousness and without drawing to any appreciable extent on their experience of the outer world.'

Jeans replies to the objection, that the mathematician finds mathematical laws in nature only because he observes nature through mathematical blinkers that he has made for himself, that the animistic interpretation of nature had been tried by primitive man and found wanting. Then the engineering interpretation had been tried, and that too had failed. But the mathematical interpretation was proving brilliantly successful. 'It would seem beyond dispute that in some way nature is more closely allied to the concepts of pure mathematics than to those of biology or of engineering.'

He views with disfavour 'the possibility of the universe having been planned by a biologist or an engineer; from the intrinsic evidence of his creation, the Great Architect of the Universe now begins to appear as a pure mathematician.'

He admits, however, that it cannot be denied that some of the pure mathematicians' concepts are taken directly from his experience of nature. For instance, there is the fundamental notion of quantity. In his opinion, the laws that nature obeys are more like those of a musician composing a fugue or a poet a sonnet than of a machine in motion. The universe can best be pictured as a 'pure thought' of a 'mathematical thinker'.

And so Jeans finds himself at one with Bishop Berkeley, who held that 'all those bodies which compose the mighty frame of the world . . . subsist in the mind of some Eternal Spirit.' Substantiality is taken to be a purely mental concept. The making of models or pictures of nature is condemned as 'like making graven images of a spirit'.

Our knowledge of the universe is merely a contact between our minds and 'a creation of mind'. We need not therefore be awed by the apparent vastness and emptiness of the universe, nor depressed by our own insignificant size. We are not terrified by the sizes of the structures created by our own thoughts.

Thermodynamics indicates very strongly that the universe must have had a beginning. If the universe is a thought, it must have been created by an act of thought.

Our new knowledge compels us to abandon our mechanist idea that we had stumbled into a universe which was hostile to life. The old dualism of mind and matter was being resolved through matter proving to be a 'creation and manifestation of mind'. The universe is 'more like a great thought than a great machine'. It shows evidence of a designing or controlling power that has something in common with our own minds, not with our emotions, morality or aesthetics, but with our mathematical powers.

Finally, he underlines that every conclusion that he has drawn is 'frankly speculative and uncertain', and perhaps wholly illusory. And after all that he has said, his last words are that 'science should leave off making pronouncements: the river of knowledge has too often turned back on itself.'

So, after the spate of speculation, we are told to curb our speculations!

On the same page, he quotes the Platonic shadows to show that we are not in contact with ultimate reality, and then tells us that we are receiving a 'whole torrent of surprising new knowledge'.

'No one except a mathematician need ever hope fully to understand those branches of science which try to unravel the fundamental nature of the universe.' Were, then, Faraday and Rutherford, the founders of electro-magnetic theory and nuclear physics respectively, for ever precluded from an insight into nature?

Jeans himself describes idealist philosophy as a reaction from mechanistic or engineering physics. Why did he remain an idealist after engineering physics had been replaced by relativistic and quantum physics? What is the philosophical reaction to them? Why had he not caught up with that, instead of lagging behind in the old philosophy produced as a reaction to Newtonian mechanics? Jeans was a prisoner of the engineering physics which he criticized so scathingly.

The enormous circulation of *The Mysterious Universe* owed much to its meretricious style and shallow philosophical thought. Jeans was pressed to write more books for the general reader, and he produced several, which were duller but written with more pains. The best of them was on *Science and Music*, which is in a thorough, entirely modest style. He was too deeply attached to music to joke about it. His *Physics and Philosophy*, published in 1942, shows him beginning to acquaint himself, in a distinctly pedestrian style, with the elements of philosophy, about which he had had such high jinks twelve years earlier.

The Mysterious Universe received fierce criticism when it was published. Rutherford was heard to say that Jeans had told him that 'that fellow Eddington has written a book which has sold 50,000 copies; I will write one that will sell 100,000.' And Rutherford added: 'He did.'

It is said that on another occasion, he told some of his younger colleagues 'that if any of them was thinking of writing a book on the Atom, and calling it the Mysterious Atom, he had better come to see him in his study for a few minutes' talk before he went to a publisher.'

Why was Jeans so successful as a popular writer? Because he was clear, vivid, confident. His classical Newtonian prejudices helped him, for he shared these with his audience, and both started from them.

Why was his influence as such a writer so unfortunate? Because he was philosophically and socially shallow. His lack of reading in his earlier popular books was striking.

But his most serious fault was his attitude towards popularization. He quite consciously undertook it because he felt that his capacity for discovery, his highest intellectual powers, were waning. He was under the vulgar impression that second-rate powers were good enough for explaining science to the general reader. This revealed a contempt for the people. It arose from his social milieu and class-feeling. He was a rich, highly individualistic bourgeois, who was primarily concerned, as a scientist, with his own personal researches. He created no school of research, and apart from a donation of £1000 for a George Darwin lectureship and £2000 to the Marie Curie Hospital, he did not bequeath much of his enormous fortune to science.

Jeans was the gifted City schoolboy, who grew up with City ideas. His appearance and businesslike manner were more like those of a cultured stockbroker than a philosopher. He was a hard-headed negotiator, for example, in contracts for books. He could be very charming when he wanted something of others and very brusque when others wanted something of him.

Milne writes that 'he was modest and unassuming in private life, though he did not exercise "charm" in the usual sense of that word outside the lecture theatre; on the lecture-platform he had a most winning delivery.' Many have thought that his voice was unattractively harsh and high-pitched.

His shyness in playing his musical instruments revealed inhibitions in his nature. In some ways he behaved like an intellectual prisoner. In his *Report on Radiation and the Quantum Theory*, he seems to take a delight in showing as clearly as possible that his own fundamental physical conceptions, those of Newton, were inadequate. And yet he never abandoned that outlook. Instead of plunging into quantum and relativity theory, he left the fields where they were necessary and sought one where the Newtonian outlook still served.

Perhaps his consciousness of being a prisoner of Newtonian conceptions helps to explain the brilliance, and the surprising conclusion, of his comparison of Rutherford with Newton. The unexpected death of Rutherford placed on Jeans the duty of reading Rutherford's posthumous Presidential Address to the Indian Association for the Advancement of Science in 1938. Jeans prefaced the Address with a tribute which is possibly the most interesting and generous comment that had yet been made on Rutherford.

He said: 'In his flair for the right line of approach to a problem, as well as in the simple directness of his methods of attack, he often reminds us of Faraday, but he had two great advantages that Faraday did not possess—first, exuberant bodily health and energy, and second, the opportunity and capacity to direct a band of enthusiastic co-workers. Great though Faraday's output of work was, it seems to me that to match Rutherford's work in quantity as well as in quality, we must go back to Newton. . . . In some respects he was more fortunate than Newton; there was nothing in Rutherford's life to compare with the years which Newton spent in a vain search for the philosopher's stone, or with Newton's output of misleading optical theories, or with his bitter quarrels with his contemporaries. Rutherford was ever the happy warrior—happy in his work, happy in its outcome, and happy in his human contacts.'

It was surprising that so loyal a Newtonian as Jeans should have compared his master with Rutherford, to the former's detriment. The emotional strength of his tribute may have been due to Jeans' admiration for one who, unlike himself, was in some ways very uninhibited, and had succeeded in escaping from the prison of Newtonian conceptions.

Jeans read many novels and detective stories, and delighted in the young children of his later years, sharing the joys of stamp-collecting with his elder son.

He published some seventy-six original papers, eight learned books and eight popular books, besides many lectures.

What is his place in the history of science? As an investigator, he will always be esteemed for his discoveries concerning the stability of liquid and gaseous bodies under gravitational and centrifugal forces, and the light they throw on the origin and evolution of the nebulae and stars. These belonged to classical Newtonian mechanics. Outside that field of ideas, of what he called 'engineering physics', he was always uncertain. He made no contribution to the two great new theories of his day: relativity and quantum mechanics. In fact, he fought to the limit of his strength against them both, though he ultimately accepted them. When he saw that they were correct, he left those regions of work in which they were the leading themes. His mind saw only the mathematical surface of things.

His theory of the origin of the solar and planetary systems, attributing them to the very rare event of one star tidally drawing matter out of another, has swiftly lost favour, and with it the support that it gave to the supposition of the possibility of the unique nature of the solar system, and the life found on its earth. Already it has been replaced by theories such as those of Fessenkov and of Lyttleton and Hoyle. Fessenkov attributes the formation of planets to the throwing-off of matter from the sun when it was rotating at a very high rate. Lyttleton and Hoyle believe that the solar system is due to the condensation of the debris of a companion star of the sun, which exploded some five thousand million years ago as a super-nova. In their view, the sun was not even the parent of the planets. Both of these theories suggest, unlike Jeans' tidal theory, that planetary systems are common in the universe. They incline towards the plurality of worlds, and thus of life like our own, rather than its uniqueness.

It was characteristic of Jeans to give almost exclusive weight to mathematical arguments; Lyttleton and Hoyle pay much attention to the chemical features of the solar system and their significance, an aspect almost ignored by Jeans.

Jeans was an example of one of the striking types of educated Englishman, in particular of Cambridge and Trinity, at the turn from the nineteenth to the twentieth century.

As a popularizer he looked at things entirely from his own point of view and as they suited his convenience. He did not regard himself as doing a job for the service of the people. He was a rich bourgeois, feeling that he had no duty to science except what he chose to perform. He was always clear, even when wrong, and wrote like a mathematical advocate.

His abandonment of mathematical physics in 1914 was a silent acknowledgement of unfitness, and a symbol of the failure of the British school of theoretical physics. This, and the character of his writing for the people, revealed the intellectual bankruptcy of the British educated bourgeoisie.

He was a very striking figure in the science of Imperial Britain during its period of decline—his gifts and achievements were classical and conservative, and his failures were radical. He was not a genius of the first rank, able to escape from the limitations of the science and society of his day.

The highest genius consists in knowing when to accept and when to reject the apparently unintelligible: Maxwell, Planck, Einstein, Rutherford had it: Jeans had not.

BIBLIOGRAPHY

Report on Radiation and the Quantum Theory, by J. H. Jeans (1914).
Problems of Cosmogony and Stellar Dynamics, by J. H. Jeans (1919).
The Dynamical Theory of Gases, by J. H. Jeans (1904).
The Universe Around Us, by J. H. Jeans (1929).
The Mysterious Universe, by J. H. Jeans (1930).
Science and Music, by J. H. Jeans (1937).
Physics and Philosophy, by J. H. Jeans (1942).
The Growth of Physical Science, by J. H. Jeans (1947).
The Stars in Their Courses, by J. H. Jeans (1931).
Through Space and Time, by J. H. Jeans (1934).
Obituary Notices: *Proc. Phys. Soc.*, **59**, 503 (1947); *Roy. Soc. Obit.*, **5**, 573 (1947).

ARTHUR STANLEY EDDINGTON

1882–1944

I. QUAKER WIDOW'S SON

EDDINGTON'S father, Arthur Henry Eddington, died of typhoid fever in 1884 when his son was only two years old. He was a Quaker schoolmaster, who became the head and owner of Stramongate School at Kendal, where John Dalton had once taught. He was of Somerset and Quaker descent, and had been a tutor in a Quaker training college in Lancashire before going to Stramongate.

His wife, Sarah Ann Shout, came from Darlington, and was descended from generations of Yorkshire Quakers. Her family name was probably of Dutch origin, a modification of Schouten.

The Eddingtons' son, Arthur Stanley, was born at Kendal on December 28th, 1882. Very soon after the death of Arthur Henry Eddington in 1884, his widow migrated to Somerset, and settled at Weston-super-mare, taking her infant son with her.

Eddington did not grow up in the Kendal atmosphere which had reared John Dalton, and superficially he matured into a very different kind of man. Dalton was notably uncouth: the reverse of a Trinity man, as Eddington became. But they possessed intellectual similarities. They both engaged in deep solitary meditation. Eddington withdrew into long meditative silences which often puzzled and sometimes embarrassed his companions and pupils. Dalton was notoriously absent-minded.

They were both competent practical men, as observers or experimenters. Yet they were both primarily interested in the development of theory, with an inclination to speculation. Both of them preferred to follow and develop, in rugged independence, their own ideas. They were less than normally receptive to intellectual criticism. Dalton's obstinacy was directly expressed in his speech and manner; Eddington's in a gentle but steely persistence.

The tendency to develop long trains of thought in meditation inclined both of them to be unhistorical in their method. This did not mean that they were not widely read. But they paid more attention to the logic of arguments from a limited number of data, than to the drift of the subject as a whole, its general character and direction, as indicated by its entire history. When they were right, they were profoundly right, and when they were wrong they were devastatingly wrong. These characteristics were no doubt in their natures, but were also fostered by the intellectual habits of the Friends, with their separation from the general social stream in their Society, and their belief that truth is to be obtained only through direct and personal inspiration.

Eddington had an austere though not unhappy childhood in Weston. His mother was poor, but not declassed by her misfortune. Eddington was reared in a middle-class atmosphere in a watering-place, which strengthened his individualistic tendencies.

In spite of his poverty, he was able to attend a private school. From 1893 until 1898 he went to the Brymelyn School at Weston, where pupils were normally boarders, though Eddington attended as a day-boy. The social separation of such a school from those for the working class was, at that time, sharp and deep.

Eddington showed arithmetical precocity when he was a child. He learned the twenty-four times twenty-four table before he could read, and he said that he was attracted to astronomy, at the age of six, by its big numbers. He received a three-inch telescope when he was ten, with which he explored the sky, and became familiar with its features.

One of the masters at Brymelyn School stimulated his interest in English literature, and another in mathematics up to the calculus. He secured a first-class in the Cambridge Junior Local Examination before he was thirteen years old, and in 1898 he was awarded a Somerset County Scholarship of £60 a year for three years.

He was allowed to enter Owens College at Manchester, though he was only fifteen years old. He studied there from 1898–1902. He arrived as a schoolboy in short trousers, much younger than his fellow-students. His first year was devoted to general study, covering what other boys had usually done at school. In this year, his Greek was good. Then he settled down to study for the Bachelor of Science degree, in which physics was his main subject. Schuster was professor of physics, and the new laboratory which he had planned, and which was subsequently so brilliantly used by Rutherford, was not yet built.

Eddington attended Horace Lamb's lectures on mathematics, and was much influenced by them. His fellow-students included many men of ability: J. W. Nicholson, Petavel, G. C. Simpson, and others. Owing to his extreme youth, he did not at first take the lead, but in his last years he won the chief honours. Many prizes and scholarships were awarded to him, which eased his financial situation.

He sat for a scholarship at Trinity College, Cambridge, in 1901, and was awarded a minor scholarship in natural science. He went into residence at Trinity in 1902. His minor scholarship was converted into a major scholarship in mathematics in 1903. In the same year he sat for a London degree of Bachelor of Science, and secured a first-class in mathematics and a third-class in physics.

His drift from physics to applied mathematics had become quite marked. He had followed the Cambridge tradition, and competed for first place in the mathematical examination, which he duly won as Senior Wrangler in 1904. No one else has achieved that in less than three years. As an examinee, Eddington was extraordinarily clever and a wonderful learner, rather than a startling genius. His Quaker self-sufficiency and perseverance, as well as native intellectual ability, made him a confident and successful individualistic competitor. He subsequently fought with the same skill and determination for his scientific conceptions.

In 1905 he was selected by Sir William Christie to be Chief Assistant at Greenwich Observatory, and in 1913, at the age of thirty, he was appointed Plumian Professor of Astronomy at Cambridge.

Ability, training, and the state of science put him in an ideal position for great scientific achievements.

He had been poor in his youth. When he died, at the comparatively early age of sixty-two, he left the very considerable fortune of £47,237. He never married, and he wrote some books which had large sales. Nevertheless, even with his simple needs, such a fortune could scarcely have been accumulated without considerable attention to money matters. Those profound silences of Eddington were, perhaps, not entirely occupied by meditations on the stars.

In his last years he became a well-to-do cultured Englishman, one of the varied personalities which constituted the English governing class. He was an outstanding intellectual figure in British imperialist society, belonging socially to the section of religious nonconformists who, by their simple living, hard-working and saving habits, contributed so much to the development of capitalism.

Eddington was not a good teacher. He was too much absorbed in his own thoughts and he had no aptitude for ready explanation. He was not a good reviewer of books, because he was apt to develop the ideas the book stimulated in himself, rather than discuss the author's ideas.

He could deliver the most brilliant prepared lectures, but often became completely and embarrassingly tongue-tied when questioned on them in the subsequent discussion. He required time to formulate his answers, and when this was granted to him, they were masterly.

It was not surprising that he had few pupils and did not create a school. He could use the most difficult weapons for his own researches, but he could not teach others his skill. The inspiration that he gave sprang directly from his achievement, and his presence as a symbol of that achievement. 'There is the man who discovered the internal constitution of the stars': the young aspirant in Cambridge could see him daily, and thus be reminded of the reality and possibility of great achievement, and thus be inspired to emulation.

Eddington's influence was less as a teacher than as an example of quality, a person whom the young man could see and not merely read or hear about. There is no educational substitute for personal knowledge of the great. Without that direct contact, the range of human possibility cannot be imagined.

In 1927, just after the publication of his *Internal Constitution of the Stars* and at the height of his achievement, the first impression made by Eddington in personal meeting was of a tall, clean-shaven, spare figure, with an overhanging brow and deep-set, speculative eyes which were the only moving feature in his quiet aspect. His face was unusually pale, and his mouth very firm, if not obstinate. He spoke rarely, and was almost incapable of speaking coherently without preparation. Only the lower half of the pupils of his eyes were visible, the rest was buried behind his tufty eyebrows. He meditated on what had been said, and only his eyes moved, as in a circular sweep of rayless searchlights. Presently his eyes came to rest on his interlocutor, but he said nothing, and then made another sweep. One wondered whether one was expected to go, and then, as one began to perspire with uncertainty, he replied to the question.

A bachelor, living entirely in the realm of thought, he paid little attention to bodily movement or instinctive presentation. He never lost the hard 'a' of the northerner's speech.

His mild and gentle presence gave no clue, except by contrast, to the inspiration of his writing on science. Though he could not marshal his personality or thoughts on the moment, he could do so superbly, given time. He was one of the very rare scientists of great distinction who could explain his own discoveries, and other people's, not only in intelligible but delightful language. He could build brilliant similes of scientific ideas out of fashionable oddities of dress and trivial details of daily life. These apt and shining similes manifested a poetic feeling. An occasional clumsiness in his prose suggested that his power of writing came from the extreme clearness with which his ideas appeared to himself. They were so clear and living that they forced themselves into brilliant expression, in spite of the natural tentativeness of their author. His literary style was, accordingly, inspired rather than cultivated. The quality of his expository books was probably related to a compensatory effort in the mental sphere of words, to integrate a personality which was not completely integrated in ordinary daily affairs.

Eddington's comparatively early death was due to cancer. He bore this terrible disease with great fortitude. In his last years his ghostly pale face was drawn with suffering, as he sat in his long reveries.

His handwriting became unsteady, and he struggled to control himself.

One of the last tasks he was invited to undertake was an expansion of a beautiful pamphlet that he wrote in 1935 on the Cavendish Laboratory, to assist in the collection of funds for new buildings. He replied that it was as impossible to add to the pamphlet as it would be to add a new part to a finished picture. He regarded his expositions as harmonious works of art, and he made them so.

Eddington was an inveterate pipe-smoker, a reader of detective stories, and an addict of crossword puzzles. He liked an occasional game of golf, and was fond of solitary bicycle rides.

He duly valued the numerous honours which were conferred on him. These included a Royal Medal of the Royal Society in 1928, and the Gold Medal of the Royal Astronomical Society in 1924. He was president of the latter society in 1921–23 and of the Physical Society in 1930–32. He was always an admirable and accomplished officer. He was elected president of the International Astronomical Union in 1938, and was awarded many honorary degrees. In 1930 he was made a knight and in 1938 received the Order of Merit.

II. STELLAR ENGINEER

In 1938 Harlow Shapley described Eddington as 'the greatest living astronomer'. It is necessary always to keep in mind this view of his professional status. Eddington had many other qualities besides his eminence as a professional astronomer. The adjective 'brilliant', generally debased by indiscriminating use, truly described his expository gift. This quality was even more striking in his most original research papers than in his popular books. Passages of difficult mathematics are interspersed by unsurpassed imaginative use of models to explain the working of cosmical physical processes.

His genius is seen in its finest form in his first paper on cosmic physics, read in 1916, in which he attempted to explain the pulsating Cepheid stars as stellar engines, in which the gaseous material of the star expanded and contracted, through periodic stoking with subatomic energy, controlled by some physical process acting as a valve.

All his great discoveries concerning the internal constitution of the stars grew out of this paper. Thus, in his best work, Eddington was a typical example of the British tradition in physical science, the combination of skill in mathematical problem-solving with a strong physical and engineering sense.

He did not find his ideal field quickly. In spite of his early academic brilliance, he was thirty-four years old before his first paper on pulsating stars was published. The ten years that had passed since his graduation could be regarded as the unconscious period of training and preparation

5. Sir Arthur Eddington

face p. 144

for his main achievement. It took all that time, as it were, for that branch of science for which he was uniquely suited to find him and claim him for its own.

The main foundation of his professional reputation is his discoveries concerning the theory of the internal constitution of the stars. This subject had been founded in 1870 by J. Homer Lane, the American astronomer, who showed that if a star consisted of a ball of perfect gas, then the internal temperature and pressure at any point inside it could be calculated if the law of the distribution of density of the material were known.

The determination of the actual distribution of density is a very difficult and only partially solved problem, but Lane showed that if the internal distribution of temperature and pressure were worked out for a star in which the density varied from the centre to the surface according to a given law, then the internal temperatures and pressures of larger and smaller stars, built on some model, or distribution of density, could at once be deduced by simple calculations. At corresponding points within such stars, the densities vary according to the mass divided by the cube of the radius, the pressures as the square of the mass divided by the fourth power of the radius, and the temperatures by the product of the mass and of the molecular weight of the gas at the point under consideration, divided by the radius.

In particular, if the radius only is altered, the density varies as the reciprocal of the cube of the radius, the pressure as the reciprocal of the fourth power, and the temperature as the reciprocal, of the radius. The latter very simple relation is known as Lane's law.

One of the prime results of Lane's investigation was his discovery that as a star contracts the internal temperature rises, as long as its material behaves as a perfect gas. During all the period of contraction, the star would, of course, be continually shining or radiating. It would be losing heat. Thus the star exhibited the paradox of getting hotter by cooling.

Lane's investigations were extended by Ritter, Kelvin, and Emden. In his own time, their significance was realized by Norman Lockyer almost alone. There was no evidence that stars could be perfect gases: the average density of the sun, for example, was greater than that of water. How, then, could it be a perfect gas?

In this case, as in others, Lockyer was not deterred by current scientific limitations. On the assumption that the stars were perfectly gaseous and obeyed Lane's laws, he arranged them in a sequence of ascending temperature with age, in contrast with the current view that their temperatures continually descended with age, the hottest stars being the youngest and the coolest stars the oldest.

Lane's results were independent of the way in which the heat was transferred inside the star. He had, however, conceived the star as blown up by the pressure of its gas-particles in the same way as a football, the pressure being produced by the temperature of motion of the particles. He supposed that heat was transferred by the stirring of the rapidly moving particles themselves, that is, by convection.

In 1894 R. A. Sampson pointed out that radiation pressure must also contribute. Heat can be transferred by waves, like those of light, and, as Clerk Maxwell had foretold, such waves would exert pressure.

Under the intense conditions inside a star, there would be strong streams of waves, which might contribute considerably to the support of the distension of the star's constituent gas. This was as far as the classical physics of the nineteenth century could go.

The theory of equilibrium, based on radiation pressure as distinguished from gas pressure, could not be developed without an improved theory of thermodynamics, that is, without the application of new physical theories.

Schwarzschild began this work in 1906, with his publication of a famous paper on the role of radiation pressure in the sun's atmosphere. His brilliant work was ended by his premature death in 1916, in the First World War.

Meanwhile, physics advanced in other directions, besides that of thermodynamics, which had culminated in the discovery in 1900 of the quantum theory. Röntgen had discovered X-rays in 1895, one year after Sampson had raised the question of the importance of intense radiation in the pressure of stellar gas.

The discovery of radioactivity in 1896 and of the electron in 1897 soon raised surmises as to the subatomic source of stellar energy and heat. Here, again, Lockyer had been before his time, and had asserted in 1900 that the stellar energy was of subatomic origin. It was Lockyer, too, who perceived the importance of the enhanced or strengthened lines in spectra, which were attributed by him to the breaking-up of atoms owing to high temperature, before the phenomenon of ionization had been explained by the discovery of the electron.

Indeed, up to the beginning of the new physics in 1895, astronomers were in advance of physicists in their premonitions of the physics of the twentieth century. The surprising thing is not that this should be so, but that the astronomers' conceptions of the nature of matter should have had so little influence. Their advantage was that they were observing matter under extreme conditions of temperature and pressure not at that time reproducible on earth, and therefore not within the range of the contemporary physicists' laboratory experiments. Their limitation was that the information coming to them from the sun and stars, of matter in states as yet unknown on earth, was fragmentary. They had to work with such data as observational astronomy gave them, not with what they would have chosen by laboratory experiment if that had been possible.

Their deductions about the possible states and properties of matter were restricted to hints and suggestions, which were vague and not susceptible of immediate clarification and proof. The physicists experimenting with matter in their laboratories on the surface of the earth paid little attention to the astronomers' ideas until they had been forced towards similar ones by their own results.

The discovery that electrons could be obtained from any kind of atom showed that atoms had parts. They could be taken to pieces and put together again. Then, in 1911, Rutherford discovered the form or structure of atoms: they consisted of a relatively heavy positively charged nucleus surrounded by a group of revolving electrons.

In 1913 Bohr published his quantum theory of the atom, which gave a comprehensive explanation of the phenomena of spectra and ionization. The spectra were due to the interaction of the electrons of the atoms with radiation. When a quantum of radiation, consisting of waves of a certain size, fell on an atom, one of its electrons absorbed it and revolved with greater energy, or flew off altogether, as in ionization. And in the reverse process, an electron in an atom might fall into a lower state of energy by emitting an energy-packet of wave-radiation.

Such were the new and main ideas of atomic physics in 1913.

Eddington was born in 1882. He completed his preliminary education in Manchester and Cambridge by 1905. He was trained in the classical school of Cambridge applied mathematics. The famous Horace Lamb, the Cambridge applied mathematician, was one of his professors at Manchester. He imbibed the essence of the Cambridge school in his intense preparation for the mathematical examination. He was first in his year, Senior Wrangler.

It was not possible to be Senior Wrangler without complete absorption in the Cambridge mathematical tradition. Among his teachers were E. T. Whittaker, A. N. Whitehead and E. W. Barnes, and his Tripos coach was R. A. Herman, who had coached nine out of the last eleven Senior Wranglers before this examination system was changed. Herman succeeded Hopkins and Routh in the tradition of great Tripos coaches. Their approach to mathematics was peculiar, and stamped on the Cambridge mathematics of the nineteenth century its most characteristic features of style.

Herman was a quiet, retiring man, sensitive to quality and elegance in mathematics. One of his pupils described him as conscientious, inspiring and methodical, an artist in mathematics, scrupulously rigorous and quick to praise beauty. 'By means of weekly problem papers he imparted all that it was possible to impart of his own facility in the problem-solving that was formerly essential to distinction in Cambridge examinations.'* He was familiar with all branches of Tripos mathematics, and especially astronomy, geometrical optics, hydrodynamics, and differential geometry.

* *Nature*, February 11th, 1928, p. 214, vol. 121.

It has been reported that Eddington once said that when he arrived in Cambridge he found that he was expected to play a game called the Tripos Examination. As it could not be avoided, he decided to play the game for all it was worth. He did so, and won. He did not feel that it had done him any harm, apart from taking some time and energy that he might have devoted to research.

Nevertheless, under Herman's gentle and perfect guidance, Eddington acquired the most typical and accomplished Cambridge mathematical style and outlook. His greatest work was done in the problem-solving Cambridge style, in which mathematics was dextrously applied to ingenious physical models and used as the servant of physical insight.

But in spite of his immense ability, he did not immediately find the kind of problems which suited this style. He graduated in 1905. His first mathematical research was on a problem in potential theory. It was not immediately successful, though eighteen years later he completed the investigation and applied the result to the calculation of the energy transmitted from a rotating cylinder by gravitational waves. In the meantime, the solution had already been found and published by A. W. Conway.

He undertook a piece of experimental research in the Cavendish Laboratory on thermionic emission, the emission of electrons by hot bodies. He soon dropped this work. In the same year, he was appointed Chief Assistant in Greenwich Observatory. This fixed his attention on astronomy.

The Cambridge training in problem-solving in a wide range of applied mathematics fitted him to turn his hand to almost any branch of physics. It encouraged versatility. His first task at Greenwich was to collaborate with C. R. Davidson in the reduction of the photographic observations of the minor planet Eros as a basis for the more exact calculation of the distance of the sun from the earth.

He initiated a programme of observation of latitude variations, which has since been constantly pursued, and he made meridian observations himself for twelve years, the last being in 1918. In 1909 he determined the latitude of Malta.

Eddington was a thorough practical astronomer. It would be a mistake to imagine that he was merely a theorist. But it is important to note that his sustained routine of personal practical work ended in 1918.

In 1906, he began a study of stellar movements. Kapteyn had discovered in 1904, and described in a paper to the British Association in South Africa in 1905, that the stars do not move at random, but appear to drift into streams, from which the shape and motion of the local universe of stars, the Galaxy, could to some extent be deduced. Eddington extended Kapteyn's examination to the data of many more stars, and discussed the mathematical characteristics of the stellar universes so revealed. This was his main work for the next half-dozen years. He submitted an essay on the subject in 1907, for which he received a Smith's Prize at Cambridge, and in the same year he was elected a fellow of Trinity College.

He devised convenient methods of analysing the systematic motions of the stars and of determining their distances. He used diagrams which showed stellar distributions in a striking way, and because of their appearance were nicknamed 'rabbits'.

He summarized this work in his first book, *Stellar Movements and the Structure of the Universe*, which was published in 1914. He chose for the frontispiece a photograph of the Great Star Cloud in Sagittarius, as if to signify both the anarchy and the order of the myriad-starred universe.

Eddington was thirty-one years old when the book was published. He expressed in it attitudes which became more and more characteristic in subsequent years. He wrote that he believed 'that excessive caution is not to be desired' in propounding hypotheses to explain the movements of the stars.

It was also outside his 'purpose to describe the steps by which knowledge has advanced; it is the present situation that is here surveyed.'

Thus at the beginning of his research career he adopted an adventurous attitude to scientific speculation and a non-historical approach. He even insisted that study of the history of a subject was a hindrance to creative research.

Typical of the first of these attitudes is his remark in this book that 'it is not necessary to suppose that the primordial matter from which (a nebula) arises is not subject to gravitation (although there is nothing inherently improbable in such a speculation).'

The second attitude is evinced in his tendency to concentrate on getting the utmost out of the latest observational and experimental data. He is inclined to try to extract the maximum from them by strong, imaginative mathematical analysis, and then squeeze from these results by speculation their last hints of suggestion.

He explained in his first chapter that 'it is to the relations between the stars—to the stars as a society—that attention is here directed. We are not concerned with the individual peculiarities of the stars, except in so far as they assist in a broad classification. . . . Accordingly it is not proposed to enter here into the more detailed study of the physical characters of stars . . . stellar chemistry and temperatures are foreign to our present aim.'

He concluded from this study of the motions of the stars that 'if we have still to leave the stellar universe a region of hidden mystery, yet it seems as though, in our exploration, we have been able to glimpse the outline of some vast combination which unites even the farthest stars into an organized system.'

Eddington was elected Plumian professor of astronomy at Cambridge in 1913, after the death of G. H. Darwin. He was only thirty years old and had not yet become a fellow of the Royal Society, though he was elected in the following year, 1914. The Plumian chair had been founded in 1704 for observational astronomy and the Lowndean chair in 1750 for theoretical astronomy and geometry.

When J. C. Adams was Lowndean professor, he had been asked temporarily to supervise the University Observatory. This arrangement was prolonged for many years, and continued under his successor, Sir Robert Ball. Observational astronomy had through this accident become detached from the Plumian chair. Ball died in 1914, and the opportunity of returning the direction of observational astronomy to the Plumian chair was taken by appointing Eddington director of the Observatory. The Lowndean chair became devoted to geometry and mathematics, theoretical astronomy being in effect detached from it. Thus the study of observational and theoretical astronomy at Cambridge were brought together, and Eddington became responsible for both. Eddington's experience as a practical astronomer, in addition to his great intellectual ability, weighed in his election at such an early age and in preference to seniors of high standing.

Eddington was settled in Cambridge when the First World War began. The researches inspired by his Greenwich interests were virtually completed, and described in his *Stellar Movements*, though he continued for a while with theoretical studies of a system of stars in a dynamically steady state, in which each star moves under the gravitational field caused by the attraction of the stars acting all together—in which each star is, as it were, a planet revolving under the central force produced by all the matter in the universe. He investigated the gradual evolution of such a dynamically steady system, in which the stars are moving in a comparatively limited number of well-defined motions, into a statistically steady state, which might be regarded as the ultimate condition of the stars.

From this point, Eddington's interest in stars as members of a 'society' rapidly gave place to his interest in them as individuals, in their mechanical, physical and chemical characters.

His change-over from the 'social' to the 'individual' approach was affected by many factors. Greenwich was essentially a centre of 'social' astronomy. It was traditionally the centre of the astronomical world. It had been founded for social purposes, in 1675, for the service of H.M. Admiralty, to which it still belongs, and for the provision of data for the use of mariners everywhere.

At Cambridge the environment was profoundly different. Observational astronomy was not conducted there for use or service, but for education and training. The serious business of Cambridge was in pure research of a theoretical or experimental character. There was an ancient tradition of mathematical investigation, and a newer and very lively tradition of laboratory research in physics.

The Cambridge professor, in contradistinction from the Greenwich astronomer, was naturally impelled to concentrate on theoretical problems which personally appealed to him. He could scarcely avoid hearing the exciting news coming almost daily from his physicist colleagues.

His studies of stellar movements had, in themselves, drawn his attention to the physical properties of stars, for it had been observed that stars of certain physical types tended to move with high speeds. Lockyer had started the classification of the stars on the basis of their physical type. In 1913 H. N. Russell and E. Hertzsprung had made a great advance by showing that the absolute magnitude, that is, the candle-power, of stars is simply related to their physical conditions, as revealed by the spectra of the light they emit.

Russell exhibited this relation in a diagram, in which absolute magnitude was plotted against spectral type. The majority of the stars formed two roughly straight lines on the diagram. A star originally became visible when it was turned by gravitational condensation from a cold nebula into a cool red star, of low density but immense size. Owing to its vast surface area, its candle-power was very great, in spite of its low temperature. As it contracted, it became hotter, according to Lane's law. The candle-power remained about the same, as the rise in temperature made up for the reduction in surface area.

When the star had contracted so much that its material no longer behaved as a perfect gas, it became a dwarf which behaved like a hot liquid ball cooling off. While the temperature fell, the surface area remained constant, so its candle-power gradually declined, until it became a faint red star. The cool red stars could therefore be either large newly-born giants or small old, worn-out dwarfs.

The candle-power of the red giants was about 1000 times as great as the red dwarfs. No red stars whatever could be found in the vast intermediate range of candle-powers. Similar, but smaller differences in candle-power were found between giants and dwarfs of other colours.

By 1914 the astronomers had succeeded in placing the stars in a startlingly simple scheme according to their absolute magnitude and spectra, or candle-power and surface temperature. It was just sufficiently clear-cut to offer the possibility of quantitative as distinguished from qualitative, or speculative, explanation in terms of physics. It inspired the hope of making a major advance towards the discovery of the way in which a star works as a machine continually pouring forth light, the nature of the materials of which it is made, and the fuel which keeps it going.

Eddington's peculiar gifts as an investigator, his mental outlook, his training, his environment, the state of physics, the stage of the subject, his personal situation, formed a harmonious combination, which enabled him to achieve his greatest work. He reviewed his progress with the stars in a remarkable presidential address to the physics section of the British Association, in 1920.

He started by referring to Sir Charles Parsons' proposal in the previous year that a deep hole should be bored into the earth to learn more of its interior. It 'may remind us that the most secret places of Nature are perhaps, not 10 to the *n*th miles above our heads, but 10 miles below our feet. . . . Distance lends not enchantment but precision to the view.'

'Science has material and non-material appliances to bore into the interior, and I have chosen to devote this address to what may be described as analytical boring devices.'

'The analytical appliance is delicate at present, and, I fear, would make little headway against the solid crust of the Earth. Instead of letting it blunt itself against the rocks, let us look round for something easier to penetrate.

'The Sun? Well, perhaps. Many have struggled to penetrate the interior of the Sun. But the difficulties are great, for its substance is denser than water. It may not be quite as bad as Biron makes out in *Love's Labour's Lost*:

> The heavens' glorious sun
> That will not be deep-searched with saucy looks;
> Small have continuous plodders ever won
> Save base authority from others' books.

'But it is far better if we can deal with matter in that state known as a perfect gas, which charms away difficulties as by magic. Where shall it be found?'

In the stars, and it is easier to investigate their interiors than that of the earth beneath our feet.

Half way through the address, Eddington says that one may be hearing whispers that his ideas are speculative. He launches into a defence of his favourite activity—ultimately his vice—speculation. 'It is often supposed that to speculate and to make hypotheses are the same thing; but more often they are opposed. It is when we let our thoughts stray outside venerable but sometimes insecure hypotheses that we are said to speculate. Hypotheses limit speculation. Moreover, distrust of speculation often serves as a cover for loose thinking—wild ideas take anchorage in our minds and influence our outlook; whilst it is considered too speculative to subject them to the scientific scrutiny which would exorcise them.'

Later on he says that 'Our model of Nature should not be like a building—a handsome structure for the populace to admire, until in the course of time someone takes away a corner-stone and the edifice comes toppling down. It should be like an engine, with movable parts.'

For the assurance of those who may have been troubled by the boldness of his ideas, he ends his address with a parable:

'In ancient days two aviators procured to themselves wings. Daedalus flew safely through the middle air across the sea, and was duly honoured on his landing. Young Icarus soared upwards towards the Sun till the wax melted which bound his wings, and his flight ended in fiasco. In weighing their achievements perhaps there is something to be said for Icarus. The classic authorities tell us that he was only 'doing a stunt', but I like to think of him as the man who certainly brought to light a constructional defect in the flying machines of his day. So, too, in Science, cautious Daedalus will apply his theories where he feels most confident they will safely go; but by his excess of caution their hidden weaknesses cannot be brought to light. Icarus will strain his theories to the breaking-point till the weak joints gape. For a spectacular stunt? Perhaps partly; he is often very human. But if he is not yet destined to reach the Sun and solve for all time the riddle of its constitution, yet he may hope to learn from his journey some hints to build a better machine.'

Here the stellar engineer wrote the apologia for his scientific mind's career, and its epitaph.

III. PEACE IN WAR

As Eddington was a Quaker, occupying a high academic position, it was not difficult for him to secure exemption from national service during the world wars. He did not participate in them and was virtually untouched by them, as his pacifist views were firmly and profoundly held.

He was not, however, unaffected by war. In 1916 a note to the *Observatory* from an Oxford correspondent raised the question as to whether German scientists could ever again be admitted to the pale of international science. It provoked Eddington into an outspoken letter on *The Future of International Science*. He wrote that 'some of the problems of our science can only be attacked by world-wide cooperation.' The permanent exclusion of any would harm the progress of astronomy. As to the behaviour of German scientists (ninety-three of them had signed an arrogant pan-German manifesto at the beginning of the war), it was not entirely bad. 'Notwithstanding official pressure, the Berlin Academy of Science has twice declined to eject its enemy members.' Their colleague Foerster, who had signed the manifesto, was a leading member of the committee for assisting aliens in Germany, 'a work which must have cost him much popularity'.

'The indictment of a nation takes an entirely different aspect when applied to the individuals composing it.

'Think, not of a symbolic German, but of your German friend Professor X, for instance—call him Hun, pirate, baby-killer, and try to work up a little fury. The attempt breaks down ludicrously. No doubt, he is a most ardent supporter of his fatherland, passionately convinced of the righteousness of its cause. Call it wrong-headed, if you will, but surely not morally debased. Far be it from me to deny his individual responsibility for his country's share in the evil that has befallen. The worship of force, love of empire, a narrow patriotism, and the perversion of science have brought the world to disaster.' The assertion that 'because we will not sit in judgment on our fellow-labourers, we are thereby condoning the faults they have committed . . . surely cannot claim the sanction of either religion or ethics.'

'Let us not imagine that the German men of science will come to us repentant, publicly repudiating the cause now consecrated by the blood of their sons. Nations repent of their folly, not by a sudden revulsion, but by the slow growth of a wiser outlook.'

He then quoted from the British Association Address of Sir Arthur Evans, the archaeologist who discovered the Minoan civilization: 'We have not ceased to share a common task with those who today are our enemies. We cannot shirk the fact that tomorrow we shall be once more labourers together in the same field.'

Eddington's attitude helped him to receive the wartime news of Einstein's great discoveries in general relativity with a receptive mind, in contrast especially with French colleagues, who, even for years after the war was ended, were blinded by nationalist passion against the theory because they refused to believe that anything good could come from Germany.

Eddington's evocation of a German scientific friend recalls his splendid obituary notices on Schwarzschild, who died in 1916 at the age of forty-three from a disease contracted on military service. In one of these notices Eddington remarked that the German authorities, with characteristic efficiency, had put Schwarzschild on to military meteorological work. There seemed to be an implied hint of criticism of British inefficiency in the use of scientists. Perhaps he was thinking of cases among his friends, possibly of himself, even if unconsciously.

But deep convictions enabled him to pursue his normal scientific work without any loss of energy through doubt. The whole of his mental force went into individual research disturbed by less even than the usual distractions of academic teaching and routine.

Eddington was admirably prepared and situated to attack a problem suited to his powers, and war gave him peace in which to pursue his work. He began his attack on the elucidation of the internal constitution of the stars in 1916, by an attempt to explain the mechanism of the pulsations of Cepheid stars.

According to his own account, he attempted to find 'some cause maintaining the mechanical energy of pulsation . . . by which it could be extracted from the abundant supplies of heat at different temperatures. . . . This might happen if the material of the star acted as the working substance of a simple thermodynamic engine, or if the radiation pressure varied in the manner necessary to perform mechanical work. The equation developed for this study naturally laid stress on the opacity (which must serve as the valve of the engine), the transport of heat by radiation, and on radiation pressure. None of these were treated in the investigations of the stellar interior then customary, so that before discussing the small oscillations the conditions of the steady state had to be investigated anew. The Cepheid problem was laid aside for a time, and attention paid to the equilibrium problem.'

In this way, he came to write his paper *On the Radiative Equilibrium of the Stars*, published early in 1917, in which he extended Schwarzschild's theory of the radiative equilibrium of a star's atmosphere to the star's interior. He found that, on the assumption that radiation pressure has the chief role in supporting the structure of the star, the resulting distribution of temperature and density in the interior follows a rather simple law.

In a giant star of low density the temperature gradient is very steep, and consequently, according to conventional theory, the outflow of the heat from the star should be very great, in fact millions of times that which is actually observed.

This difficulty can be removed if the intense beams of radiation do exist in the star, but are unable to escape. In that case, they would radiate from the centre, where the temperature is highest, and beat upon the superincumbent layers, pushing them outwards by radiation pressure. The weight of the upper layers of the star, due to the gravitational attraction on them, would then be mainly borne by the radiation pressure.

If the pressures due to radiation and gravitation balanced, then it could be shown that the relation which would hold between mass and density of the star and its effective or surface temperature roughly corresponded with the observed facts.

His theory enabled him to estimate the average densities of the giant stars of various colours. It showed that the average luminosity or candle-power would be roughly the same for the various types. It showed how the candle-power of these giants could be compared with that of the sun. It determined the maximum effective or surface temperature which a star could attain, and it indicated by how much the masses of individual stars are likely to deviate from the mean mass.

This paper exhibited Eddington's powers at their highest. There was a close correspondence between the many deductions from the new theory and the observed properties of stars. The results were not of a special nature applying only to some peculiar or freak star—they applied to all, or the majority of, stars, and they dealt with the major features, their mass, luminosity, etc.

'It is clear,' wrote Eddington, 'that we cannot arrive at much certainty with regard to the conditions in a star's interior, except in so far as the treatment can be based on the most general laws of nature. There are some physical laws so fundamental that we need not hesitate to apply them even to the most extreme conditions, for instance, the density of radiation varies as the fourth power of the temperature, the emissive and absorbing powers of a substance are equal, the pressure of a gas of given density varies as its temperature, the radiation pressure is determined by the conservation of momentum . . . the weak link in the present investigation is that I have assumed without much justification that a certain product . . . is constant throughout a star . . . I plead to be allowed provisionally one rather artificial assumption.'

In this paper Eddington had his great powers under complete discipline. With the allowance of not more than one 'rather artificial assumption', bold imagination of how physical processes were occurring, and masterly use of the methods of mathematical physics, he obtained profound and general results.

He added much to these results in the next half-dozen years, and his body of work on the internal constitution of the stars formed his greatest achievement. In his other fields of work he failed to discipline so successfully his tendency to speculation. One 'rather artificial assumption' was not enough; he invoked two, three, and more and more, until in his final work undisciplined speculation ran riot.

His greatest investigations, in which speculation was strictly controlled by fact and calculation, were completed by about 1925, when he was forty-two years old. In his youth he worked under supervision. At Greenwich he was set tasks. He formed habits which enabled him to carry on the control of his speculative demon for some ten years after he had become completely independent, at the age of thirty, as Plumian professor. But the demon gradually got loose, until it rampaged through all his work and left a tangle of desolation. His speculative tendency was innate, but it is possible that it would not have latterly been so wasteful if he had lived and worked in a different kind of scientific society, in which there was more serious mutual criticism, not from the point of view of intellectual debate, but with the serious intention of advancement by the group in the solution of profound problems.

The Cambridge of the 1920s encouraged Eddington's speculative weaknesses and did not help to overcome them. The latter part of his work was ruined by unrestrained individualistic speculation.

Eddington collected and described the harvest of his researches on the *Internal Constitution of the Stars* in his greatest book, published with this title in 1926.

In his original paper of 1917, he had assumed that his star behaved as a perfect gas and that therefore his results could apply only to the giant diffuse stars, which were certainly gaseous. For such a star, he had shown that the absolute magnitude (including all its radiation, visible and invisible) was dependent only on its mass. This was a startling result. It seemed incredible that the candle-power of a gaseous star could be directly measured merely by weighing the star.

He did not suppose that this applied to dwarf stars such as the sun, whose mean density was greater than that of water, and therefore presumably did not behave as a perfect gas.

Eddington's extension of Schwarzschild's theory in 1916 was still largely in the style of thermodynamics. He had not clearly visualized the mechanism by which radiation exerts pressure. He had calculated that the intense radiation in the interior must be of about the length of X-rays, and he had tentatively considered J. J. Thomson's theory of the scattering of X-rays as the mode by which the radiation exerted pressure, and he had examined Bragg's measurements of the opacity to X-rays of matter in the laboratory.

After 1917 Eddington's attention turned more closely to the study of the physics of atomic processes, in order to discover more exactly how radiation and matter interacted inside the stars. He was assisted in this study by the swift growth in atomic physics, especially after Rutherford's arrival in Cambridge in 1919, and the stimulating discussions which this progress inspired. Rutherford's colleague, C. D. Ellis, guided him in the literature of X-ray absorption processes, experimental and theoretical, based on the new quantum atomic physics.

In 1921 he realized that the radiation pressure is exerted by the photoelectric process, by which wave-radiations push electrons out of atoms, and *vice versa.* In Eddington's 1917 paper he had assumed that the interior of a star consisted of free atoms, released from all chemical combinations by the high temperature. He supposed that the atoms of different kinds were mixed together and had an average atomic weight of about 50. He was working with the nineteenth-century idea of the atom as a small compact ball.

Newall, Jeans, and Lindemann pointed out to him that the twentieth-century conception of the atom as a nucleus surrounded by satellite electrons would suggest the existence of a different condition. It was known that the number of satellite electrons in an atom was roughly equal to the atomic weight. Hence, if all the satellite electrons were stripped away in the violent conditions inside the star, the average atom of atomic weight 50 was replaced by twenty-six particles, consisting of the nucleus and twenty-five electrons. Hence the average weight of the independent particles in the star was reduced to approximately 2, the extra weight of each heavy nucleus of atomic weight 50 being, as it were, shared out among the twenty-five very light electrons.

Hence the average weight of the constituent particles inside a star is approximately 2 whatever their chemical composition. It became possible to calculate the temperature, density and pressure inside a gaseous star of known mass and radius.

Furthermore, owing to the breaking-away of the satellite electrons, the atoms would, in effect, become very much smaller, and hence be capable of behaving like perfect gases when they had been so compressed that the stellar material was denser than water.

It appeared that even the dwarf stars, like the sun, whose average density is higher than that of water, might nevertheless behave as perfect gases.

By 1924 Eddington had shown that all the masses of stars which were accurately known from astronomical observation and measurement could be correctly calculated from his theory. Dense dwarf stars, such as the sun, did indeed behave like perfect gases.

Eddington had now found the complete general explanation of the Russell-Hertzsprung relation of 1913. There was, however, a group of stars which did not obey his mass-luminosity law. Typical of these was the faint Companion of Sirius, which was very small and yet greatly perturbed that mighty star. Gravitational theory indicated that, in order to have its observed effect, it must be 55,000 times denser than water. Eddington pointed out that that was not unreasonable if its constituent particles consisted of the parts of almost completely ionized atoms, which, in spite of being so closely packed together, still behaved like those of a perfect gas owing to their very small size.

He explained that if the Companion of Sirius had a density of 55,000 the light emitted from the star would be sensibly pulled back, owing to the size of the Einstein relativity attraction of the dense matter for the issuing radiations. He calculated that the retardation would be about 20 kilometres a second, and his result was confirmed shortly afterwards by Adams at the Mount Wilson Observatory.

As the intensity of radiation increases as the fourth power of the temperature, the radiation pressure correspondingly increases with temperature. At one million degrees it is equal to about 2550 atmospheres, at ten million degrees it is about 25,500,000 atmospheres. It begins to take an important part in resisting the gravitational pressure. 'The outward flowing radiations may thus be compared to a wind blowing through the star and helping to distend it against gravity.' Eddington calculated the proportion of the outward pressure due to radiation and that due to the ordinary gas-pressure exerted by the very hot particles.

The proportion was about the same throughout the star, and did not depend on the density or the opacity of the material. It depended only on the mass of the star and the average weight of the particles.

'We can imagine a physicist on a cloud-bound planet who had never heard of the stars calculating the ratio of radiation-pressure to gas pressure in a series of globes of gas of various sizes, starting say, with a globe of mass 10 grams, then 100 grams, 1000 grams, and so on, so that his nth globe contains 10^n grams.' He showed that for globes up to 10^{33} grams in mass, the gas pressure will overwhelmingly exceed the radiation pressure but for globes over 10^{35} grams in mass, the reverse occurs. For globes of mass within the range of 10^{33} to 10^{35} grams 'we may expect something interesting to happen. What "happens" is the stars'.

He remarks that the lightest known star weighs about 3×10^{32} grams, while the heaviest is about 2×10^{35} grams. 'It is as though nature had a standard model before her in forming the stars.' Heavy stars are very rarely more than 100 times heavier than light stars.

It appears that gravitation draws matter together to form stars. The radiation pressure raised inside, as the temperature rises through contraction, ultimately prevents the star from growing beyond a certain mass. 'The significant fact that stellar masses congregate just at the point where radiation pressure is beginning to endanger the star . . . leads us to think that radiation pressure is the agent which has cloven chaos into stars.'

Eddington devoted much ingenuity and effort to the elucidation of the mechanism of the Cepheid stars. Shapley had suggested that their periodic variations in luminosity were due to pulsations in the ball of stellar gas, and Eddington worked out the theory.

In a steady star the gravitational pressure would be balanced by gas and radiation pressure. Suppose that the star were caused to contract. Its temperature would rise owing to the increase in pressure, and the star would then expand. Its surface would not, however, rest at its original position, but sweep beyond, in an over-expansion. This would sweep back, passing again through its original position, and an oscillation or pulsation would be started. Eddington calculated that such a pulsation in a ball of gas of the same mass as a Cepheid star would have a period approximately equal to what is observed. He deduced that δ Cephei should have a period of 3·57 days, compared with the observed period of 5·37 days.

If the pulsation were started by accident and not replenished, it would die down in about 8000 years. Only a comparatively small liberation of energy was needed to keep the pulsation going. In δ Cephei, which emits 160 ergs of energy per gram per second, 0·5 ergs per gram would be sufficient.

'This heat can only be made available as mechanical work if the star behaves as a thermodynamic engine; that is to say, excess heat must be added to matter when at a high temperature and withdrawn at a low temperature. We require, in fact, something corresponding to the valve-mechanism of a heat engine.' The obvious position for placing the 'valve' was 'at the point of entrance of sub-atomic energy into the engine'. There must be a periodic variation in the output of sub-atomic energy in the centre of the star, in phase with the pulsations, and the rate of liberation of sub-atomic energy must increase nearly proportionally to the square of the temperature or to the two-thirds power of the density in order to keep the pulsations going.

Eddington's theory of the Cepheid pulsations was incomplete and contained many difficulties, but it was one to which he was particularly attached. It was the beginning of his best work, and he enjoyed devising ingenious physical models, which might be made to work by dextrous juggling with plausible postulates.

Eddington gave serious consideration to the problem of the origin of stellar energy in his first researches in 1916 on radiative equilibrium. But he shelved the question, as he found that radiative equilibrium could be analysed without any special hypothesis as to the origin of the energy.

In 1919 H. N. Russell published a suggestive review of the more precise conditions which a sub-atomic source of energy for stars must satisfy. Then, in 1920, Eddington's colleague F. W. Aston showed that the mass of a helium atom was considerably less than that of four hydrogen atoms, so that if four hydrogen atoms could be combined into one helium atom, there would be a large output of sub-atomic energy.

Later in 1920, J. Perrin in France and Eddington both suggested that the synthesis of helium out of hydrogen might be the source of stellar energy.

In 1926 Eddington had no clue as to how that synthesis might occur. The process seemed so mysterious that he distrusted all predictions. 'The attention paid to temperature, so far as it concerns the *cookery* of the helium atom, seems to neglect the adage "first catch your hare." . . . How the necessary material of 4 mutually repelling protons and 2 electrons can be gathered together in one spot, baffles the imagination. One cannot help thinking that this is one of the problems in which the macroscopic conception of space has ceased to be adequate, and that the material need not be at the same place (macroscopically regarded) though it is linked by a relation of proximity more fundamental than the spatial relation.'

Bethe showed in 1938 how helium could be synthesized at stellar temperatures out of hydrogen through the intermediation of carbon atoms, without any transcendental hypotheses.

Helium synthesis as the source of stellar energy emphasized the importance of hydrogen as the raw material and fuel of the material universe. Eddington, simultaneously with B. Strømgren, calculated the abundance of hydrogen in the stars and showed that there was a plentiful supply for cosmic needs.

IV. SIDELINES

Eddington served as Secretary of the Royal Astronomical Society from 1912 until 1917. Through this position he was, for that period, officially in charge of the Society's correspondence. Holland was neutral during the First World War, so Eddington received on behalf of the Society from Dutch correspondents and associates their customary communications. Among them was W. de Sitter, who had followed, and mastered, Einstein's development of relativity. During 1914 and 1915, Einstein had completed his General Relativity Theory, which comprehended gravitation. He published the first complete account in a series of papers in November, 1915. De Sitter sent a copy of one of Einstein's 1915 papers to Eddington. Owing to the interruption of direct communications between Germany and England by the war, this remained for some time the only copy in England. De Sitter sent a series of papers expounding the new theory, which were published in the Royal Astronomical Society's *Monthly Notices* in 1916 and onwards. These excellent papers gave the main principles and results, and were the first exposition of them in the English language.

Eddington at once saw the tremendous significance of Einstein's extension of relativity theory. He had the ability and adaptability of mind to understand it thoroughly and quickly, in spite of its revolutionary novelty, unfamiliar mathematics, and German origin.

Against his will he undertook to explain it to his colleagues. He delivered many lectures and wrote many articles on it, both learned and popular, at the expense of time and energy that he had intended to devote to his own original researches.

The new theory, with its profound implications, its paradoxical results in terms of old ideas, its flights of thought through the universe, penetrating and embracing everything in it, appealed to the Icarus in him. Eddington not only restated Einstein's work and de Sitter's expositions, he joyfully took wing in flights of physical and mathematical thought and fancy of his own.

This tendency appeared even in his first technical book on the subject, his *Report on the Relativity Theory of Gravitation*, made for the Physical Society of London, and published in 1918. It contains only 91 pages and is a masterpiece of concise and elegant exposition. He explains that 'the main difficulty of this subject is that it requires a special mathematical calculus which, though not difficult to understand, needs time and practice to use with facility.' In Britain in 1916 Eddington was perhaps the only youngish man (he was thirty-three) of the requisite ability, who had the time to master the new mathematical technique, and the detachment to study the new conceptions with the minimum of prejudice. Indeed, his Icarus spirit prompted him to engage in exquisitely elegant and accomplished intellectual baiting of the prejudiced, which seemed to leave them delightfully and ridiculously out-of-date and yet without making absolutely clear why.

The Report began with an exposition of the Restricted Principle of Relativity, starting from the Michelson–Morley Experiment, which had been announced by Einstein in 1905, and had been accepted by English physicists as a mathematical device for describing such phenomena as the increase of mass with velocity, but had not been regarded by them as a fundamental theoretical basis of physical nature superseding Newton's.

The second chapter dealt with the relations of space and time, and Minkowski's conception of Space-Time and its implications. The obsoleteness of the old conception of 'force' was explained. The third chapter contained a beautiful short lesson on the Theory of Tensors and how to use them. The fourth expounded Einstein's new law of gravitation; and the fifth contained an account of the crucial phenomena by which the new theory could be tested: the exact explanation of the rotation of the orbit of Mercury in the direction of motion, which proceeds at the rate of 43 seconds a century. The observational fact was known with an accuracy of more than thirty times the probable error. The Newtonian theory of gravitation could account for only one-half of this precession of the perihelion of Mercury, and the anomaly was the most outstanding one in astronomy; for the conflict between observation and theory was sharp and much more fully established than in any other unsolved astronomical problem. Included in the chapter also were the theory and calculation of the magnitude of the deflection of rays of light by neighbouring masses and the displacement of lines in the spectra of stars due to the retarding effect of their masses on the rays of light emitted by them.

The sixth chapter contained an account of the gravitation of a continuous distribution of matter, and the seventh an elegant lesson on the Hamiltonian Principle of Least Action and its use. The Report concluded with a chapter on the curvature of space and time.

Eddington discussed the properties of de Sitter's 'world', or conception of space-time, which he had developed from Einstein's theory. He also included a bold but inadequately founded speculation of his own: the existence of a fundamental unit of length in the universe. It was out of harmony with the straightforward expository nature of the bulk of the Report, but characteristic of his own original and future work on relativity and other fundamental physical theory.

Eddington's next technical book was on *The Mathematical Theory of Relativity*. It was published in 1923, and was based on his university lectures in Cambridge. He put them into a consecutive form in 1921, as a mathematical appendix to the French translation of his semi-popular book, *Space Time and Gravitation*, first published in English in 1920.

The tendency to speculation noticed in the *Report* became far more pervasive in the *Mathematical Theory* of 1923. He started from the fundamental hypothesis that 'everything connected with location which enters into observational knowledge—everything we can know about the configuration of events—is contained in a relation of extension between pairs of events.'

He concluded his treatise by saying that 'we have been engaged in world-building—the construction of a world which shall operate under the same terms as the natural world around us. The most fundamental part of the problem falls under two heads, the building material and the process of building.'

'In one sense deductive theory is the enemy of experimental physics. The latter is always striving to settle by crucial tests the nature of the fundamental things: the former strives to minimize the success obtained by showing how wide a nature of things is compatible with all experimental results.'

The language and attitude of his later speculative philosophical works is already very clearly expressed—the tendency to invent the world mentally, and the preparation of a position in which deduction is exalted over experiment, and finally intellectual fancy over material fact.

The most important original contribution in the book was an interpretation of Einstein's law of gravitation and a generalization of Weyl's combined theory of electromagnetism and gravitation.

Eddington's other contribution to relativity theory was made with G. C. McVittie in 1930. They showed that Einstein's mathematical equations for the universe implied that it was unstable. This led to his investigations of the theory of the expanding universe.

Eddington and McVittie were, however, forestalled by Lemaître, who independently discovered a brilliant solution of the same problem and published it in an obscure journal in 1927.

Presently it was realized that Eddington and McVittie, and Lemaître, had long been forestalled by the Soviet theorist A. Friedmann, who had published as early as 1922 a demonstration that the Einstein universe was unstable. The solutions were independently discovered yet again by the American theorist H. P. Robertson.

Eddington's original contributions to the theory of relativity were considerable, but, compared with those of the founder, not at all of the first rank. He probably came to the subject too late (at the age of thirty-three) to have the chance of exerting his full mental power on it, and his English and Cambridge training, with its thinking in terms of physical models, and the application of mathematical problem-solving methods, was unsuitable for discovering a profound systematic all-embracing theory. Apart from Newton and Clerk Maxwell, Hamilton and Dirac, the British have rarely been able to leave their traditional method and enter the highest realm of general theory. When, like Eddington, they tried, they went astray. In Eddington's case contact with physical reality was lost. It seems he could not keep it without using physical models in the traditional British way. He had not got Einstein's sense of concrete physical logic, never lost even in the most abstract mathematics. Einstein was always a theoretical *physicist*, with the strongest sense of the reality of nature. He laboriously sought for and painfully acquired the right mathematics, and only just enough, to handle the physical ideas that he had very strongly and vividly in his mind.

Without physical models, Eddington was adrift in the sea of mathematics, and ultimately lost in reverie.

The greatness of Einstein and the magnitude of his discoveries become all the more impressive when it is seen that Eddington contributed no major fundamental addition and only two or three considerable applications.

Thus, like Jeans, Eddington as a theoretical physicist did not belong to the rank of Einstein and Planck. Perhaps his enormous abilities were intrinsically as big as theirs, and he was only unable to make a similar contribution to fundamental theory because his mental habits had been fixed in the British tradition, with its weaknesses as well as its strengths.

Eddington's effectiveness as an exponent of relativity was reinforced by his competence as a practical astronomer. This enabled him to take a leading part in the famous solar eclipse expeditions in 1919, which proved that Einstein's theory of the bending of rays of light by matter was correct.

After Eddington had received the news in 1916 of the general theory of relativity, F. W. Dyson, who was then the Astronomer Royal and an authority on eclipses, reviewed the opportunity for verifying Einstein's theory of gravitation at the forthcoming eclipse on May 29th, 1919. He found that the Hyades, and hence an unusual number of bright stars, would be near the sun, so that, if their rays were deflected on passing, there would be many examples of displaced images on a photograph of the eclipse. It might easily have happened that an equally good eclipse for the purpose would not have occurred again for thousands of years.

Energetic and thorough preparations were made for two expeditions, one, led by Eddington and E. T. Cottingham, to the island of Principe in the Gulf of Guinea off West Africa; and the other, led by A. C. D. C. Crommelin and C. R. Davidson, to Sobral in Brazil. At both places the eclipse would last for more than five minutes.

The astronomers left for their distant stations in an atmosphere of world attention. The centres of science waited for news in tense excitement. The eclipse came, and then two telegrams. One from Crommelin at Sobral with the two words: 'Eclipse splendid'; and the other from Eddington at Principe: 'Through cloud hopeful.'

At Sobral the sky was clear for four out of the five minutes of the eclipse. Nineteen good photographic images of stars near the edge of the sun's disc were obtained. At Principe two plates with measurable images of five stars were obtained, in spite of the cloudiness.

Comparison of the plates with photographs of the same spot in the sky in the absence of the sun showed that, for the Principe plates, the images were displaced $1''{\cdot}61 \pm 0''{\cdot}30$ from their normal position, and for the Sobral plates the displacement was $1''{\cdot}98 \pm 0''{\cdot}12$.

The observations triumphantly confirmed the Einstein prediction of $1''{\cdot}74$ as against the Newtonian prediction of one-half that amount—$0''{\cdot}87$.

On his return, the fascination of Eddington's exposition was reinforced by concrete astronomical facts and the romantic details of the expeditions. *Nature* reported from Cambridge that 'On Tuesday, December 2nd, 1919, in the hall of Trinity College, a lecture open to the University was given by Professor Eddington on the theory of relativity. Apart from the interest of the lecture, which attempted—sometimes lightly and sometimes almost dramatically—to present a popular account of the subject, the most striking thing about it was the enormous attendance. Fifteen minutes before the lecture began there was a queue half-way across the Great Court of men anxious to obtain admittance, and during the lecture the hall was entirely filled with dons and students listening breathlessly to hear an intelligible account, if one could be given, of the new theory. The keen interest was due, no doubt, largely to curiosity stimulated by the newspaper accounts of the subject, but also partly to the feeling, to which at least some hope of satisfaction can be given, that a further great unifying principle is needed in natural philosophy. Whatever the reason, however, the size and appreciation of the audience were no less extraordinary than the subject of the lecture and the brilliance of its exposition.'

While he was following all these sidelines on relativity, his own major original discoveries on the internal constitution of the stars were pouring forth in full spate, and with equal expository brilliance.

V. GUIDE TO SPACE, AND THE INTERIOR OF THE STARS

The First World War dammed up the burst of scientific discovery in the first quarter of the twentieth century, so that it pressed with all the more violence on the uninitiated. The artificial obstacles to its steady flow to them increased their desire to know and understand it.

Eddington found himself equipped with talent and appointed by circumstance to be their guide to the new frame of nature revealed by Einstein, the constitution of the stars revealed by himself, and the related work of others.

One of his earliest lectures in this capacity was his address to the Royal Institution, on February 1st, 1918, on relativity. There is a century-old Royal Institution ceremony of opening the door of the auditorium, to admit the lecturer, who takes a few steps to the lecture table while the audience look at him with anticipatory interest. Eddington skilfully seized the audience's attention by proceeding at once to develop his esoteric subject out of the little ceremony with which most of those present had long been familiar.

'There were many difficulties to encounter in entering the room just now,' he commenced. 'To begin with, we had to bear the crushing load of the atmosphere, amounting to 14 lb. on every square inch. At each step forwards it was necessary to tread gingerly on a piece of ground moving at the rate of twenty miles a second on its way round the sun. We were poised precariously on a globe, apparently hanging by our feet, head outwards into space. And this acrobatic feat was performed in the face of a tremendous wind of aether, blowing at I do not know how many miles a second through us. We do not claim much credit for overcoming these difficulties—because we never noticed them. But I venture to remind you of them, because I am about to speak of some other extraordinary things that may be happening to us of which we are quite unconscious.'

So he set the tone of wonder, and aroused alert anticipation, and prepared the audience to become aware of the Fitzgerald–Lorentz contraction, which they have hitherto been performing unconsciously all their lives.

He told them of Minkowski's conception of space-time, that had the great advantage of getting 'rid of all idea of a conspiracy', which the Fitzgerald-Lorentz proposal seemed to suggest: a kind of shortening or lengthening trick by nature to evade observation. By regarding space-time as a unity, different appearances from different points of view became quite natural. 'You cannot have a conspiracy of concealment when there is nothing to conceal.'

'We cut Minkowski's space-time world in a certain direction so as to give us separately space and time as they appear to us. We have been imagining that there exists some direction which would separate it into a real and absolute space and time. But why should there be? We do not attempt to cut the space-world in a particular direction so as to give us the *real* horizontal and vertical. The words 'horizontal' and 'vertical' have no meaning except in reference to a particular spot on the earth. So for a particular observer the space-time world falls apart into its four components, up-and-down, right-and-left, backwards and forwards, sooner-and-later, but no observer can say that this division is the one and only real one.' In this way, he gently loosened his listeners' conventional beliefs in absolute space and absolute time.

Then he explained that the notion of gravitational force was a clumsy description of the mutual behaviour of masses of matter. 'The earth moves in a curved orbit not because the sun exerts any direct pull, but because the earth is trying to find the shortest way through a space and time which have been tangled-up by an influence radiating from the sun.'

Eddington ordered his many general expositions into the first and most brilliant of his popular guide-books to the new science—*Space Time and Gravitation*, an outline of the General Relativity Theory. The first edition was published in 1920. Professor F. A. Lindemann (Lord Cherwell) in his review wrote that it was very proper that 'the first original work on the subject in our own language should come from the pen of the man who not only was the first on this side of the Channel to study the theory and see its fundamental importance, but who also had such a large share in its triumphant experimental verification at the last solar eclipse.'

Eddington wrote that Einstein had provoked a revolution of thought in physical science by his theory. The achievement consisted essentially in separating far more completely than hitherto the share of the observer and the share of external nature in the things that we see happen.

He asks to be excused for the limitations of non-mathematical presentation, for that which is peculiar to the limited imagination of the human brain can only be abstracted from phenomena with the aid of mathematical symbolism. 'As the language of a poet rings with a truth that eludes the clumsy explanations of his commentators, so the geometry of relativity in its perfect harmony expresses a truth of form and type in nature which my bowdlerized version misses.'

The exposition is introduced by a prologue in the form of a Socratic discussion on the nature of geometry between an Experimental Physicist (whose tones echo Rutherford's), a Mathematician (who even speaks with the words of Bertrand Russell: mathematics is 'the subject in which we never know what we are talking about, nor whether what we are saying is true'), and a Relativist, who speaks with the voice of Eddington himself.

The discussion starts with an analysis of the meaning of Euclid's proposition that any two sides of a triangle are together greater than the third. The Physicist says he will prove it by measuring the sides with a scale. The Relativist replies that they seem to be talking about different things. 'I was speaking of a proposition of geometry—properties of space, not matter. Your experimental proof only shows how a material scale behaves when you turn it into different positions.'

The laws of the behaviour of material scales were just as much physics as were those of electromagnetism.

The Physicist admits he is hazy about definitions, but there are so many interesting experiments to get on with.

The Relativist replies that he is not naturally inclined to be rigorous about these things, but 'although I appreciate the value of the work of those who are digging at the foundations of science, my own interests are mainly in the upper structure.'

They become involved in a discussion of the meaning of length, and hence of space. The Relativist says that 'natural geometry is the theory of the behaviour of material scales.' The Physicist observes that the space described by natural geometry must be 'a sort of abstraction of the extensional relations of matter.'

The Relativist replies that he does not ask for any violent effort of the imagination when he says that space is warped, that is, non-Euclidean. It seemed 'logical to conclude that space as known to us must be the abstraction of these material relations, and not something more transcendental', and he goes on to admit that 'space is conventional'. Following Poincaré, he chooses, for describing nature, that kind of space which is the most advantageous. But he differs from Poincaré in his choice, for whereas Poincaré preferred to use Euclidean space, the Relativist prefers non-Euclidean.

'I call the space thus chosen *physical space*, and its geometry *natural geometry*, thus admitting that other conventional meanings of space and geometry are possible', and if they were transplanted into an extremely intense gravitational field, they would directly perceive that space was non-Euclidean.

The Physicist says that he cannot imagine himself perceiving non-Euclidean space, so the Relativist tells him to 'look at the reflection of the room in a polished door-knob, and imagine yourself one of the actors in what you see going on there.'

Then they get on to the difficulties of understanding the meaning of simultaneity, and discuss whether any particular moment of time on the star Arcturus can be the same as the present instant on the earth. They are dependent on light for any conceivable experimental test. But light takes years to come from Arcturus, and consequently any test through it may be years in error. The Relativist says that there is in fact 'no such thing as absolute simultaneity of distant events', and it is therefore better not to base physics on this notion. 'The fundamental measurement is not the interval between two points of space, but between two points of space associated with instants of time.'

So the Relativist blithely remarks that their 'natural geometry is incomplete'; they must add to it a dimension of time. The Physicist queries whether it is right to speak of *geometry* if one is including time. So the Mathematician is called in. He doesn't find any difficulty in incorporating time, it's just one more variable to the three he was dealing with before: the three dimensions of Euclidean space. But then, of course, he is only a Mathematician, who doesn't know what he is talking about, nor whether what he is saying is true.

The Relativist says that 'space without time is as incomplete as a surface without thickness.' The space-order of natural objects cannot be discussed without reference to their time-order as well.

The Relativist sums up by remarking that we have believed in a Euclidean space for 2000 years because measurements favoured it. He sees no reason for changing the rules of the game now. When he speaks of space, he means 'the space revealed by measurement, whatever its geometry'.

The conservative physicists, 'frightened by the revolutionary consequences of modern experiments', have begun to play with the metaphysical idea of a pre-existing space to which they assign Euclidean properties, and which cannot be explained by experiment. The Relativist, on the other hand, is much more modest, he defines space by measuring it, and 'declines to consider anything more transcendental.'

The tables are turned on the Physicist: he is shown to be a metaphysician, while the Relativist is a realist.

After this stimulating Prologue, Eddington gives a clear conventional account of the Fitzgerald Contraction, and how its explanation leads to the restricted Principle of Relativity: it is impossible by any experiment to detect uniform motion relative to the aether.

He discusses how an aviator flying past at 161,000 miles a second would appear. If he was lying flat in his machine, in the line of flight, 'We would see a figure about three feet high, but with the breadth and girth of a normal human being. And the strange thing is that he would be sublimely unconscious of his own undignified appearance. If he looks in a mirror in his conveyance, he sees his usual proportions. . . . But when he looks down on us he sees a strange race of men who have apparently gone through some flattening-out process. If the reader has watched a cricket-match through a pair of prismatic binoculars, he will have seen this effect exactly.'

It was the reciprocity of the appearances which was so difficult to realize, a paradox beyond even the imagination of Dean Swift. 'If the Lilliputians had appeared dwarfs to Gulliver, and Gulliver had appeared a dwarf to the Lilliputians—but no! that is too absurd for fiction, and is an idea only to be found in the sober pages of science.'

Certain hypotheses had insensibly crept into all physical theories some two thousand, and others two hundred, years ago. Einstein had brought them to light. 'The most dangerous hypotheses are those which are tacit and unconscious.' He had shown that they were unnecessary, and stood in the way of understanding the simplicity of nature. Their presence in our physical thought, especially as unconscious preconceptions, had prevented us from seeing, and from finding, the simplest way of conceiving space and time.

'Sometimes by instinctive habit, sometimes by design, we attempt to eliminate our own share in the observation, and so form a general picture of the world outside us, which shall be common to all observers. A small speck on the horizon of the sea is interpreted as a giant steamer. From the window of our railway carriage we see a cow glide past at fifty miles an hour, and remark that the creature is enjoying a rest. We see the starry heavens revolve around the earth, but decide that it is really the earth that is revolving.'

Einstein discovered the physical conception and logical argument by which these appearances could be brought into consistence. Minkowski gave them form in his famous lecture to the Eightieth Meeting of German Scientists and Doctors, in 1908: 'The views of time and space, which I have set forth, have their foundation in experimental physics. Therein is their strength. Their tendency is revolutionary. From henceforth space and time each separately vanish as shadows, and only a union of the two preserves reality.'

We are not fully equipped by our senses for forming an impersonal view of the world. Eddington then compared *Alice in Wonderland* and *Gulliver's Travels* as two essays in relativity. Lewis Carroll, who was always making Alice change in size, took the ordinary scientific view, that the observer changed, rather than the observed. But 'Dean Swift took the truer view of the human mind when he made Gulliver attribute his own changes to the things around him; it never occurred to Gulliver that his own size had altered.'

In the nineteenth century intense efforts had been made to explain physical phenomena in terms of an aether. Light consisted of waves, atoms were vortices, in this aether, which was assigned a density and rigidity, and sometimes even 'a place in the table of the elements'.

The real death-blow was given to the materialistic conception of the aether when attempts were made to explain matter as a state of the aether. 'For if matter is vortex-motion or beknottedness in aether, the aether cannot be matter-some state of itself.'

Eddington analyses the notion of gravitational force, and concludes that 'the relativity view is that a field of force can, like length and duration, be nothing but a link between nature and the observer'.

'There are absolute things in the world—not only matter, but certain characteristics of empty space.' Nevertheless, 'the absolute world is of so different a nature, that the relative world, with which we are acquainted, seems almost like a dream'.

Eddington gives the elementary mathematics of the Minkowski conception, and concludes that 'the material structure of the four-dimensional world is fibrous, with the threads all running along time-like tracks; it is a tangled warp without a woof'.

He explains that gravitational energy presumably has mass, and calculates that the mass of the 'gravitational energy of the earth is of the order *minus* a billion tons'.

He begins his discussion of fields of force with Lucretius' statement of the first law of motion, in his *De Natura Rerum*, some 1700 years before Galileo and Newton: 'all things must be moved and borne along with equal velocities though of unequal weights through the unresisting void.'

Newton gave apparent precision to the notion of force, in order to prescribe an agent which could change motion. But the nature of gravitational force remained very mysterious, because it was all-pervasive. 'More than 200 theories of gravitation have been put forward', but most of them were too materialistic, 'filling space with the hum of machinery'.

Einstein was the first to discover a true physical model of the gravitational field—the conditions in a lift. Its upward acceleration exactly produces 'an additional gravitational field superimposed on that normally present'.

We never feel the force of gravitation. When we are falling freely under gravitation we feel nothing. Normally, we feel the resistance of the earth on which we are standing, against gravitation, and the pressure is exerted by 'the bombardment of the soles of our boots by the molecules of the ground'. The aviator is aware of fields of force, which occur when he changes speed or direction, and which cannot be discriminated from gravitation. It is one of the causes of his difficulties in navigation.

Force turns out to be what the observer perceives when he is following a curved track through the four-dimensional space-time world.

Newtonian mechanics assumes that there is a super-observer. 'If *he* feels a field of force, then that force really exists.' It is quite possible that there *is* such a super-observer, 'whose views have a natural right to be regarded as the truest or at least the simplest.'

'A society of learned fishes would probably agree that phenomena were best described from the point of view of a fish at rest in the ocean.'

The equivalence of the field of gravitational force, and that which is experienced in the lift leads to Einstein's Principle of Equivalence, that 'a gravitational field of force is precisely equivalent to an artificial field of force, so that in any small region it is impossible by any conceivable experiment to distinguish between them.' Hence, Eddington concludes, 'force is purely relative'.

After discussing force, Eddington outlines the elementary geometry of space. He begins by quoting W. K. Clifford, 'that the hypothesis that space is not flat, and again that its geometrical character may change with time may or may not be destined to play a great part in the physics of the future; yet we cannot refuse to consider them as possible explanations of physical phenomena, because they may be opposed to the popular dogmatic belief in the universality of certain geometrical axioms—a belief which has arisen from centuries of indiscriminating worship of the genius of Euclid.'

Eddington concluded that 'there is no *shape* inherent in the absolute world', and he preferred to 'think of matter and energy, not as agents causing the degrees of curvature of the world, but as parts of our perceptions of the existence of the curvature.'

These views had been marvellously anticipated by Clifford, who wrote in 1875: 'The reality corresponding to our perception of the motion of matter is an element of the complex thing we call feeling. What we might perceive as a plexus of nerve-disturbances is really in itself a feeling; and the succession of feelings which constitutes a man's consciousness is the reality which produces in our minds the perception of the motions of his brain. These elements of feeling have relations of *nextness* or contiguity in space, which are exemplified by the sight-perceptions of contiguous points, and relations of succession in time which are exemplified by all perceptions. Out of these two relations the future theorist has to build up the world as best he may. Two things may perhaps help him. There are many lines of thought which indicate that distance or quantity may come to be expressed in terms of *position* in the wide sense of the *analysis situs*. And the theory of space-curvature hints at a possibility of describing matter and motion in terms of extension only.'

In 1873 Clifford had written that 'the geometer of today knows nothing about the nature of actually existing space at an infinite distance; he knows nothing about the properties of this present space in a past or a future eternity. He knows, indeed, that the laws assumed by Euclid are true with an accuracy that no direct experiment can approach, not only in this place where we are, but in places at a distance from us that no astronomer has conceived; but he knows this as of Here and Now; beyond his range is a There and Then of which he knows nothing at present, but may ultimately come to know more.'

What does the Relativist have to say today about the magnitude of forces, real or apparent, as degrees of curvature in space-time? Eddington relates a fable.

'A race of flat-fish once lived in an ocean in which there were only two dimensions. It was noticed that in general fishes swam in straight lines, unless there was something obviously interfering with their free courses. This seemed a very natural behaviour. But there was a certain region where all the fish seemed to be bewitched, some passed through the region but changed the direction of their swim, others swam round and round indefinitely. One fish invented a theory of vortices, and said that there were whirlpools in that region which carried everything round in curves. By-and-by a far better theory was proposed; it was said that the fishes were all attracted to a particularly large fish—a sun-fish—which was lying asleep in the middle of the region; and that was what caused the deviation of their paths. The theory might not have sounded particularly plausible at first, but it was confirmed with marvellous exactitude by all kinds of experimental tests. All fish were found to possess this attractive power in proportion to their sizes; the law of attraction was extremely simple, and yet it was found to explain all the motions with an accuracy never approached before in any scientific investigations. Some fish grumbled that they did not see how there could be such an influence at a distance, but it was generally agreed that the influence was communicated through the ocean, and might be better understood when more was known about the nature of water. Accordingly, nearly every fish who wanted to explain the attraction started by proposing some kind of mechanism for transmitting it through the water. But there was one fish who thought of quite another plan. He was impressed by the fact that whether the fish were big or little they always took the same course, although it would naturally take a bigger force to deflect the bigger fish. He therefore concentrated attention on the courses rather than on the forces. And then he arrived at a striking explanation of the whole thing. There was a mound in the world round about where the sun-fish lay. Flat fish would not appreciate it directly because they were two-dimensional; but whenever a fish went swimming over the slopes of the mound, although he did his best to swim straight on, he got turned round a bit. (If a traveller goes over the left slope of a mountain, he must consciously keep bearing away to the left if he wishes to keep to his original direction relative to the points of the compass.) This was the secret of the mysterious attraction, or the bending of the paths, which was experienced in the region.'

The effect of the gravitational field, on the old view, was to slow down the speed of light. When the light-rays from a star passed near the sun, they consequently behaved as if they were passing through a medium of varying density, decreasing with distance from the sun. The front of the waves would be slewed round in the same way that the front of sea-waves is slewed round when they roll on to a beach that is shelving across their direction. The magnitude of the deflection of the stellar rays by the sun can therefore be calculated by the same formulae which govern the degree of the slewing of the wave-front on the beach. Hence the deflection can be calculated by a simple formula of the geometrical optics of refracting media.

Eddington then quotes Newton's First Query. 'Do not Bodies act upon Light at a distance, and by their action bend its Rays, and is not this action (*caeteris paribus*) strongest at the least distance?'

After discussing Momentum and Energy, Eddington refers to the notion of *Action*, which is not to be confused with Newton's 'Action and Reaction'. It is equal to mass or energy multiplied by time, and 'is more fundamental than either. Action is the curvature of the world.'

The laws of gravitation, mechanics, and electromagnetism have all been summed up in the Principle of Least Action. 'Action is one of the two terms from pre-relativity physics which survive unmodified in a description of the absolute world. The only other survival is entropy. The coming theory of relativity had cast its shadows before; and physics was already converging to two great generalizations, the principle of least action and the second law of thermodynamics.'

Eddington completed his exposition with a review of the Nature of Things. He starts with the physicists' belief in the existence of a real external world. He repeats the conventional idealist criticism of the common-sense belief in the real existence of an object, for example a penny. But this appears differently, according to the aspect from which it is seen, so, Eddington says, 'Physical reality is the synthesis of all possible physical aspects of nature.' Consequently, as it is natural to explain the complex in terms of the simple, 'it carries with it the necessity of explaining the familiar in terms of the unfamiliar.'

The search for physical reality is not necessarily utilitarian, but it has turned out to be by no means profitless.

The statement that the world is four-dimensional is not so clear as it appears at first. An aggregate of a large number of things need not have any particular number of dimensions, an element of *order* is necessary.

'Consider, for example, the words on this page. To a casual glance they form a two-dimensional distribution; but they were written in the hope that the reader would regard them as a one-dimensional distribution.'

The law of gravitation is 'merely the definition of a vacuum; . . . matter is a symptom and not a cause.' Acceleration is not a change of velocity but an independent entity, 'much simpler and more universal than velocity'.

The mind perceives those features which have some degree of permanence. 'When the eye surveys the tossing waters of the ocean, it is the waves that strike the attention', which are a 'motion of the wave-form, which is not a motion of the water at all.'

It is the mind which builds up permanent substance out of relations, and has 'imposed these laws on an indifferent world. . . . It is structure, not material, which counts. . . . Mind filters out matter from the meaningless jumble of quantities, . . . it exalts the permanent and ignores the transitory. . . . Is it too much to say that the mind's search for permanence has created the world of physics?'

'Give me matter and motion,' said Descartes, 'and I will construct the Universe.' But the mind reverses this, and says, 'Give me a world—a world in which there are relations—and I will construct matter and motion.'

But there is one most inconvenient fly in the ointment. Atomicity and quanta exist. 'Whence arises this discontinuity?' There is at present no ground to believe that it is due to the mind. Perhaps in the future it will be discovered that atomicity and the other laws of physics are identical. But so far as can be seen, atomicity definitely has its seat in external nature, and perhaps is the only physical law grounded there and not in the human mind.

Physical science is a knowledge of structure, not of content. The unknown content which runs through all the physical world 'must surely be the stuff of our consciousness'.

So the Guide, after taking us on a thrilling tour through the new Universe, brings us to the place of embarkation for home. Like Virgil bidding farewell to Dante on returning to the shades, or the genius of Defoe speaking to himself through Robinson Crusoe, he calls to us from the landing-stage, as our ship weighs anchor for the murk and mud of reality:

'We have found a strange foot-print on the shores of the unknown. We have devised profound theories, one after another, to account for its origin. At last, we have succeeded in reconstructing the creature that made the foot-print. And Lo! it is our own.'

Eddington's *Space Time and Gravitation* was the first, and most brilliant, if not the soundest of the non-technical expositions of the General Theory of Relativity in the English language. It was imitated by many, and translated into ten languages.

Brilliant though it was, it was not the best of his books of general exposition. It was surpassed by his *Stars and Atoms*, published in 1927. This expounded the substance of his solid contributions to the theory of the constitution of the stars. In it imagination and speculation were strictly disciplined by the realities of what was known about stars, and were not allowed to stray beyond their proper functions.

The great success of this book, which was translated into eleven languages, started the competition from Jeans for the attention of the wide public, and to the spate of books from both which made them, for better and for worse, the chief spokesmen of science to the Anglo-American populace in the 1930s.

The story that Eddington had to tell is the one that has already been outlined in the account of his contribution to the theory of the constitution of the stars. Here we will look at his technique of presenting the story to the general reader.

The idea of the book arose out of his Evening Discourse to the British Association at Oxford in 1926. He delivered it in the packed hall of the University Union. He had a pale and faded appearance, and his voice was not strong, but he fascinated his audience by the wonder of his similes. The most dramatic of these was invented, apparently extempore, as he paused in his meditative manner in the darkened and silent hall, and surveyed a photograph of a Wilson cloud-chamber track of an α-particle, on which there was a very large and fine thumb-mark, inadvertently made on it by the photographer: 'I wonder if there is an undercurrent of suspicion in your minds,' he said, 'that there must be something of a fake about this photograph. Are these really the single atoms that are showing themselves—those infinitesimal units which not many years ago seemed to be theoretical concepts far outside any practical apprehension? I will answer that question by asking you one. You see a dirty mark on the picture. Is that somebody's thumb? If you say Yes, then I assure you unhesitatingly that these streaks are single atoms. But if you are hypercritical and say "No, that is not anybody's thumb, but it is a mark that shows that somebody's thumb has been there", then I must be equally cautious and say that the streak is a mark that shows where an atom has been. The photograph, instead of being the impression of an atom is the impression of the impression of an atom, just as it is not the impression of a thumb, but the impression of the impression of a thumb. I don't see that it really matters that the impression is second-hand instead of first-hand. I do not think we have been guilty of any more faking than the criminologist who scatters powder over a finger-print to make it visible, or a biologist who stains his preparations with the same object.'

A long silent thrill held the audience, which presently broke into an applause rarely heard in scientific meetings.

When Eddington was asked to expand his lecture into a book, he said that he would rather add to it some others that he had written, and he would hold them for a while, in order to save himself the trouble of writing new ones for a number of forthcoming engagements.

Eddington was a model author in the preparation of his manuscripts and his dealings with publishers.

In his approach to the presentation of his theory of the constitution of the stars, he explained that the universe provides space for their accommodation on the most lavish scale. 'The stars are scattered through space about as sparsely as thirty cricket balls flying about in a volume equal to that of the earth.' Space is ample enough for millions of island universes, each consisting of thousands of millions of stars lodged on that scale.

What is a drop of water to this? Yet each one contains thousands of millions of millions of millions of atoms, each about one hundred-millionth of an inch in diameter. 'Here we marvel at the minute delicacy of the workmanship.'

Midway in scale between the atom and the star is the human body. Man is slightly nearer to the atom than to the star. His body contains about 10^{27} atoms; whereas 10^{28} human bodies would be required to constitute a star.

Man is in the central position to look both ways, 'the road to a knowledge of the stars leads through the atom, and important knowledge of the atom has been reached through the stars.'

The most familiar star is the sun. What is it like? Its surface is covered with cyclones of flaming gas. One could always safely forecast weather on the sun; cyclone or anticyclone, it will be '*very warm*', and the temperature will be about 6000° C.

What is it like within? As the architect can work out the stresses in his building without boring into the foundations, so the astronomer can work out the pressure inside without boring a hole. If the surface manifestations are to be understood, if the reason why 'one star differeth from another star in glory', is to be discovered, 'we must go below—to the *engine-room*'.

One can calculate the temperature by working out a model of the way in which the hot gases and radiation resist the gravitation of the upper layers, and then choose a central temperature which leads to the surface temperature of 6000°. The central temperature giving the result is about 40,000,000°. Such a temperature is not meaningless. Molecules in the room fly at 500 yards a second, and at a temperature of 40,000,000° the speed would be 100 miles a second. That is a speed to which the astronomer is quite accustomed, and as for the experimental physicist, he is quite contemptuous of it, he handles atoms shot off from radium at 10,000 miles a second. To him the speed of atoms inside the stars is only a jog-trot.

The atom consists of a heavy nucleus with 'a girdle or crinoline' of electrons. Each kind of atom has 'a different quorum' of electrons. On the earth we generally meet atoms fully attired in their complete crinolines of electrons, but in the stars it is different.

Then Eddington expounds ionization. Atoms moving at high speed may knock electrons out of the crinolines of other atoms with which they collide. They may leave a trail of electrons and positively charged nuclei along their tracks; the evidence of furious driving.

The high temperature causes the atoms to strip each other, and reduce the physical conditions to unusual simplicity. It is at low temperatures, such as those on the earth, that matter has complicated properties. 'Stellar atoms are nude savages innocent of the class distinctions of our fully arrayed terrestrial atoms.'

The stars are of remarkably uniform mass, that of the sun is—'I will write it on the blackboard:

2,000,000,000,000,000,000,000,000,000 tons.

I hope I have counted the 0s rightly, though I dare say you would not mind if there were one or two too many or too few. But Nature *does* mind.' Then he explains that the correct counting of the 0s for the stars is done by radiation pressure.

So there is a wild hurly-burly inside the stars—the atoms tearing each other apart, and being pierced by radiation, and recombining all in thousand-millionths of a second.

Modern atomic physics has a knockabout comedy aspect. Stellar evolution is not stately, but more like hair-breadth escapes on the films. 'The music of the spheres has almost a suggestion of—jazz.'

A wave of radiation picks its way from the centre of the star to the surface in about ten thousand to ten million years, according to the mass of the star.

The radiation sets up such a great pressure and takes so long to pass through the star because of its intensity. It ionizes the atoms as quickly as they can recapture the electrons, and thus makes the material more opaque than it would be on earth. 'It is like an army of mice marching through your larder springing the mousetraps as fast as you can set them. Here it is the time wasted in resetting the traps—by capturing the electrons—which counts.'

Eddington has to expound his own great discovery that dense stars are nevertheless perfect gases. It is due to the contraction in size of atoms when they are highly ionized, and have lost many outer electrons. Their diameters are only one-hundredth of their size at terrestrial temperatures. They had overlooked the fact that in the stellar ball-room 'crinolines are no longer in fashion', and hence the atoms can get closer together.

Then he discusses the Companion of Sirius, the first star whose existence was deduced from perturbations before it had been seen. Eighteen years later, an instrument-maker focused a telescope on Sirius, not for investigation of the star, but to test the optical perfection of a new object-glass. When the observer, Alvan Clark, saw a point of light close to Sirius he was probably disappointed and tried to polish it away. However, it could not be removed, and was found to be the hitherto unseen Companion.

As the Einstein effect of retardation of light is proportional to the mass divided by the radius, and the radius was very small, though the mass was normal, the effect for the Companion was about thirty times that for the sun.

Some of the stars are very rarefied. Betelgeuse has a density of about one-thousandth that of air, which in earlier times would have been regarded as a good vacuum.

Eddington made important discoveries concerning the cloud of atoms in space. The vast regions between the stars are not entirely empty. Every cubic inch contains about one atom. This is extremely few, but space is so enormous that the total mass of the dispersed atoms for the whole universe is very great.

An atom in the stellar spaces will rarely be more than an inch away from its next atomic neighbour.

The system of the stars is floating in a placid ocean of rarefied atoms, in which there is little relative motion and only minor slow currents.

He described E. A. Milne's beautiful explanation of why calcium is so prominent in the sun's chromosphere. This consists of a few kinds of atoms that 'are able to float—float, not on the top of the sun's atmosphere, but on the *sunbeams*. The art of riding a sunbeam is evidently rather difficult, because only a few of the elements have the necessary skill. The most expert is calcium.' This is odd, because calcium atoms are rather heavy.

The atoms are like tiny shuttlecocks, which are kept up by the pressure of the sun's light. They acquire momentum from the light by absorbing it. But in most atoms, when this absorption occurs, an electron is lost, and the atom is then in a condition where it cannot absorb any more of the sun's light, and drops like a stone. Calcium however happens to have an electron which is not thrown off its atom by the sun's light, but only raised to a higher orbit. It revolves around this, and after a time drops back into the old orbit, where it may again get a kick up into the higher one. The calcium atom is able to absorb 20,000 such kicks a second, and so receive sufficient momentum to keep it permanently floating on the sun's rays.

The special skill demanded is the tossing of an electron 20,000 times a second without dropping it.

In most other atoms, the kicks are too dangerously big, and liable to detach the electron altogether, and hence leave the atom to fall back into the sun. Only the calcium atom has an electron which can do the required juggling trick with rays of the wavelength of those emitted by the sun.

Milne's theory enables the mass of the chromosphere to be calculated. It comes out at about 300 million tons. 'One scarcely expects to meet with such a trifling figure in astronomy. It is less than the tonnage handled by our English railways each year.'

As for the absolute brightness of the stars, you cannot tell it merely by looking at them. You must know their distance. Otherwise, you may 'mistake Arcturus for a ship's light' and be out by a factor of a quintillion.

With regard to their source of energy, he writes in 1927: 'I suppose that many physicists will regard the subject of subatomic energy as a field of airy speculation. That is not the way in which it presents itself to the astronomer.'

For him, the measurement of subatomic energy is one of his commonest operations, for it is nothing more than the measurement of the energy radiated by the stars, which is subatomic in origin.

The great advantage of the view that 'the energy of a star is derived by the building up of other elements from hydrogen' is that the process is certainly possible.

'To my mind the *existence* of helium is the best evidence we could desire of the possibility of the *formation* of helium.'

Though the synthesis of helium out of hydrogen seemed to Eddington the appropriate source of energy for the ordinary stars, he suspected that the far more drastic process of the total annihilation of matter was the only one which would provide a sufficiently long life for the complete process of stellar evolution. Thus, in the end, perhaps in the white-dwarf stage, 'the ultimate particles one by one yield up their energy and pass out of existence. Their sacrifice is the life-force of the stars which now progress on their high adventure:

> Atoms or systems into ruin hurl'd
> And now a bubble burst, and now a world.'

Eddington ended his first Guidebook on an arrogantly majestic note of superiority: man shaping nature by his mind. He ended his second in an exactly contrary spirit.

'I should have liked to have closed these lectures by leading up to some great climax. But perhaps it is more in accordance with the true conditions of scientific progress that they should fizzle out with a glimpse of the obscurity which marks the frontiers of present knowledge. I do not apologize for the lameness of the conclusion, for it is not a conclusion. I wish I could feel confident that it is even a beginning.'

How differently the two books end! and how significantly different! *Space Time and Gravitation* was brilliant but derivative, its lack of creative originality was bolstered by a charming but arrogant mentalism. *Stars and Atoms* needed no philosophical defiances: its author was explaining what he had himself discovered, he arranged his discoveries in beautiful and fascinating words and images, but he modestly left them to speak for their dazzling selves.

His lame conclusion might be compared with T. S. Eliot's

> This is the way the world ends
> Not with a bang but a whimper,

but Eddington's came from the strength of the successful investigator, pausing before his next advance into the unknown, whereas Eliot's came from despair.

It is evident that Eddington was master of the fable, the simile, the paradox, the epigram, the pun. He used the anthropomorphic method of allowing processes to live and speak as beings. He was master of the dialogue and the natural conversational style. He had a dramatic sense, and created beautiful entries and exits for his scientific characters. He had a gentle, poetic fancy. He had humour.

He was apt at choosing quotations, but these were nearly all well known and from classic writers. He rarely, if ever, revealed in a phrase significance that had not been seen before. He was conventional in his interests, which were very similar to those of many Cambridge scientists of the period, though more gracefully expressed.

His general view of the new discoveries was that of the social group to which he belonged, and he communicated this attitude to a very much wider public through his books. As a writer for the general reader, Eddington must be seen not only as the explainer of new scientific ideas, but also as the communicator of the attitude of a social group towards them.

As a Trinity man, Eddington was an intellectual aristocrat, and belonged to the society of Balfour and Baldwin, Bertrand Russell and Whitehead, as well as the tradition of Newton and Clerk Maxwell. He was an associate of those who ruled England. There was nothing in his expository works which could ever seriously have upset them.

On Relativity he was inclined to be naughty. He sought to shock the conventional without departing from the criteria of the conventional, and he often stretched his analogies beyond their legitimate correspondence. But on the Stars he was wonderfully and effortlessly precise. It was his discoveries about them that made him the greatest astronomer of his day, and his writings about them which exhibited riches of expository technique for making the discoveries of science intelligible to mankind in general.

VI. SOPHIST

With the publication of the *Internal Constitution of the Stars* in 1926, at the age of forty-three, Eddington's attention diverged more and more to philosophy and speculative theorizing. The last chapter of *Space Time and Gravitation*, in 1920, with the title, 'The Nature of Things', had contained a sketch of a philosophy of physical science. He extended this in a course of Gifford Lectures at Edinburgh University in 1927, which were published in his book, *The Nature of the Physical World*, in 1928. He developed his arguments further in *New Pathways in Science* in 1935, and *The Philosophy of Physical Science*, in 1939.

Eddington begins his *Nature of the Physical World* with a description of the ordinary table at which he has been writing his MS., and the description of it in terms of contemporary physics. The ordinary table has extension, it is comparatively permanent, it is coloured, and above all it is *substantial.* By this he did not mean merely that it did not collapse when leant on, but that substantiality was part of its intrinsic nature.

By contrast, the scientific conception of the table was mostly of emptiness, containing numerous sparsely distributed electric charges rushing about at great speed, whose combined bulk was less than a million-millionth of the table itself. In spite of its strangeness, it is an efficient table, the pages of his MS. are supported by millions of particles bombarding it from beneath, so that it does not fall.

Eddington says there is nothing *substantial* about this table, it is mostly empty space, though admittedly pervaded by fields of force, which, however, are not 'things' but only 'influences'.

In dissecting matter into electric charges, the picture of it which first gave rise to the notion of substance has been lost on the way. 'Whether we are studying a material object, a magnetic field, a geometrical figure, or a duration of time, our scientific knowledge is summed up in measures.'

'We feel it necessary to concede some background to the measures—an external world; but the attributes of this world, except in so far as they are reflected in the measures, are outside scientific scrutiny. Science has at last revolted against attacking the exact knowledge contained in these measurements to a traditional picture-gallery of conceptions which convey no authentic information of the background and obtrude irrelevancies into the scheme of knowledge.'

Modern physics, by 'delicate test and remorseless logic' has proved that the second table is the only one that is really there. But modern physics will never succeed in exorcising that first table, 'strange compound of external nature, mental imagery and inherited prejudice', which is visible to the eyes and tangible to the grasp.

But are not the two tables merely two different aspects of the same thing? Yes, they are no doubt ultimately to be identified in some fashion; it is true that science starts from the familiar world and must return to it; but 'the part of the journey over which the physicist has charge is in foreign territory.'

Until recently the physicist used to borrow the raw material of his world from the familiar world, but now his raw materials are electrons, quanta, Hamiltonian functions, etc. There is a familiar table parallel to the scientific table, but there is no familiar electron parallel to the scientific electron.

After the physicist has put together his scientific table it is permissible to compare it with the familiar table, but premature attempts at linkage are entirely mischievous.

'Science aims at constructing a world which shall be symbolic of the world of commonplace experience.' It is not necessary that every individual symbol used should correspond to something in the familiar world.

'The external world of physics has thus become a world of shadows. In removing our illusions we have removed the substance.' We have seen that substance is one of our greatest illusions. However, we should be careful to investigate whether 'in our zeal to cut out all that is unreal we may not have used the knife too ruthlessly.'

Perhaps 'reality is a child which cannot survive without its nurse, illusion.' 'The frank realization that physical science is concerned with a world of shadows is one of the most significant of recent advances.'

It is difficult to treat the physical world as purely symbolic, and there is always a tendency to relapse into 'incongruous conceptions taken from the world of consciousness'.

When he thinks of an electron, the picture of a little hard red ball rises in his mind. He knows he is incorrigible, and that the younger men are already finding this picture too concrete and are constructing the new scientific world out of symbols that do not even obey the laws of orthodox arithmetic. He finds it difficult himself to rise to that plane of thought, but recognises that it has got to come.

He illustrates the nature of exact science by analysing the procedure in the investigation of an elephant sliding down a grassy hillside. The ponderous object is described in scientific language as of mass two tons. Then the steepness of the hill is described as an angle of 60°, and the slipperiness of the soft, yielding earth is replaced by a coefficient of friction. 'And so we see that the poetry fades out of the problem, and by the time the serious application of science begins we are left with pointer-readings.'

The whole subject-matter of exact science consists of pointer-readings, and the very definite conceptions that we seem to have of objects in the external world do not enter into it. 'From the point of view of exact science the thing that really did descend the hill can only be described as a bundle of pointer-readings.'

Eddington says that the limitation of the scope of physics to pointer-readings is not a craze of his own, but is the outcome of a tendency discernible a century ago, and only brought out completely by relativity theory.

The description of the physical properties of a body is merely got by applying various measuring indicators to it, '*and nothing more*'.

Eddington's sharp separation of the familiar world and the physical world leads him to various conclusions. His conception of the physical world has an 'idealistic tinge', which is very different from his earlier philosophical views, in so far as he 'had any'.

He says in *Science and the Unseen World* that 'it is probably true that the recent changes of scientific thought remove some of the obstacles to a reconciliation of religion with science; but this must be carefully distinguished from any proposal to base religion on scientific discovery. For my own part I am wholly opposed to any such attempt.'

The exploration of the external world leads, he says, to a shadow-world of symbols. The enquiring mind feels there must be more behind this, and so returns to its own starting-point in human consciousness. There he finds other stirrings, other revelations (true or false) than symbols. We can and must believe we have an inner sense of values, otherwise we cannot have the faith to begin scientific investigation.

We do not believe that our discoveries are final truths, but we truly believe that we are on the road. 'If our so-called facts are changing shadows they are shadows cast by the light of constant truth.'

'We need not turn aside from the measure of light that comes into our experience showing us a Way through the unseen world.'

Eddington seemed determined to cut the familiar world off from the world of symbols, so that he could leave his own consciousness to engender such beliefs as it wished. It was not surprising, then, that it brought forth the beliefs of the Quakers, in whose Society he had been reared, and that he was sympathetic to mysticism. He conceived the world as a kind of crude mind-stuff, an undeveloped mixture of life and matter in varying proportions. The mind-stuff was continuous, and yet differentiated. 'In this background we must find, first, our own personality, and then perhaps a greater personality. The idea of a universal

Mind or Logos would be, I think, a fairly plausible inference from the present state of scientific theory.'

Eddington remarked that 'if it were necessary to choose a leader from among the older philosophers, there can be no doubt that our choice would be Kant.' Ritchie comments that Kant's view of the definite limits of scientific knowledge and its separation from the extra-scientific realm was sympathetic to Eddington's outlook. Kant had been deeply influenced by the East Prussian Pietists, as Eddington had by the Quakers.

Possibly the early beliefs that he had acquired prevented him from making the simple philosophical corrections that would have enabled him to see that symbols are not sharply separated from the things they deal with, but are abstractions of certain aspects, and come out, and are part, of them. As R. B. Braithwaite commented, he never seems to have fully realized, that things, events, have properties which physics can study, and also properties that it cannot. 'Physics is a treatment of a selection of the properties in the world, not of a selection of the things.'

Eddington appears to have been aware that his treatment of the origin of symbols was inadequate, and excused himself by saying that Whitehead and he had tunnelled through the mountain from opposite sides, Whitehead starting from the familiar world, in which he showed that the symbols of science were derived from familiar things by abstraction, whereas he had started from science, in which a symbolic world was synthesized and compared with the familiar world. It was important 'not to confuse the two entrances to the tunnel'.

A. D. Ritchie refers Eddington's total separation of the familiar and the scientific worlds to confusion of *subjective* and *objective*, which 'Eddington has done nothing at all to clear up.' The comparison of the familiar and the scientific tables 'may pass muster as a joke, but is not one of his best jokes. It depends on a way of imagining that makes imagination obscurantist.'

Eddington's metrical view is misleading because, says Ritchie, it 'cannot be used except on the basis of previous non-metrical notions and operations.' These are qualitative, not quantitative. Euclid arrived at his geometrical symbols by moving his measuring instruments or tools, Galileo added symbols which would include the description of motion besides shape.

M. Cornforth points out that Eddington's pointer-readings are not 'things which just happen in the experience of physicists' from which scientific laws may be formulated, but things 'produced by the scientists themselves' for a definite purpose. The pointer-readings are not the subject matter of science. This consists of the various aspects of the world which the pointer-readings are used to record. Science is essentially experimental, and hence, ultimately, social in origin. Eddington's mysteries arise through failing to formulate, and forgetting, the ultimate basis of science in social activity.

Eddington described laws as primary and secondary. Primary laws apply to single individuals, while secondary laws are statistical and apply to crowds. The Second Law of Thermodynamics is of the second kind. He regarded it as possibly the most important law in nature.

Formerly, primary laws were supposed to follow the rule of cast-iron determinism, but the quantum theory had shown that even the description of individual particles obeyed statistical laws. As he believed that statistical laws were not deterministic, he supposed that the discovery that primary laws were statistical had abolished determinism.

'It is difficult to see now any justification for the strongly rooted conviction in the ultimate re-establishment of a deterministic scheme of law, except a supposed necessity of thought.'

Schrödinger's wave-theory of the electron was interpreted by Eddington as indicating the probability that, at any instant, an electron would be in a certain place. He regarded it 'not as a physical theory but a dodge, and a very good dodge, too'. Dirac used 'basal entities inexpressible by numbers'. Eddington said that he 'liked to think that we may have not yet heard the last of the idea'.

Heisenberg's principle of uncertainty indicated that if an electron has a narrowly defined location, then its velocity will fall within certain limits. Thus the laws governing the properties of the electron are fundamentally statistical. Eddington says that: 'One would not be surprised if in the reconstruction of the scheme of physics which the quantum theory is now pressing on us, secondary law becomes the basis and primary law is discarded.'

Braithwaite starts a penetrating criticism of Eddington's views on statistical laws with the remark that 'the importance of entropy, like entropy itself, is always increasing'. This law, the Second Law of Thermodynamics, holds, in Eddington's words, 'the supreme position among the laws of Nature', and, as Braithwaite comments, 'a law with such pretensions should have a pedigree above suspicion.'

Eddington introduced his discussion of the idea of entropy, which may be interpreted as a measure of degree of 'randomness' in a physical state, by considering the shuffling of cards. He says that if a pack of new cards, in the order arranged by the maker, is shuffled for a few minutes, the order disappears and 'will never come back, however long you shuffle. Something has been done which cannot be undone.'

He qualifies this by saying that there is a 'ghost of a chance' that if the cards are shuffled long enough, the cards will come back to the original order. This is due to the comparatively small number of cards in a pack.

But when collections of atoms are shuffled, the chance is enormously less still, because the numbers of atoms are so very large.

Braithwaite points out that this is not correct: 'If only you shuffle long enough it becomes not indeed certain but as probable as you like that any specific arrangement will appear.' The frequency of the operation of shuffling increases the probability of the return to the original order. If a man started to shuffle a pack of cards as frequently as a molecule collides in air, and if 'a snail started to crawl round the universe (finite but unbounded) at the rate of one centimetre during the life' of the system of the stars, then it would have 'got round the universe many million times before it would become at all likely that the man would have got the pack back to the original order.'

Eddington repeats the illustration that 'if an army of monkeys were strumming on typewriters they *might* write all the books in the British Museum', and says that the chance of this happening is 'decidedly more' than of molecules in a vessel of air concentrating by chance all in one half of the vessel.

But given time enough, both of these things will happen, to any degree of certainty that you may specify.

Of course, says Braithwaite, the truth of the Second Law of Thermodynamics, regarded as dealing with the shuffling of atoms, may be very probable during the next 10^{100} years, but as the supreme physical law, applied to all time, 'it is not even approximately true'.

Eddington related the notion of time to the increase of entropy or randomness of the shuffling of the particles in the world. The space-time map of the world gave no sign as to the direction of its change, whether from 'past to future, or from future to past'. The 'moving-on of time is a condition of consciousness. We have direct insight into "becoming" which sweeps aside all symbolic knowledge as on an inferior plane. If I grasp the notion of existence because I myself exist, I grasp the notion of becoming because I myself become. It is the innermost Ego of all which *is* and *becomes*.'

Eddington does not identify 'becomingness' with entropy, but regards the latter as the physical criterion of the former. Hence man's insight into becoming and time is, in Braithwaite's words, 'derived from a reading from an entropy-clock in the brain.' H. W. B. Joseph then asks: 'How can I acquire my consciousness of (time) by a consultation with brain-cells that must be useless until I have acquired it?'

As Braithwaite has shown, entropy does not always increase, and therefore cannot be used, as Eddington attempts to use it, as a criterion for the passage of time. The sense of time, Braithwaite considers, need not be found in physical theories, but may 'quite well arise out of our consciousness of the physical world.'

Eddington asserted that the replacement of primary laws by statistical laws in the description of physical nature had destroyed strict causality and provided a basis for the freedom of the will. However, statistical mechanics indicates that crowds of atoms must behave with a very high degree of determinism. As Braithwaite puts it, the individual soldiers described in *War and Peace* are only 'able to move as they liked on the condition that the army as a whole followed its destined course.'

Eddington's theory of 'world-building', of the supremacy of the mind as the synthesizer of the scientific theory of the physical world, attributes simple and complete priority to consciousness, 'the mind is the first and most direct thing in our experience, and all else is remote inference.'

G. Dawes Hicks commented that 'Eddington is very confident that we have an intimate acquaintance with our own minds, that they are much less mysterious to us than either our bodies or the external world.' Hicks says that well-nigh every psychologist would deny that, and the 'plain man' must be very surprised to find what kind of a mind is revealed by 'direct self-knowledge'.

Eddington said that 'we picture the mind like an editor in his sanctum receiving through the nerves scrappy messages from all over the outside world, and making a story of them with, I fear, a good deal of editorial invention.'

Can this, asks Hicks, by any stretch of imagination, be said to be the information which direct and infallible self-knowledge affords? 'Could any account of the mind and its operations be more diametrically opposed to what self-scrutiny reveals?'

Hicks then considers the process of world-building which leads to a representation according to Eddington, of 'a physical world which will give a shadow-performance of the drama enacted in the world of experience.'

Eddington represents the 'world' by an equation containing 265 coefficients, which covers all physical relations, and 'gives a numerical measure of the structure'. He eliminates most of these as 'useless lumber', that is, as Hicks says, 'of elements that will not form part of a building which will shadow the things of common experience.' Finally, sixteen coefficients are left, ten for the construction of geometry and mechanics and six for electromagnetism.

Eddington starts from the world of common experience in his 'world-building', and also uses it as the criterion as to whether his attempts have been successful. As Hicks says, his proceeding seems 'to presuppose the very thing it is professing to account for'. And what is the outcome? A 'representation of external nature from which all reference to the qualitative has been rightly excluded.'

This may be necessary for the mathematical representation of the world, but the philosopher, says Hicks, 'must insist that the more the quantitative aspect of the world is emphasized, and rightly emphasized, for certain purposes, the more the importance of the qualitative aspect of the world must be emphasized. Otherwise, we fall into the dilemma on which Lotze once commented, that the more abstractly our symbols are conceived, the more they pass into general notions which apply fairly well to every thing, but give us no adequate knowledge about anything.'

Eddington tells us that we bring to physics a conception of space which must have originated many million years ago and has become embedded in human thought. 'But the space of physics ought not to be dominated by this creation of the dawning mind of an enterprising ape.'

Hicks remarks that, according to Eddington, the mathematician is not the only 'world-builder'. No small portion of it has, apparently, been built by our ape-like ancestors. In fact, 'we, or our ape-like ancestors, have put into nature well-nigh everything that we discern there. Colours, sounds, temperatures—the warmth of the air, the scent of the grass, the stir of the breeze, the solid substance of things—these are "fancies" which the mind has projected into the external world.' Hicks concludes that without reference to some kind of model based on the 'familiar world' the whole account of that world becomes purely subjective, and in fact, however much the physicist may protest that he knows nothing about the structure of the electron, he is all the while assuming that it has certain physical qualities.

To what general conclusions are we led by our glance at Eddington's philosophy? One difficulty is that his conclusions were just what those of a man like him would be. The brilliant arguments seem superfluous.

In the *Nature of the Physical World*, he wrote as an amateur, with the amateur's courage and absence of responsibility. His wide knowledge and his facile command of metaphor enabled him to make many startling remarks and suggestions. To the professional philosopher, his fundamental conceptions appear naïve and over-simple. Nearly all of his lively phrases, which give a first impression of being illuminating, prove on closer examination to be inexact and often false.

As a great scientist entering another field, he used his scientific knowledge to plough up the ground, and provided the professional philosophers with the opportunity of refreshing themselves by harrowing his furrows. His value to them was as a stimulus to exercise and activity, rather than as the provider of seed. It is not easy to see any important new philosophical idea in Eddington's philosophical works.

The *Nature of the Physical World* was much more interesting, and less aggravating, to the general reader than the philosopher because of its entertaining style. It enabled him to glimpse some of the problems with which both the scientific and philosophical professionals were struggling.

His later book, the *Philosophy of Physical Science*, is less light-hearted and gay, and more sober, than the *Nature of the Physical World*. It is directed more to the examination of the nature of physical knowledge rather than the nature of the physical world, as if he felt the need to answer the strictures made by the professional philosophers on his earlier work, and especially on his cavalier treatment of the origin of scientific concepts.

Eddington was in his personal life a very modest man, yet his speculativeness was rash to the degree of arrogance. E. A. Milne complained that his obstinate refusal to recognize mistakes in his theory of stellar opacity was ostrich-like.

The 'quiet effrontery' of his literary style, which amused scholars in the same Cambridge intellectual circles, often angered those without, who were earnestly trying to understand the crises in modern life reflected in science.

He was as metaphorical as Bergson, but wrote as a philosophical scientist, whereas Bergson wrote as an artistic philosopher.

His attitude as an amateur in philosophy may be compared with that of his colleague at Trinity College, the great mathematician G. H. Hardy, when he made an excursion from pure mathematics into mathematical logic, in a discourse on the nature of Mathematical Proof. Hardy asked to be excused 'if, in a desperate attempt to be interesting, I lose myself in discussions where I am admittedly an amateur.'

As a professional pure mathematician he was, in his own subject, 'quite as intolerant of amateurs as a self-respecting professional should be'. He had no difficulty in understanding that mathematical logic 'is also a subject for professionals, that it demands a detailed knowledge which I do not possess and, so long as I am active in my proper sphere, have hardly leisure to acquire; and that I am certain to be guilty of all sorts of confusions which would be impossible to a properly qualified logician. Indeed there is only one thought which gives me courage to proceed, and that is that I may be concerned less with strictly logical questions than with questions of general philosophy. However treacherous a ground mathematical logic, strictly interpreted, may be for an amateur, philosophy proper is a subject, on the one hand so hopelessly obscure, on the other so astonishingly elementary, that there knowledge hardly counts. If only a question be sufficiently fundamental, the arguments for any answer must be correspondingly crude and simple, and all men may meet to discuss it on more or less equal terms.'

He was inclined to apologize with less sincerity to those mathematicians who dislike all discussions savouring of philosophy. He felt that this distaste was based on 'an unreasoning shrinking from anything unfamiliar, the distaste of pragmatists for truth, of the engineers for mathematics, of the pavilion critic at Lords for the in-swinger and the two-eyed stance'. It is reasonable to ask that 'this dislike of the fundamental for its own sake' should be put aside.

Eddington's excursion into philosophy disturbed not only the professionals in his own social class, in whose colloquial idiom, like Hardy, he wrote. From the standpoint of those outside the middle-class professionals providing the executive of English early twentieth-century society, his philosophy seemed narrow and shallow, besides being littered with brightly expressed but inexact conceptions. It appeared to be a product of ideological disintegration, iridescent with decay.

Caudwell found the source of the disease in the mechanistic philosophy of capitalism. Such a society, said Engels, 'based on commodity production has the peculiarity that in it the producers have lost control of their social relationships.' Man becomes the slave of machines owned by a few, and finally of mechanistic philosophy. From that, like Eddington, he conceives the world in terms of a mechanics of pointer-readings. Man naturally tries to escape from such a world, and in Eddington's case, he bolts directly from science into naïve idealist philosophy and mysticism. As Ritchie said, he found 'the transition from science to religion short and easy'. He believed in contemplation for the sake of understanding, in contrast with those who studied philosophy in order to change the world as well as understand it; and such, so Ritchie says, are the exponents of modern Black Magic. Unfortunately, however, it is not easy to see why, according to the contemplative view, 'the realm of physics is worth contemplating. Indeed, Newton thought it somewhat trivial.'

The source of the unsatisfactoriness of Eddington's philosophy is to be found in the break-up of the mechanistic philosophy created during the last three centuries. Caudwell quotes Max Planck's view that 'we are living in . . . a moment of crisis . . . the iconoclast has invaded the temple of science. There is scarcely a scientific axiom that is not nowadays denied by somebody.' Planck is aware of a 'general feeling of collapse of the old order, together with a complete helplessness and lack of understanding as to its cause'. Specialization and efficiency increase in each domain, but lead 'to an increasing anarchy and contradiction between the domains'. A synthetic world-view, 'a living theory in touch at all points with practice', becomes less and less possible.

In such a situation, one must go forward or backward. Eddington, says Caudwell, has followed the conservative path, 'a mystical positivistic attitude to all spheres of ideology outside one's little garden'.

It is the result of the break-up of the old bourgeois world-view. Nature has become a game of symbols, where 'the physicist can go round and round in circles without encountering anything real. Nature has become unknowable.'

And Man, 'dragging with him all the rich qualities of interesting life, has entered arid regions of *theology*.' Because a table consists of sparsely distributed electric charges, it is no longer necessary to accept unitarianism, 'we can go right back to the Virgin Birth and the miracle of the loaves and fishes.'

Eddington tried to smuggle the ghost of mechanistic philosophy from the old world of capitalism that is breaking up into the new world of socialism that is being constructed. He acted thus because, socially, he was entirely within the old class system, through his professional post and his residence in Cambridge society. The completeness of his social circumscription is reflected in the character of his style, sprightly but donnish, and the narrow range of allusion, in his very numerous illustrations and metaphors, to social customs and knowledge outside his own class.

The new socialist society must create its own new philosophy. Eddington did not participate in this, and consequently had no clue to a comprehensive philosophy consistent with the new creative life.

VII. MAGICIAN

'I believe there are 15,747,724,136,275,002,577,605,653,961,181,555, 468,044,717,914,527,116,709,366,231,425,076,185,631,031,296 protons in the universe, and the same number of electrons'. So Eddington begins the chapter on 'The Physical Universe' in his *Philosophy of Physical Science*. The total number of particles in the universe, the sum of the protons and electrons, is, in briefer notation, $2 \times 136 \times 2^{256}$. Eddington called his number, which he labelled *N*, the *Cosmical Number*. He regarded his discovery of it as his greatest achievement.

He devoted the main effort of his later years to an attempt to derive the chief numerical constants of the universe from this fundamental number *N*. His calculations consisted of a mixture of abstruse mathematics, some of which could be recognized by experts as elegant, and of many obscure scientific principles. No one claimed to understand more than a little of this work, chiefly expounded in his *Relativity Theory of Protons and Electrons*, and his *Fundamental Theory*, published posthumously in 1946, and even those who understood a little, accepted less.

The most intelligible summary of this aspect of Eddington's work has been given by E. T. Whittaker and H. C. Plummer, which will be followed here.

In 1928 Dirac published his relativistic quantum dynamics, in which he showed how the equations for describing the quantum behaviour of the electron could be modified so as to satisfy the requirements of relativity. It was one of the great achievements of the first half of the twentieth century in theoretical physics.

Dirac's discovery revived Eddington's interest in relativity theory, and the completion of his work on the internal constitution of the stars, marked by the publication of his book in 1926, left him with time and energy for new mathematical investigations.

Eddington set out to construct a comprehensive theory of Nature, which, following and going beyond Dirac, would combine the quantum and relativity theories and transcend them both. These two theories had separately achieved a remarkable completeness, and power of predicting in their own fields the result of any experiment or event. However, to give a complete account of the structure of physical reality, the two theories would have to be combined.

Eddington set out to discover a theory which would enable the ratio of the masses of a proton and electron to be determined by calculation. He sought for a common basis for the quantum and relativity theories, and held that this could be found in the state of equilibrium of a radiationless self-contained system of a very large number of particles. Einstein was the first to solve this problem for large bodies. The presence of matter produces a closed curved space of radius R, which is called an *Einstein* Universe.

In quantum mechanics a radiationless steady system is described as being in its *ground state.*

Eddington therefore sought to determine the conditions which must be satisfied by a system of N particles, on the one hand, as an Einstein universe at zero pressure and temperature; and on the other, as a quantized system in its ground state. The relativity solution will be expressed in terms of a gravitational constant and a cosmical constant, and other microscopic constants.

A comparison of the two solutions will show how these various fundamental constants of nature are connected.

When the system is in the ground state, its particles will be in the N states of lowest energy. The total energy of the Einstein universe is expressed in terms of N, R—the radius of the universe—and the quantum constant. The result is equated to the energy or mass of the Einstein universe, as deduced from relativity theory.

Hence it is possible to obtain two equations from which N and R can be calculated separately. When R has been calculated, the cosmical constant and the limiting speed of the recession of the spiral nebulae can be calculated from it. The whole theory can then be checked by comparing the calculated speed of recession of the nebulae with that determined by astronomical observation. A completely satisfactory agreement is obtained.

N, the number of particles in the universe, is deduced by the calculation of the number of independent wave-functions of quadruple form with the necessary relativistic property, in spherical space. The result is $N=\frac{3}{2}.136.2^{256}$. (The first factor in an earlier result was amended to $\frac{3}{2}$.) From this cosmical number N, Eddington deduced the ratio of the electrical and gravitational forces between a proton and an electron and the range and magnitude of certain forces in the atomic nucleus, and obtained results which agreed with the observed values.

N is the number of particles in an Einstein world composed of atoms of hydrogen, each of which can be separated into a proton and an electron.

Eddington obtained all these results from theory, and does not make use of any numbers derived from the experimental measurement.

The proper mass or proper energy of an observed particle depends on the highest energy-level of the Einstein world, for which he had a formula. The proper energy of n particles is less than n times the energy of one particle. The difference is equal to the gravitational potential energy, and gravitation is a consequence of the exclusion principle, which forbids an energy-level to have more than one occupant.

Exclusion energy is represented by interchange energy arising from the indistinguishability of particles.

Thus the various forces of nature, hitherto regarded as distinct and independent, are shown to be manifestations of the same force.

Eddington deduced the existence of a force which would bind a proton and electron together to form a neutron. He explained the forces between heavy particles in the atomic nucleus without invoking Yukawa's particle. He did not believe in this particle, and reserved the term 'meson' for the particle found in cosmic rays, for which he also gave an explanation. He deduced from his theory the masses of the neutron and the cosmic ray meson, the forces between the heavy particles in the nucleus, the mass-defects of deuterium and helium, the magnetic moments of the hydrogen atom and the neutron, and other constants.

Plummer comments that it was not without justification that Eddington described this work as *Fundamental Theory*. It is not deduced logically from existing theory, and Eddington did not claim to do this. He began the development in December 1928, by his assertion that the fine-structure constant, which occurs in the quantum theory of spectra, must be the reciprocal of a whole number. Two years later he concluded that this was 137. He deduced his cosmical number N, soon afterwards, and presented his results in 1936 in his *Relativity Theory of Protons and Electrons*. In his Dublin lectures of 1943 on the *Combination of Relativity Theory and Quantum Theory* he expounded a new principle which enabled him to calculate all the fundamental constants except the cosmical number.

What is to be made of Eddington's procedure? He said that 'An intelligence, unacquainted with our universe, but acquainted with the system of thought by which the human mind interprets to itself the content of its sensory experience, should be able to attain all the knowledge of physics that we have attained by experiment. He would not deduce the particular events and objects of our experience, but he would deduce the generalizations we have based on them. For example, he would infer the existence and properties of radium, but not the dimensions of the Earth.'

Eddington presents his calculations as the proof that it is 'clear that in claiming to determine *a priori* the number of elementary particles in the universe we are not usurping a prerogative which has usually been ascribed to the Creator of the Universe.'

Milne has pronounced some of the lucid patches of Eddington's final 'unified theory' 'beautiful' and other patches 'palpable fudge', but takes refuge by declaring that it has 'All the marks of the production of a man of genius, not excluding tiresomeness; whether or no it will survive as a great scientific work, it is certainly a notable work of art.'

Dingle describes the impression of scientific insight and profundity which Eddington made in personal contact. 'He might be profoundly wrong, but he would inevitably be profound.' He exchanged letters with Eddington on his *Fundamental Theory* a few weeks before his death. He pointed out that Eddington appeared to give words and scientific terms a special meaning of his own, different from those of accepted usage.

Eddington replied: 'You are a bold man to set any limits to what a "particle" may mean since the dear old classical particle became obsolete.' This attitude seems to give a clue to his manufacture of particles and principles in order to make his equations come out, and give the desired results.

Finally, Eddington wrote to Dingle that 'I am interested in your detailed criticism because I am continually trying to find out why people find the procedure obscure. But I would point out that even Einstein was once considered obscure, and hundreds of people have thought it necessary to explain him. I cannot seriously believe that I ever attain the obscurity that Dirac does. But in the case of Einstein and Dirac people have thought it worth while to penetrate the obscurity. I believe they will understand me all right when they realize they have got to do so—and when it becomes the fashion "to explain Eddington".'

Dingle writes that he would not care to dissent from that view.

Eddington was perhaps conscious of the shadow of death when he expressed these remarkable opinions. But considered with detachment, does it not suggest that Eddington had an intense intellectual passion to equal and surpass Einstein and Dirac, whereas his gift was not in that direction? Einstein and Dirac have each made major contributions to theoretical physics, not once but several times, and in each case with general experimental confirmation and acceptance. Eddington's chief accepted contribution, his generalization of Weyl's combined theory of the gravitational and electromagnetic fields is considerable but not major.

Eddington was essentially an enormously gifted student, with a strong, cultivated imagination. He could learn anything. He was the exponent and applicant first of classical physics, then of relativity, and then of quantum theory. He had not the independence of a theoretical creative genius such as Einstein or Dirac, and his letter to Dingle suggests that he was conscious of it. His creativeness lay in application, not in the creation of fundamental theory. In this, his typical British quality of practical judgment in application was seen to best advantage. His greatest successes were in the application not even of the newest theory, but of classical thermodynamics and early quantum theory.

Eddington's 'unified theory', apart from its obscurity, explains too much; indeed it explains everything, and hitherto such theories have generally been found ultimately to explain nothing. At best, it is a fragmentary work, which contains flashes of insight that will be appreciated by future generations, like Leonardo da Vinci's scientific researches, probably after what is significant in them has been more completely discovered by very different paths and methods.

At worst, they will prove to be nothing more than an advanced form of number-magic, the working-out of arithmetical coincidences and then the claiming for them of special significance.

Eddington invokes numerous fictitious particles to assist him—'the standard carrier, extracules, intracules, hydrocules'. He tacitly uses scientific terms in new senses, and produces new principles to make arguments fit. Is it genius or megalomania? It looks not unlike a return, in an abstruse form, to the interest in numbers as such which he manifested as a child. Many scientific authorities have been cautious not to criticize this work, and to protect themselves by saving clauses on the possibility that it may contain discoveries of the profoundest character.

After all, it may prove to be the boldest, the most imaginative, the most revolutionary work of the twentieth century.

But the most remarkable numerical concidences may have no scientific meaning. Chandrasekhar has revealed startling numerical coincidences among the magnitudes of the universe and the constants of nature, purely as a demonstration of the meaninglessness of such coincidences. Is Eddington's 'unified theory' nothing more than an enormous, misguided essay in this direction?

On the whole, it seems probable that it will prove to be an extreme individualist aberration, a product of the weak side of Quaker solitary quietism and obstinacy, which remained uncorrected through the absence of an adequate system of mutual scientific criticism, because of a defect in social organization. Owing to lack of correction it grew into an imposing but spurious excrescence, and proved, in the end, to be a descent, if not into black magic, into number magic.

BIBLIOGRAPHY

Stellar Movements and the Structure of the Universe, by A. S. Eddington (1914).

Report on the Relativity Theory of Gravitation, by A. S. Eddington (1918).

Space Time and Gravitation, by A. S. Eddington (1920).

Mathematical Theory of Relativity, by A. S. Eddington (1923).

Internal Constitution of the Stars, by A. S. Eddington (1926).
Stars and Atoms, by A. S. Eddington (1927).
The Nature of the Physical World, by A. S. Eddington (1928).
Science and the Unseen World, by A. S. Eddington (1929).
The Expanding Universe, by A. S. Eddington (1933).
New Pathways in Science, by A. S. Eddington (1935).
Relativity Theory of Protons and Electrons, by A. S. Eddington (1936).
The Philosophy of Physical Science, by A. S. Eddington (1939).
Fundamental Theory, by A. S. Eddington (1946).
Reflections on the Philosophy of Sir Arthur Eddington, by A. D. Ritchie (1948).
'Professor Eddington's Philosophy of Nature,' G. Dawes Hicks, *Proceedings of the Aristotelian Society* (1928–29).
'Professor Eddington's Gifford Lectures,' R. B. Braithwaite, *Mind*, N.S. 38 (1929).
Science versus Idealism, by M. Cornforth (1946).
The Crisis in Physics, by C. Caudwell (1939).
Obituary Notices: *Roy. Soc. Obit.*, **5**, 113 (1945–48); *Proc. Phys. Soc.*, **57**, 244 (1945).

FREDERICK GOWLAND HOPKINS

1861–1947

I. PIONEER

HOPKINS had one quality at least which was not possessed by the four scientists we have so far considered. Unlike them, he had no British predecessor in his own branch of science. Rutherford was the pupil of J. J. Thomson; Eddington and Jeans were trained in the Cambridge school of applied mathematics. All of them were intellectual heirs, through Stokes, Rayleigh, Clerk Maxwell, and J. C. Adams, of Isaac Newton.

Hopkins was not the heir of a great tradition. He had to create British biochemistry, and secure its position and his own, in the face of many difficulties. The gifts by which he accomplished this achievement were of the most surprising and fascinating kind. One might have expected that he must have blasted his new path by elemental scientific energy, like Rutherford's. This was not so; he never bullied nature or human obstructors. His chief weapon was a diplomacy of the scientific intellect. He captured his positions not by frontal attack, but by flanking and undermining movements. He had the supreme, and very rare, merit of being simultaneously diplomatic and honest.

His scientific advances were protected by remarkable aesthetic qualities. He expressed bold ideas with exquisite modesty and grace. He wrote with superb literary style and taste. Many of those who were not ready for his new vision were disarmed by his noble humility and intellectual elegance. No one ever advocated radically progressive scientific conceptions with such tact.

Some people imagined that because he was almost infinitely patient, understanding and kind, he must be weak. No one of perspicacity could have believed that, and his long life and work give the most striking contradiction of it.

His first note was published in 1878 at the age of seventeen, and his last paper in 1948, seventy years later, after his death in 1947 in his eighty-sixth year. During his long life he published more than 130 papers, in which he pursued his biochemical philosophy and various experimental investigations with sustained tenacity. His later papers were not less distinguished than his earlier papers, published more than fifty years before.

As would be expected from his intellectual qualities, he was sensitive, and was sometimes subject to despondency. But he enjoyed good basic health. He told Dale that he could not remember having a common cold until he was over fifty years of age. He kept well without physical exercise.

Hopkins was small and slight in figure. His head was long, and big for his frame, and his face was even and handsome. His hair and drooping moustache emphasized the distinction of the vertical lines of his face. He usually stood with his head slightly inclined forward, pondering. This made his brow yet more striking.

His courtesy was almost Chinese in perfection. He received students, colleagues, and visitors with extraordinary kindness, and listened politely and with genuine interest to all. Many thought that he allowed his time to be wasted by everyone except themselves. He was always affectionately known among his colleagues as 'Hoppy'. He was accused, in his kindness of heart, of regarding his geese as swans. But through his solicitous inspiration his geese were liable to be transformed into swans.

How could a man of such gentleness and delicacy of feeling triumph over prejudice, blank incomprehension and thick-skinned opposition? By imaginative intellectual subtlety practised with indefatigable pertinacity.

Hopkins was primarily a biochemical philosopher with an inspired vision of the chemical aspect of living processes. He saw in his imagination the molecules at work in the living cell. He knew what they were doing as if they were his personal friends. The biochemist, he said, 'strives to obtain a mental picture of that invisible dance and transformation of molecules which in every organ or tissue underlies its visible activities.'

An American pupil of Hopkins once opined that he was not a particularly good experimenter, because he knew what was going to happen before he started. His experiments were merely practical work to confirm conclusions at which he had already arrived in his intellectual imagination. The criticism was not without point, though in fact Hopkins was an experimenter of exceptional gift.

He persistently explained and taught and lectured, for half a century, on his vision of life as a polyphasic system, whose myriad molecular interchanges were in dynamic equilibrium. What Rutherford's vision of the nuclear atom, with its ever-whirling parts, was to physics, Hopkins' dynamic vision of the living cell has, in a degree, been to biochemistry.

Hopkins was a great imaginative theorist, atomistic and specific in his thought. He was convinced of the importance of *simplicity* in living processes. This emphasis of the simple gave a tremendous intellectual relief to those confronted with the complexity of living processes.

He was a kind of Galileo of biochemistry, who made the advance from the static to the dynamic conception of the processes of the living cell. He began the analysis of the dynamics of the cell on the principle that its complexities arose from the activities of units which in themselves were simple and relatively easily knowable.

His vision of the gossamer interweaving of the atoms would have been nothing more than a fancy if he had not confirmed features of it by experiment and been led by it to new experiments.

He brought to this side of his work a particular skill in chemical analysis. This was due to inborn abilities which had been specially trained and sharpened by peculiar circumstances. As a boy he had already become a naturalist and a student of insects. His note of 1878, published in the *Entomologist*, when he was seventeen years of age, was on the beetle *Brachinus crepitans*. He discovered early in the 1880s that the wings of the common British brimstone butterfly contained a yellow pigment which could easily be dissolved. He found that it was related to uric acid, a result that he published in 1889.

The discovery that the colour of certain butterflies, and hence their mimicry, might be related to excretory substances such as uric acid, established a relation between excretory mechanisms and evolutionary significance which was of considerable biological interest and attracted Ray Lankester's attention. Hopkins prepared a memoir on this investigation, which was published, through Lankester, in the *Philosophical Transactions* of the Royal Society in 1895. His concern with uric acid had led him, in 1893, to the publication of an improved method of estimating the amount of uric acid in urine, which remained the standard method for many years and established his reputation.

After this achievement, he was invited, in 1898, to become a lecturer in chemical physiology at Cambridge University. In the same year he introduced the first easy and reliable method of crystallizing proteins, an advance of great technical importance. His next experimental achievement was the discovery in 1901, with the collaboration of S. W. Cole, of tryptophane, one of the most important of the score of amino-acids from which proteins are built.

Feeding experiments with proteins lacking tryptophane led him to the conception of accessory food factors, later called vitamins, whose presence in minute quantity is essential for normal health. He published his first announcement of the existence of vitamins in 1906 and produced the first conclusive and convincing experimental proof in 1912.

Meanwhile, in 1907, with the collaboration of W. Morley Fletcher, he founded the modern quantitative theory of the chemistry of muscular contraction.

After the interruption by the First World War in 1914–18, he discovered, in 1921, at the age of sixty, the sulphur-containing substance glutathione. This is one of the most important of the agents, or catalysts, effecting the oxidations taking place in living tissues. Its mode of operation throws light on the part played by catalysts in the molecular dynamics of the cell.

In his long struggle to establish the status of his science, the position of Reader in Biochemistry was created in Cambridge for him in 1902. This was not the first recognition of 'biochemistry' in Britain, but it was the first associated with the modern dynamical conception of the science.

Hopkins was given the title of professor in 1914, but he was still without a modern, or even a decent, laboratory. He did not receive this until 1924, when he was sixty-three years old, after the trustees of the well-known hat-manufacturer, Sir William Dunn, had provided £210,000 for the purpose.

He soon set extended and varied biochemical researches in movement in his new Institute of Biochemistry. He did not organize them on traditional lines, like a general giving orders to his officers on various sectors in a campaign or an industrial manager organizing a research team, but rather after the pattern of his vision of the biochemical dynamics of the living cell. The whole institute became a kind of cell, in which many of the multifarious molecular activities underlying life were being investigated—simultaneously, and all held together by the fine thread of Hopkins' imaginative intellect.

He pictured enzymes, or ferments, as having a major, or the major, role in the organization of living processes, and he created in his institute an atmosphere of intellectual ferment. He regarded each colleague in his institute as a kind of human enzyme, and he regarded each enzyme in the living cell as a kind of human colleague. He was himself the Grand Enzyme, who catalysed all.

Thus his method of directing research was based on his conception of the mode of the organization of processes in the living cell by enzymes. He directed research by, as it were, the method of life itself. There was a mutual relation between his insight into human beings and into enzymes.

H. Wieland, one of the masters of the classic German tradition in organic chemistry, described Hopkins' Institute, in 1938, as a *universitas* of biochemistry, where scientists of every land and every discipline were welcome and appreciated, and could learn how to engage in biochemical investigation and pursue any aspect of the subject.

Some considered that Hopkins did not organize research at all—the great German chemist regarded Hopkins' organization of research as 'exemplary', and noted that more than 600 papers had already, by 1938, been published from Hopkins' laboratory.

He did not retire from the direction of the Institute until 1943, at the age of eighty-two. He was determined to struggle to the last for the prestige of his new department and subject, and desired to accomplish all that was humanly possible to guarantee that they should continue their development in undiminished brilliance.

Hopkins shared a Nobel Prize in 1929 and was elected President of the Royal Society in 1930. These honours came to him when he was nearly seventy years old, but they did not cause the least slackening in his mission. He utilized his increased authority to advance with undiminished determination his vision of life as 'the expression of a particular equilibrium which obtains in a polyphasic system'.

In his speech at Stockholm on receiving the Nobel Prize, he said that he was not sure whether honours and rewards are the more desirable in the days of full vigour or after most of one's work is done. He was not sure that the most obvious answer was the right one. 'Youth, it is true, needs sustenance, but should need no tonic. The pulse of age may be quickened, by the recognition of past efforts and unsuspected capacity for further effort thus revealed.'

After this speech in 1929, Hopkins published more than fifty papers, and continued his scientific activity until his death, at the age of nearly eighty-six, with a sustained intellectual vigour and enthusiasm which is almost unparalleled.

As President of the Royal Society, of the British Association, and of the British Association of Scientific Workers, Hopkins supported progressive proposals to utilize science for the benefit of mankind. His experience as a chemical analyst and medical student had acquainted him with human problems and suffering, and his discoveries in the science of nutrition had given him the fundamental knowledge by which he knew how much could be done to improve human health and happiness through the proper application of science.

Hopkins contributed more than any other Englishman of his generation to the proof of the possibilities of science for human betterment. At the age of seventy-five he was still giving perfect and profound expressions of his belief in the power of scientific method to acquire insight into nature and to improve the state of man.

In 1936 he delivered a wonderful oration at Birkbeck College, in which he spoke of the Pace of Science. Among many remarkable things, it contained a magnificent appeal for the new profession of science writing, as a bridge between the scientists and the people, so that the advance of science should be continually explained, with the aim of securing understanding and support for its prosecution and its harmonious absorption and utilization by society.

He concluded his address with the words: 'The pace is great, and the pace is growing greater.' And as time passes, it is seen that Hopkins' own great achievement appears ever greater.

Hopkins was the master of a superb literary style, and wrote much, but never wrote a book. This is not altogether surprising. With his strong sense of literary discrimination, he was primarily an essayist, a Montaigne of biochemistry. Experimentalists' books are usually compilations. Hopkins' taste was too severe to allow him to undertake such a work.

Every paragraph that he wrote was deeply pondered and elegantly expressed. It is scarcely possible to find a logical sequence of experimental data sufficiently long to form a book, which admitted throughout of such an exacting standard of literary expression. Scientific books are nearly always uneven, because the author must include for the sake of balance and logical development a number of topics with which he is not directly acquainted and in which he is less interested.

Hopkins was necessarily restricted by the perfection of his style to the choice of particular aspects, for which he could find satisfying expression. So he treated the same subjects over and over again from varying aspects, making, as it were, a series of perfect miniatures, each subtly different.

Hopkins came in touch with government as a chemical analyst who had assisted in the detection of crime, and then as an authority on the science of nutrition.

He was an original member of the Medical Research Committee established in 1913, to administer funds for medical research, which had been set aside from the contributions of national insurance. In that position he made the proposals that W. Morley Fletcher should be appointed the first Secretary of the new Medical Research Council and that H. H. Dale should be appointed to the staff of the National Institute of Medical Research. He exerted influence in committees through his scientific quality and transparent sincerity, rather than by forcefulness or adroit manœuvring.

He was knighted in 1925 and awarded the Order of Merit in 1935.

He became ill in 1938, and had an operation the effects of which troubled him for his remaining years. But his mental vigour continued, even when his eyesight declined almost to blindness. He was taken daily to his laboratory, where he directed experiments that he could not see, until a year or two before his death on May 16th, 1947, a month before attaining the age of eighty-six.

He had lived simply all his life, and left the modest fortune of £14,021.

His achievement in carrying out the tasks of the great pioneer of British biochemistry, not with roughness but with grace, is unique.

The quality of Hopkins which is most likely to be overlooked is his strength. His fragile body and infinitely deferring personality carried an intellect of undeviating tenacity, subtle yet clear and beautiful.

His enthusiasm and quality of activity continued with extraordinary power to his end in advanced old age. Even as death approached, he did not pause in his search into life.

II. LONE SCIENTIFIC YOUTH

'The first ten years were spent alone with my mother at Eastbourne': so, with characteristic literary skill, Hopkins started his brief *Autobiography*, and communicated the tone of his childhood. His father died in his infancy; so Hopkins, like Eddington, was a widow's only son reared in a seaside resort. The Hopkins' home was named Cavendish House. Eastbourne pier was built just in front of it, and Hopkins describes, as one of the most vivid memories of his childhood, the screwing of the first pillar of the new pier into its base. Six horizontal bars were attached to the top of the pillar, and a coastguard pushed at each bar, so that the vertical pillar was rotated. On the top of the pillar sat another coastguard, wearing a tarred straw hat, and giving the time to the men with a fiddle in the fashion of an old salt driving the team on a capstan.

Hopkins wrote that the building of the pier made a profound impression on him, and he decided to himself that he would be an engineer when he grew up.

He had, however, no mechanical skill with his hands. He could not remember ever making anything in his childhood, and he felt the lack of constructive skill in his subsequent life in the laboratory, admiring it very much in others.

He read his first novel before he was ten years old. This was *David Copperfield*. He revelled in its sentimentality, and was not touched by its humour. He recorded that his taste for sentiment in literature and drama remained with him through his life.

He wrote much doggerel verse, which, when he examined it again in his old age, seemed to him to be precocious, but without sign of genius.

When he was about eight or nine years old, his mother revealed to him a telescope and microscope which had belonged to his father, and 'allowed limited contact with them'. The telescope was incomplete, and in that state not of much use, but the microscope functioned properly. With this, Hopkins examined all kinds of curious things, animate and inanimate, that could be found in profusion, especially on the seashore. He says that he felt that the powers of the microscope revealed to him 'something very *important*', the most important thing that he had yet met in life, and far more significant than anything that he had yet heard of at school.

Unfortunately, however, there was no one with whom he could discuss the revelations of the microscope and his sense of their importance. 'I had the kindest of mothers, but she was a complete stranger to the world I was dimly visualizing'. He had no friends of his own age, apart from a cousin whom he saw only occasionally.

He was sent at the age of six to a dame's school, where, again like Eddington, he was a day-boy among boarders, and in his case the only day-boy. The boarders were not allowed out of the school except for walks in formation.

Hopkins gradually learned that one of the boys, about thirteen or fourteen years of age was 'a sadist of unbelievable ingenuity in method'. Another of the older boys truckled to him, and the pair tortured a number of the others until, in several cases, their spirit and health were utterly broken. Hopkins said that hitherto he had met with nothing but kindness in his relations with human beings, and his gradual revelation of the existence and nature of human cruelty filled him with horror.

The two maiden ladies were entirely ignorant of what was going on under their roof, until they were informed about the time Hopkins left the school. Though he escaped from the torturers himself, because he was a day-boy, he became suspicious of other human beings, and his self-centredness, which had already been engendered by loneliness, was increased. This discovery of cruelty had a profound impression on Hopkins, lasting far beyond his childhood.

The solitary child by the seashore, the kind and uncomprehending mother, the maiden ladies whose incarcerated pupils took revenge on them, for loss of freedom, by torturing the smaller boys in a manner peculiarly horrifying to unmarried women, make a sombre picture.

Hopkins left the school in 1871 because his mother had decided to settle with her mother and an unmarried brother, James Gowland, in a house at Enfield near London. The district at that time was still more in the countryside than the suburb, and the Hopkins family had quite a nice house there.

Hopkins' childhood was very lonely, but not oppressed by poverty. His ancestors and relatives included some notable and one great man, Gerard Manley Hopkins, the poet. Hopkins' literary ability and sensitive subtlety were not unique in his family.

His great-grandfather, Captain Hopkins, was said to have been one of Nelson's captains at Trafalgar. Captain Hopkins had a large family. One of his sons, Manley Hopkins, became the British Consul-General in the Hawaiian Islands. He wrote a history of Hawaii, a poem not without merit under the title 'Spicilegium Poeticum', and several valuable treatises on marine insurance. Manley Hopkins was the uncle of the father of Frederick Gowland Hopkins, and his eldest son was Gerard Manley Hopkins.

Hopkins could remember that he had once, as a boy, accompanied his mother on a visit to Manley Hopkins and that he had there seen Gerard.

Another of the captain's children, Louisa, married Johann Leopold

Abel, a German by birth, who settled in England, and one of whose sons was Fritz Augustus Abel, the noted authority on the chemistry of explosives, who became a Fellow of the Royal Society and was raised to the baronetcy as Sir Frederick Abel.

Some of Manley Hopkins' brothers had gone into business in London, and one of these, the grandfather of Hopkins, became a bookseller and occasional publisher in Bishopsgate. Hopkins' father Frederick was born there. He was educated in the district, and duly entered the business, where he continued until he married. Shortly after his marriage, he took his wife to settle in Eastbourne, and unfortunately died soon after their first and only child was born.

Frederick Hopkins was a charming and able man, with many friends, and an enthusiastic amateur of science. His particular friend was his second cousin Frederick Augustus Abel, with whom he was associated in research on tides and currents. For several years before his marriage he had been president of a City *Scientific and Literary Society*. He left notebooks and abstracts which exhibited a lively interest in science. He would have probably been a most sympathetic father to his gifted son, and have been able to help him to unfold and educate his genius. His premature death threw that genius into isolation, out of which it struggled through long years and at length emerged in a unique form.

On his mother's side he was descended from a Sunderland shipping family named Stafford, one of whose daughters eloped with a jeweller named Gowland. The couple settled in Leadenhall Street, where Hopkins' mother was born and lived until her marriage.

She returned to the roof of Gowland relatives very soon, owing to her husband's death, so Hopkins was brought up in the atmosphere of the Gowlands rather than the Hopkins. Enfield remained his home for twenty years.

After discussion, Hopkins was sent to the City of London School, because of the high reputation of the headmaster, Abbott. His mother did not like to send him to a boarding school, and he, with his morbid memories of Eastbourne, hated the prospect. So, at ten and a half years of age, he began the daily train journeys from Enfield to the City, where the school was housed in old buildings in dismal surroundings in Honey Lane. He arrived in the morning just before school began, and, on the instructions of his mother, returned home on the earliest possible train after school. Apart from a rowdy half-hour's break in the morning, where all of the pupils mingled and milled around a tuck-shop in an underground corridor, there was no occasion for him to meet other boys, and this was not a good one for making friendships.

So, following lonely Eastbourne, Hopkins did not make a single friend at the City of London School. The environment of a city grammar school in the nineteenth century was of a black and inorganic character. The soot from the city dyed the walls of the classrooms and the boards of the floors. Surrounding warehouses shut out the light. Artificial illumination was feeble. The pupils' school-life was sharply cut off from the colours and sights and sounds of the green fields, and the social division of the middle-class boys from the ragged working-class urchins of the streets was equally sharp. The demands of middle-class segregation, as well as the premature loss of a father and hence of possible brothers and sisters, contributed to Hopkins' loneliness. Middle-class London, by descent and education, marked Hopkins. He never lost an agreeable form of the Londoner's accent.

Hopkins took the train to school, sat through the lessons, and duly returned home early, day after day and week after week, for three and a half years. The masters in the lower forms of the school were old, dull, and perfunctory. But he received good teaching in chemistry from Henry Durham. He entered a school examination, which was open to all the boys in the school, in his third year, when he was about thirteen years old. Three were awaded first-classes: he was one, and the other two were W. H. Perkin and A. G. Perkin, the sons of the great discoverer of synthetic dyestuffs, who both became eminent professors of chemistry.

Hopkins remembered that he was startled at the result, for he had not received any special preparation and did not even possess a chemistry textbook. And well might he have been startled! The two Perkin boys lived in a big house with a fine private chemical research laboratory in the grounds. They had unsurpassed opportunities for acquiring an interest in chemistry, and satisfying it. To equal two such able boys with every opportunity, when one had so little opportunity of one's own, was indeed a remarkable performance. Yet Hopkins did not realize at the time that it was a sign of special aptitude for chemistry.

In the following term he won the form prize for English and an essay prize for a dramatization of Scott's *Old Mortality* in blank verse. But he never received these prizes, owing to a strange 'untoward incident'. The prizes were awarded at the speech-day after the summer term. In the middle of this term Hopkins suddenly became a truant. For several weeks he left home each day in the morning and returned home at the appropriate hour in the evening. During the day he went for walks in the country and the city, visited the docks and read in libraries. He became thoroughly weary of these ways of passing the time, but could not bring himself to confess to his mother and guardian-uncle. The end of term came, and he accompanied his mother to Wales weighed down by feelings of guilt. Inevitably the truth became known, and was revealed at the beginning of autumn term. His mother had a painful interview with Dr. Abbott, to the latter part of which Hopkins was called. He was not expelled, but his mother was advised to take him from the school.

Hopkins has tried to remember the causes of this event, which took him from the school before he had received his prizes, at the age of fourteen.

He thought that it was due to sheer boredom. He recorded that during the three and a half years that he had spent at the school, none of the masters had ever spoken a word to him individually, except in class. The only occasion on which Abbott spoke to him was during that miserable interview preceding his removal from the school.

Abbott, who had a great reputation as a teacher of the senior boys, was not interested in the junior teaching. This was conducted by a group of masters who were old and near their retirement, and, as Hopkins discovered long afterwards, were badly underpaid. They did not bother even to keep careful attendance sheets, for Hopkins was away for weeks without his absence being noticed. If he had stayed a little longer, he would have received his prizes from the headmaster, and he might have attracted attention. He was about to be moved up into a class that was taught by Abbott himself, who would no doubt have quickly perceived his ability during personal instruction. Under Abbott's influence he might have read classics, and like many other of his pupils, have won a scholarship at Oxford or Cambridge.

But his tie with the school was broken, and without the acquisition of a single friend in all those years. Now, at fourteen, he was as lonely as ever, a shy, frail outcast, who had manifested to hurt and shocked parent and guardian signs of apparent moral turpitude.

They chose for him a new school of a different kind, which was supposed to build character rather than instil learning. The headmaster, who was also the proprietor, was a man without educational qualification, recognized or unrecognized. He and his inadequate staff were ignorant. There was virtually no teaching of classics, that of mathematics was elementary, and the science consisted of reading from Huxley's *Elementary Physiology*. There was a chemical laboratory, but it was never opened in Hopkins' time.

The pupils at this school were sent to take examinations at the College of Preceptors. In spite of the wretched teaching, Hopkins managed to win one of the science prizes in this examination. It surprised himself, and also his headmaster who nevertheless included a notice of his success in his future advertisements of the school.

Though Hopkins received poor teaching, he was not unhappy at this school, and made some good friends. He left before the age of seventeen, and had to begin an occupation.

A friend of the family found him a place as a clerk in an insurance office in the City. After six months the same family friend advised that he should be taken away and found another job which would give more opportunity to his brains. Hopkins said that he did not know whether this was a sign of success or failure, but he held this family friend in grateful memory, for somehow or other he managed to convey to his mother and uncle that he had scientific abilities.

Apparently it was for this six months' work as clerk that Hopkins received a cheque for £40. He jokingly remarked that he thought of framing it, and his employer, who took the suggestion seriously, warmly supported the idea.

Hopkins himself was anxious to enter a scientific calling, so his father's old friend and cousin, Fritz Abel, was consulted, and immediately advised that he should go to Cambridge.

But his uncle happened just then to meet at lunch a City acquaintance who was connected with the firm of Allen and Hanbury. This acquaintance was of the opinion that a Cambridge training would be useless for a man who would have to earn his living, and advised that Hopkins should be articled to an analytical chemist. He happened to know of one with a vacancy for an apprentice. So Hopkins found himself in a very busy commercial laboratory, engaged entirely on pressing routine work. He remained in this laboratory for three years. He picked up empirical methods from the assistants and learned to perform the analyses quickly. But he learned no theory. He was virtually an extra assistant, working from 9 a.m. to 6 p.m., and, in the apprenticed period, paying fees to his employer in addition.

After the journey home in the evening, Hopkins had little strength left for private study.

His employer was a clever but unattractive man. He made frequent visits to one of the shelves, where there was a bottle ostensibly containing a reagent, but in fact containing whisky. Hopkins subsequently felt that those important years between seventeen and twenty, which he spent in this laboratory, were wasted, apart from acquiring some manipulative skill. He probably did not realize how much he had learned, especially of a technique with which academic chemists had little acquaintance.

Then, at the age of twenty, he inherited a small legacy from his parental grandfather. Abel again advised that he should go to Cambridge. But Hopkins decided to use his opportunity to attend the chemistry course of Frankland at the Royal School of Mines. During the term that he spent there, he never met Frankland in the practical laboratory. On one occasion the professor was making his rounds in the practical class, and got as far as the next man to Hopkins, but then an attendant came and called him away. He delivered two or three lectures, but for some reason the series was interrupted. These three lectures were the only formal instruction in chemistry which Hopkins received in his life.

At the end of the term, Dr. Percy Faraday Frankland, who was only three years his senior, and the professor's son, invited him to work in a private laboratory that he had established in a house in Pembridge Square for the analysis of samples of steel rails and creosote for one of the railways. Hopkins was left to work alone for long hours, and found the dark winter days grim. After six months the contract ended; Frankland married and Hopkins was out of a job.

He decided that he must obtain a professional qualification, so he entered for the examination for the Fellowship of the Institute of Chemistry. He took some technical classes at University College, which he found of little use. But in the examination he did exceptionally well. His performance came to the notice of Dr. Thomas Stevenson (later knighted for his services), the Medical Jurist at Guy's Hospital and Home Office expert witness, who was connected with the Institute of Chemistry. Stevenson gave evidence in many notorious trials, especially in cases of poisoning. His laboratory at Guy's Hospital consisted of one room in the clinical department. This was an unusual arrangement in those days. It gave the laboratory scientists an access to the interesting cases which was then almost unique, for the separation of medical and non-medical functions was generally very complete.

Stevenson invited Hopkins to join his staff as the second of two assistants. This marked the turning-point in Hopkins' fortune. His lonely youth ended, and he found himself, though a humble assistant, in a new, appreciative, and more distinguished world.

III. BUSY ANALYST

Hopkins entered Stevenson's laboratory just after the conclusion of the Lawson case of aconite poisoning. Thereafter, he conducted the analyses and prepared the scientific evidence in numerous cases for Stevenson's brief.

He particularly remembered two cases. In one of these, the Bartlett case, a wife had poisoned her husband by pouring chloroform down his throat. Hopkins analysed the contents of the victim's stomach and found there was plenty of chloroform in it. Stevenson's evidence was strong, and completely withstood cross-examination. But Sir Edward C. Clarke, who was the woman's counsel, made a marvellous speech in her defence, the profound impression of which Hopkins never forgot. He thought it 'must have been the most wonderful example of forensic eloquence ever heard in court'. The jury was mesmerized and declared the woman not guilty, a popular verdict because a man ought to have been in the dock with her.

The other case which Hopkins particularly remembered was the murdering of a woman by a man called Lipski, who poured nitric acid down her throat. There is no doubt that the woman had been murdered with nitric acid, but there was only circumstantial evidence that Lipski had administered it. The verdict of guilty was unpopular, and there was a press campaign and petition to the Home Secretary for his reprieve. After the trial, however, Hopkins noticed that nitric acid stains on Lipski's clothes, alleged to have been acquired in the course of his trade as a cleaner of silver mounts on umbrellas and walking sticks, had an unusual colour. Investigation showed that this was due to the presence of sulphuric acid. The acid stains on the floor near the woman's bed revealed the same characteristic. The police tried to find a specimen of the acid used by Lipski in his workshop, but did not succeed. However, they secured specimens of nitric acid from all the places of supply in the neighbourhood. Some ten samples were brought to Hopkins, who analysed them, and found that one contained sulphuric acid. This came from the shop where Lipski always obtained his supply. The Home Secretary resisted the pressure for the reprieve and Lipski ultimately confessed his guilt.

Hopkins' service under the able and distinguished Stevenson gave a remarkable extension to his experience of life, all the more remarkable after his solitary youth. He learned that life or death, and the actions of Ministers of State, depended on the accuracy of his chemical analyses. For a man of conscience and ability, such knowledge strongly developed the sense of responsibility and the need for the utmost technical skill. He became expert in the identification and analysis of poisons, especially of those of plant origin, such as strychnine and the alkaloids.

Hopkins made good friends in these days. The senior assistant, Bodmer, was an excellent musician, and taught Hopkins much about the musical life of London. Also at Guy's he became friendly with John Wade, a chemical lecture assistant, like Hopkins without academic qualifications. Wade's family was cultivated, his father being an excellent amateur painter and his sister a musician.

Under Wade's influence, the pair decided to study for degrees at London University, as external students. They had first to pass the stiff matriculation examination. Hopkins had left school six years ago, and found the task of re-learning Latin hard. His only opportunities for study were in the train between Enfield and London, for his family expected him to be social when he returned home after work. He developed a technique of concentration and avoidance of interruption by other passengers in railway carriages.

Hopkins recorded his opinion that this cramming did him some permanent moral and mental harm. He had a receptive but untenacious memory, which made him a good examinee. The learning of knowledge without reference to one's taste and innate abilities increased this tendency towards rapid acquisition and equally rapid forgetting, and Hopkins felt throughout his life the temptation to get things up for the moment, and not assimilate them permanently so that he was furnished with what he called 'sound mental storage'. This, and the lack of early systematic academic training, left him with the sense of his ever being 'an amateur intellectually', an opinion of himself that he frequently expressed to the end of his long life.

Of course, an essential element in his greatness was that, unlike his peers, he was not burdened with certain conventional knowledge and experience which could turn 'sound mental storage' from an acquisition into a hindrance, and he was free from the repetitions and narrowness of vision of the intellectual professionals.

As his training had been not in ideas, but in the routine of analysis, his professional skill was a technique subordinate to, and at the service of, his unregimented imagination. He was able to think more freely and creatively than others, and yet had a thorough technical training at his command.

At the age of nearly twenty-eight, having acquired a little more capital, he decided to work for a medical qualification at Guy's Hospital. Simultaneously with his entry as a student, he was the first recipient of the newly founded Gull Research Studentship. This was a unique circumstance, and he was also about five years older than the usual student.

Hopkins believed that he owed the studentship to Stevenson's recommendation, for at that time he was almost unknown to the rest of the staff. It gave him a special status and privileges from the start of his medical course, but he found the task of justifying his holding the studentship by research, at the same time as he was following his routine medical studies, a very hard one. He gained the gold medal for chemistry and honours in *materia medica*, which he actually crammed in two days before the examination. He found this facility 'no desirable gift', for he forgot it all as quickly. Later, when he had to teach anatomy and *materia medica*, he had once more to cram them all over again.

Hopkins' transference from professional chemical analysis to the study of medicine changed his social perspective and status. The chemical analyst was a scientific craftsman, whereas a medical doctor was a species of gentleman. Hopkins began to make friends and acquaintances in this class.

He assisted the talented Washbourne, then junior physician at Guy's, to introduce Koch's bacteriological technique. In 1890 he and Washbourne went to Switzerland to stay with the Hale Whites, and enjoyed the winter sports. On another occasion they visited the Pasteur Institute, saw Pasteur, and made the acquaintance of Roux and Bordet. They were struck by the 'extraordinarily bad nursing arrangements'. Hopkins recalls that at this time he was very keen on Henry James.

He tried, as Gull Student, to make a number of researches on clinical cases, but these generally gave no publishable result, being too complicated for quick scientific advance.

His first published research as Gull Student arose out of the fashionable medical interest in uric acid, to which gout and many other ailments were attributed. The physician Newton Pitt believed it might be the cause of migraine, so Hopkins estimated the amount of uric acid appearing in the urine on various kinds of diet. The existing methods of estimating uric acid were either inaccurate or took much time, so Hopkins was led to improve the technique. Within a week he worked out an improved method that was not bettered for many years, and he published an account of it in the Guy's Hospital Report of 1891.

He received his medical qualification in 1894, having been appointed to the hospital staff just before.

Hopkins first noticed among the outpatients at Guy's that patients who apparently had enough to eat could still be under-nourished. He formed the opinion that the importance of nutritional errors in the causation of disease was underrated, and that the current treatment for scurvy and rickets ignored the significance of the old observations. He had the ambition of studying these diseases from the nutritional standpoint, but was fated to lose contact with the clinical material.

The typewriter was bringing a stream of young women into London, who were replacing the old male writing clerks. These girls went to the new teashops for a cup of tea and a bun for lunch, whereas the men went to the pubs and had a cut from the joint, two vegetables, and a glass of beer.

These girls used to call at the Hospital with a vague malaise, but no identifiable disease. An old surgeon was convinced that they could be cured with lemon juice. Hopkins made a mental note of the existence of unexplained nutritional effects. But he did not start an investigation then.

He became one of the demonstrators in practical physiology, under direction of the lecturer, E. H. Starling; he gave classes in toxicology; and was busy with many other tasks. He started a private research laboratory, together with his friend Wade, in an inexpensive room in Denman Street, and even engaged a private assistant. He collaborated with A. E. Garrod in 1895–96.

He obtained very valuable clinical experience from the system then peculiar to Guy's, by which six students for a period of six months acted as clinical assistants, and had first choice, and hence the most interesting and instructive of the cases coming into the hospital. The ward in which these cases were treated contained Stevenson's laboratory. This system enabled good men to learn much in a short time.

The hospital *Gazette*, which was being edited by Hopkins, ran a column on pathology, to which practitioners were invited to send interesting specimens for report. It was a popular feature, and one of the hospital clerks perceived that it might be commercially exploited in a journal specially run for the purpose. Hopkins suggested that it would be better to send reports by post for a fee.

An organization called the 'Clinical Research Association' was set up for the purpose, and Hopkins vacated his Denman Street laboratory to provide it with a room. His private assistant joined the staff of the new undertaking, and he also found its first secretary. The enterprise, which started virtually without capital, was a great and immediate success. Hopkins himself looked after the chemical side, doing all his work after five in the evening. He had started with a decent salary, and was under the impression that he would have the opportunity of becoming one of the proprietors when further capital was required. But as the firm succeeded, the original proprietors did not need to call in Hopkins. He was not given the opportunity of investing in the firm, and so could not share in the large profits that were made. He felt that he had not been well treated.

Starling improved the physiological laboratory and added an excellent room for chemical work. Hopkins pursued such research as he could find time for in this room, together with J. B. Leathes. He met Bayliss, who came often to collaborate with Starling.

In the midst of this intensely busy life, Hopkins met his future wife, Jessie Stevens. Their courtship lasted for five years, from 1893. Hopkins recorded in a noble reference to his wife his indebtedness to her, especially in those qualities in which he felt himself lacking: the courage to face life's difficulties and practical skill in coping with them. They had three children, a son and two daughters. After their marriage in 1898, they took a top flat in Lincoln's Inn Fields, finding on the floor below them a young socialist propagandist and his wife, the Ramsay MacDonalds.

Hopkins spent a few months in this flat before being invited to Cambridge. He told Dale that the only occasion on which he met MacDonald at this time was when the latter rushed upstairs to ask him to come and look after his wife while he went to get the doctor to attend to her first confinement. Thirty-five years passed before Hopkins met MacDonald again, at a Royal Society dinner. Hopkins recorded that Mrs. MacDonald was a remarkable woman. She was the daughter of the physicist Gladstone, and brought to her husband, besides herself, a useful private income.

Hopkins had the prospect of appointment to the clinical staff of Guy's and a prosperous medical career. His sympathy and skill would probably have given him great success as a physician. Stevenson had indicated that he hoped that he might be his successor.

But in spite of his very strenuous life he had already established his reputation as a chemical physiologist by the work on the estimation of uric acid. He had followed this by the publication in 1895 of the application of his technique to the investigation of the uric acid pigments in the wings of butterflies. His work showed that there may be a close association 'between pigments which function in ornament and substances of excretory nature'.

This paper completely broke away from the medical background, and exhibited the use of chemical technique for the investigation of a phenomenon of fundamental interest in evolutionary biology. It was a pure essay in biochemistry, and by its kind and significance added a new dimension to Hopkins' intellectual personality.

In 1898 the Physiological Society met in Christ's College at Cambridge. After dinner, as Hopkins was walking through the great gate of the College, Sir Michael Foster caught him up, took his arm, and invited him to come to Cambridge to teach and undertake research in chemical physiology. It is possible that, among other influences, Starling had suggested Hopkins to Foster.

If he accepted the post, he would be expected to start work in Cambridge in two months' time. The basic salary was to be only £200 a year, but he would be able to increase his income considerably by supervising the medical students at Emmanuel College.

Hopkins had only just married and settled down in London; he was earning much more than that, with very good prospects. He suffered agonies of indecision, but with the firm and unwavering support of his wife in favour of Cambridge, he decided to accept Foster's invitation. So he went to Cambridge almost at once, and entered an atmosphere to which he had hitherto 'been so complete a stranger'.

IV. CAMBRIDGE

Hopkins came to Cambridge with a unique background of an entirely different character. This was of profound assistance to his originality, but put an immense new burden on him. He found that he had to teach the medical students anatomy as well as chemical physiology. He had the severe task of cramming anatomy once more. He sat up until two in the morning, swotting the details, with his wife beside him, to see that he did not inadvertently allow the fire to go out. Like other scientists who have come to Cambridge from a different intellectual sphere, he found that a tradition had to be learned, besides the teaching of science. Cambridge men had been brought up in the tradition, and naturally repeated it when they became teachers. Hopkins had never been an internal university student and had had virtually no higher academic training in his life. It was hard to learn at short notice many of the things that Cambridge men wanted, and additionally hard to learn to present them in the form to which they were accustomed.

The strenuous teaching of the Emmanuel students continued for twelve years, leaving not very much time and energy for research. His laboratory accommodation was unbelievably bad. Until 1914 he had only one small ill-lit laboratory. His research began, and continued for twenty-five years, in a building described by the Needhams as 'extremely old and uncomfortable . . . in Corn Exchange Street, many of the rooms of which were semi-cellars.'

In the period 1923–31, the Institute of Biochemistry produced a lighthearted journal called *Brighter Biochemistry*. Hopkins wrote an amusing dialogue for one of its issues, in which he referred to the laboratory conditions in 1898. W. B. Hardy under the name of Collophilus borrows a test-tube from Hopkins, under the name of Aminophilus, because his own is under repair. The laboratory assistant, 'Charles', has just caught a mouse in the cellar. Hardy has purloined an egg from the pantry before leaving home, and offers Hopkins the yolk, because he needs only the white himself. The laboratory has no polyvalent ions, so domestic common salt is used, and ammonium sulphate is obtained from the Professor of Agriculture. Metallic ions of high valency are prepared by dissolving pen-nibs in hydrochloric acid. Hopkins suggests that Hardy might use alum instead of cerium or thallium salts for coagulation.

Hopkins then compares these old conditions with the multiple resources of the new laboratory of 1925, and finally with a prophetic laboratory of 2000 A.D. By then, buildings have become superfluous, and 'it is with Brains and the Biocalculus that progress is now made'. However, there are still experiments worth making. The Junior Super-investigator replies that he 'found, yesterday, for instance, that gastric hydrochloric acid contains exclusively the heavier isotope of chlorine.'

Quoting again the personal description of the Needhams: 'Much of the equipment was antique, too; for example, a centrifuge which, mounted on boards and having no arrangements for attachment to a base, used to ride all over the floor of a cellar when in operation.'

Hopkins kept the rats, with which he made his classic vitamin investigations, in these rooms, and he carried out their entire feeding and upkeep by himself. 'Upstairs was the shelf on which had stood in the sunlight the bottle of acetic acid which had led to the discovery of tryptophane. Below were the ancient benches on which 'Charles' (William J. Jolley, the laboratory assistant) had cut up the quarters of horses towards the end of World War I for the extraction of tryptophane.'

Hopkins' university appointment was raised to the status of a University Readership in 1902, and from that time he had an adequate income. In spite of his heavy teaching duties, scanty time for research, and bad conditions, he discovered the first easy method of crystallizing proteins in quantity in 1898. He discovered tryptophane, with the collaboration of Cole, in 1901. He carried out the feeding experiments on rats which led him to enunciate the existence of vitamins in 1906.

During all this time, eight years, he had not yet been elected a fellow of a college, a considerable social and academic hindrance in Cambridge. However, in 1905, he was elected a Fellow of the Royal Society, so he could not very well be passed over any longer. He was elected Fellow and Science Tutor of Emmanuel College in 1906. His supervisory duties then became heavier than ever, yet he managed to publish in 1907, together with Morley Fletcher, their classical paper laying the foundation of the chemistry of muscular contraction, and in 1910 his famous paper in which he conclusively and convincingly proved the existence of vitamins.

However, his health and strength could bear no more. He happened to knock his head against an iron spiral staircase in the laboratory. This was followed by 'uncanny cardiac symptoms' and 'months of extreme mental and bodily discomfort'.

He wrote that only those who have experienced 'neurasthenia at its worst can properly appreciate the misery involved'.

He made a complete recovery within six months, largely owing to the enormous relief and encouragement given to him by Trinity College, which elected him to a fellowship, and Praelectorship in Biochemistry, which carried a handsome salary and virtually no teaching duties. Not until the age of forty-nine, and after a serious breakdown in health, was Hopkins given an appointment which relieved him of financial anxiety while providing him with the leisure for research.

His friend Morley Fletcher initiated the proposal that he should be elected a Fellow of Trinity, and Hopkins was profoundly grateful for the support he received from that college and the inspiration and encouragement that he gained from its fellows.

In 1914 the University established a separate department of biochemistry and appointed Hopkins the first professor. His income, however, remained much the same as before.

The old Physiological Laboratory, in which Hopkins had worked since 1898, was vacated by the physiologists, who received a new building in 1914. The whole of the old building was then given to Hopkins for his new department. Improvement was delayed by the First World War, but afterwards progress began. The University allowed Hopkins to occupy the Balfour Laboratory, which had been used by students from Girton and Newnham before they were admitted to university classes, and he persuaded the Beit Research Fellowship Trust to pay the rent. Then, in 1923–24, on the advice of Morley Fletcher, the trustees of Sir William Dunn provided £210,000 for the support of biochemistry at Cambridge. The modern Institute was built, replete with the many facilities to which he humorously referred in 'Brighter Biochemistry'.

As a lecturer, Hopkins was at his best with advanced students, but his greatest influence was in the inspiration of research. He often suggested problems, but he preferred to encourage research workers to follow their own bent. He liked to have many different lines of investigation proceeding simultaneously. Bacterial chemistry, plant biochemistry, chemical embryology, enzyme chemistry were welcomed, or introduced by himself, at various times.

Hopkins held weekly or fortnightly tea club meetings in his laboratory, for nearly fifty years. Here his genius was seen at its finest. One of the laboratory workers would read a paper on some aspect of his research. Sometimes the paper would be very dull and the assembled workers would wonder how anything interesting could possibly be said about it, but Hopkins never failed. He would make suggestive comments about some aspect, and if the paper were completely without interest, he would take some commonplace phrase and develop an argument of his own, with historical allusions and philosophical imagination.

If there were anything in the work, it would be subjected to the politest but severest criticism. Hopkins created an atmosphere in which almost anything could be said without fear of offence, because of his high and disinterested character, his perfect manners and taste, and his splendid intelligence.

Hopkins was the most accessible scientist of his distinction. He was willing at almost all times to discuss problems with his research staff, and, as E. Mellanby has said, one invariably went away feeling happier than when one came. Many felt that he wasted much energy in these interminable discussions, but he was genuinely modest and interested in other people's ideas, and, as Marjory Stephenson said, he probably found that these discussions helped him to work out his own ideas.

Hopkins characteristically never kept a list of his papers or a set of reprints of them. He forgot about the remarkable note that he had published as a boy in 1878.

He worked entirely with his own mind and hands. Dale once remarked to him that he himself felt that he would not be able to work in isolation in a private laboratory. Hopkins replied, 'Oh, I am sure I could.' Dorothy Needham and Marjory Stephenson have described his 'intuitive habits of thought and his uncanny and incommunicable skill in handling material'. When he was at work he concentrated intensely. His gift of separating substances of great physiological importance from complex messes, which he never succeeded in communicating to others, seemed, 'when one was privileged to see it in operation, to be due to an intensely acute observation and a retentive memory for important detail'.

He used to go to the Laboratory on Sunday mornings to make experiments entirely by himself, which he described as too foolish to be seen by his assistants.

He posed as being ignorant of organic chemistry, and derived pleasure from producing chemical 'rabbits out of the hat'. These were protections against his deep, permanent feeling of limitation which his early lack of academic training had imposed on him.

Dale has described the actual occasion which led to the discovery of tryptophane. He was himself a research student in physiology at the time. Hopkins had been taking a practical class, in which the students had been using the test for proteins introduced by Adamkiewicz in 1874.

The presence of proteins is detected through the production of a violet colour on the addition of strong sulphuric acid to a solution of the suspected material in acetic acid.

One of the students, John Mellanby, who subsequently became professor of physiology at Oxford, found that the acetic acid from his bottle would not give the reaction. Hopkins was brought to the bench, and verified the fact. The acetic acid in all the other bottles in the laboratory gave the usual reaction.

The failure of the reaction in a particular case appeared to have important significance.

Dale met Hopkins in the corridor after the class, and Hopkins described what had happened, and added, 'with that upward jerk of his shoulders, jig on his feet, and whimsical grimace, with which he so often signalled a reaction of excitement to something new or puzzling'. 'I have a feeling, Dale, that if we can find the meaning of a colour-reaction like that, it may open up a way to new knowledge of the old protein molecule itself.'

In the later Cambridge years, after the building of the Institute, forty or more research workers were to be found in the laboratories. The programme in such large research institutions is usually organized on some managerial plan. The professor may organize the whole of the staff to work by his own methods on different aspects of an important problem. He may reserve a considerable department for his own personal work and act as administrative manager for the other departments.

Hopkins' highly individual, personal, intimate method of research could not be applied in those ways. He held his institute together not by managerial organization but by personal intellectual inspiration. Under his control, he gave great freedom to others to try their wings, and quite a number failed to fly. But he did not ruthlessly throw them out of the nest because they had not obviously found their place in some wheeling flock or formation.

Hopkins had created British biochemistry and had given it a unique philosophy. He became anxious about the future of his creation, for he was not sure who could carry it on. He had to have an operation in 1938, when he was seventy-seven. He unfortunately contracted an infection, and suffered from a localized paralysis, arthritis in the hip, and an increasing blindness.

But still he would not retire, for he could not plainly see his successor. Finally, however, he could no longer continue, and in 1943 at the age of eighty-two, he retired from his chair, though he still insisted on coming to pursue his personal researches in his own room. These last researches were of extraordinary interest, for they were on the pigments of butterflies, which he had begun nearly sixty years before. Latterly, the noted German chemist Wieland had greatly extended these researches with modern methods and resources. Hopkins in his old age returned to the subject, and, in spite of his years, and modest resources, succeeded in adding to Wieland's remarkable results.

Hopkins' departure inevitably produced a change in his institute. He had held it together by an idea rather than a method. Without him, the cell tended to disintegrate, and its materials had to be reassembled and reanimated in another way.

V. EXPERIMENTAL ELUCIDATION OF THE CHEMICAL PROCESSES OF LIFE

Hopkins' first published scientific note, his letter on the bombardier beetle in the *Entomologist* in 1878, was descriptive. He reported that the beetle had been very plentiful near his Enfield home, and he had also found a number on the South Downs near Eastbourne, where he had not seen them before. The plentifulness of the beetle seemed to be an exception to the general scarcity in the order in this year. 'It is a very sociable insect, and I have seldom seen one without finding others close by. The beetles are very partial to my sugar compound, and have swarmed on trees prepared for moths.'

Hopkins' style was already characteristically well balanced. Fifty-seven years later, in his address as Honorary President of the London Natural History Society, Hopkins recalled that in his youth he had been an ardent collector of butterflies, moths, and beetles. He first made the acquaintance, 'with infinite pleasure', of the little bombardier beetle on a day in March 1878. He was not previously aware of their means of attack or defence, so when he discovered for himself that they emitted a jet of violet vapour when disturbed, he became intensely curious as to how the volatile substance could be made and stored by the insect, and what it might be. So he placed one of the beetles in a test-tube, and encouraged it to discharge its jet of vapour against the side of the tube. Then he replaced it with others, until he secured a splash of the condensed vapour. He made experiments with it, but the quantity was far too small for his chemical skill, and they failed. From that time, however, his 'fate was sealed. Though the designation was not yet invented, I became there and then a biochemist at heart.'

His note of 1878 contained no account of his unsuccessful experiments, probably because of the shyness of youth. Hopkins made experiments with butterflies in his mother's scullery at Enfield, and may in these have made his original observation that a pigment found in the wings of many different kinds of butterflies was soluble in hot water. He became convinced in the early 1880s that the wing pigments of *Pierid* butterflies were derivatives of uric acid, while that of the common white butterfly was uric acid itself.

His next publication appeared eleven years later in the *Proceedings of the Chemical Society*, after he had learned his analyst's technique. He applied it to the chemical identification of the pigment. He found that it resembled 'mycomelic acid', a yellow derivative of uric acid which had been prepared by Wöhler in 1853 by heating uric acid and water in sealed tubes.

In 1891, he suggested in a short note to *Nature* that it might be named 'lepidotic acid'. In this note, he referred to E. B. Poulton's work on animal coloration as being 'somewhat limited in standpoint'.

After he had introduced at about this time his improvement in the method of estimating uric acid, he again examined the chemistry of the butterfly pigments, publishing his results in his notable memoir of 1896 in the *Philosophical Transactions*. He concluded that the opaque whiteness of the wings of the *Pieridae* is due to uric acid. The yellowness in the wings of related yellow butterflies was probably mycomelic acid. The excreta of these insects also contained the yellow pigment, so that both in white and in yellow insects 'a normal excretory product subserves the purposes of ornament'.

These pigments found in butterflies were not investigated again until 1925, when H. Wieland began a notable series of researches on them. With the aid of modern chemical technique, he showed they were considerably different from what Hopkins had supposed, and they were renamed *pterins*. They have since been found to have remarkable biological significance.

Wieland found that the white pigment was not, at least in major part, uric acid, as Hopkins had believed, but a related substance with very similar properties. He renamed it *leucopterin*, and the yellow pigment from the brimstone butterfly *xanthopterin*. A substance similar to xanthopterin has been found in human urine. It appears to be connected with tissue oxidation. A yellow pterin has been found in cells of the walls of the intestine, and appears to be connected with pernicious anaemia.

It has been claimed that artificial anaemia in rats is arrested by injections of pterin obtained from butterflies.

After fourteen years of work the Wieland school believed they had unravelled the constitution of the pterins, and then in 1940 they discovered that their results were illusory. They succeeded in synthesizing leucopterin and xanthopterin, and proved their actual constitution to be much simpler than had been supposed.

Hopkins had meanwhile returned to the field after forty-five years, with the hope of helping to clear up the constitutional question; but Wieland's latest results, of which he had heard during the Second World War, suggested that this was now unnecessary. Nevertheless in 1942, when he was eighty-one years old, he gave a fascinating account of what he had recently been able to contribute himself.

Hopkins had shown in 1895 that the yellow pigment gave a purple derivative when heated with sulphuric acid. He had named it 'lepidoporphyrin', but he now proposed to call it 'rhodopterin'. The German chemists had paid little attention to it, and he thought it worth noting that it could be obtained in considerable amounts. He made his extracts from 3000 of the full orange yellow *Colius crocea*; he was not able to obtain more than about 100 specimens of other species.

It had seemed in 1939 that the purple rhodopterin was worthy of further study. Dealers in butterflies informed him that sufficient quantities should be obtainable from abroad. But soon afterwards the Second World War started and completely cut off the source of supplies. He made enquiries in Britain, and many helped him, but the number of specimens which could be spared from private collections and museums could only be small. He was unable to follow up several interesting lines.

The Germans, in contrast, had been able to work with remarkable quantities of material. In 1933 Wieland and his collaborators used 250,000 white cabbage butterflies for their source of leucopterin. The Bavarian Minister of Education instructed schoolmasters to supervise the catching of cabbage butterflies during the summers of 1934 and 1935. Xanthopterin was obtained from 16,000 *Catopsilia rurina*. Tartter extracted the wings of 850,000 cabbage butterflies.

Hopkins happily noted that in 1940 Tartter announced that he had found uric acid, besides leucopterin, in the white pigment. Hopkins remarked in 1935, after Wieland had shown that the pigments were not quite what he had originally suggested, that he would have regarded such a correction, when younger, as a tragedy. But now he could bear it with equanimity. The pigments were indeed related to uric acid, and Wieland had himself long believed that the white substance was uric acid, until he used a discriminatory test which was unknown in Hopkins' earlier days.

Hopkins' knowledge of uric acid analysis was acquired in his long experience as a routine analyst, for an important part of such work was the analysis of patients' urine for medical consultants. Through uric acid his technical analytical skill, the needs of human physiology and the medical profession, and the general biological role of uric acid in nature, as exemplified by butterflies, came together.

He published a note on his improved method of uric acid estimation in 1891, and a short and confident paper on it in the *Proceedings of the Royal Society* in 1893. The method depended on the absolute insolubility of ammonium urate in ammonium chloride. 'The crucial fact that *saturation* with ammonium chloride renders the separation rapid and absolutely complete does not seem to have been previously observed.' The method enabled the uric acid to be estimated to within 1 per cent, some ten to twenty times more accurately than by other quick methods then in use; and it was very much quicker than other known exact methods.

Hopkins' method remained the standard for many years. It was of much practical importance because of the universal need for urine analyses, and interest in it was heightened by the medical profession's fashionable interest, at the end of the nineteenth century, in uric acid as an alleged cause of many ailments, especially gout. After the 1890s, however, the incidence of gout in England rapidly decreased, probably owing to changes in dietary and living habits, and increase in general abstemiousness.

Hopkins while at Guy's collaborated with A. E. Garrod on the application of the spectroscope to the study of haematoporphyrin, which is the cause of the dark red colour of the urine of persons suffering from sulphonal poisoning. They also examined the colouring matter urobilin found in urine, and showed that contrary to the suggestions of some investigators, the kinds which appeared in health and in illness were the same, and the urobilin found in faeces was also the same as that in urine.

Hopkins became the leading authority on the chemistry of urine, and was invited by Schäfer to contribute a section on the subject to his noted *Textbook of Physiology*, published in 1898. This is the only systematic text that Hopkins ever wrote, and it is not difficult to see why. It contains a very rare combination of technical mastery with philosophical vision. The extension of such a treatment to other fields of physiological chemistry would have been an enormous intellectual task, which could scarcely have been accomplished in addition to all his other activities. He would have had to leaven the whole crushing bulk of conventional medical chemistry with his new vision.

The section contains sixty-eight full and illustrated pages, dealing with the general physical and chemical properties of urine. Then he describes the chief constituents: urea, ammonia, uric acid, xanthin bases, creatinin, hippuric acid, proteins, amino-acids, aromatic substances, carbohydrates, glycuronic acid, oxalic acid, fatty acids, pigmentary substances in normal and pathological urine and salts.

He opens his treatment with the significant sentence: 'The chemical study of the urine gains its chief importance from the light which it throws upon the processes of metabolism.' Thus he regarded the study as a method of elucidating the processes going on in living tissue and how it is built up. The contents of the urine were end-products of the metabolic processes occuring both in health and disease, removed from the body in the highly complex excretion from the kidneys.

The study of individual end-products had come first because it was the simplest approach and could utilize the existing methods of analysis. This had led, however, to the overshadowing of the study of the urine as a complex whole. This could not be foretold from a knowledge of the individual constituents, but only from a study of their mutual interactions, and it was necessary if the exact processes and interactions occurring in the kidney itself are to be understood. This depended on the application, not of the 'earliest and best understood of the weapons of chemistry', analysis, but on the new theory of 'the condition of equilibrium in salt solutions'.

If the urine were regarded merely as a 'final chapter in the history of metabolism', the actual states in which the individual acids and bases appear in the urine would be of little interest for the understanding of the normal or abnormal functioning of the body.

Hopkins concluded his detailed technical account with an equally philosophical conclusion on the comparative study of uric acid secretion from the various orders of animals, and its evolutionary significance. He writes that a series of analyses is needed in order to form a judgment of the line by which the simple urinary excretions of the invertebrates have evolved into 'that most complex of physical fluids—mammalian urine'. Birds, reptiles and arthropods excrete uric acid and spiders guanin. The mammals, amphibia, fishes and certain molluscs excrete urea. The waste products from the tissues of birds and mammals are the same. The uric acid found in mammalian urine may be a vestigial relic of the earlier type of excretion, and it may be paralleled by the minute amounts of urea found in the urine of birds and reptiles.

Most of the contributors to Schäfer's textbook were already eminent men, so the appearance of Hopkins, then a demonstrator at Guy's, in their company sealed his reputation, and this was confirmed by his remarkable contribution.

Following his own philosophy, Hopkins proceeded from the study of the technique of urinary analysis to its use as a tool for elucidating the course of chemical processes in the body. The medical interest in gout had stimulated the study of the effect of diet on the uric acid in the urine. Hopkins set out to deduce from the effect of diet on uric acid the chemical processes in the body by which the uric acid had been produced. What are the precursors of uric acid?

With the collaboration of W. B. Hope, he confirmed that after a meal containing protein, the uric acid in the urine increased immediately, but for a shorter period than the increase in urea. They pointed out that this was inconsistent with the belief that the uric acid was derived from the nucleins into which proteins are broken down by preliminary digestion, because the process of digestion takes time. When protein was broken down by pepsin into nucleins, which were then fed to the animal, no increase in uric acid was observed.

Hopkins and Hope concluded that the increase in uric acid immediately after a meal did not come from nucleins, 'but from some more soluble constituent of the diet acting either as a direct precursor or as a factor in a synthetic process.'

So, in 1898, Hopkins had begun his search into the specific roles of parts of proteins in nutrition, and had already made progress. In order to learn more of the role of proteins, it was necessary to know more of their properties and constitution, so he extended research into their preparation and chemistry.

Hopkins and Brook had noticed by chance that albumen in urine is strongly precipitated by bromine, and they found that chlorine had a similar effect, and also iodine. The compounds so produced contained a 'strikingly large' percentage of halogen. Hopkins found that the chlorination of proteins had been carried out by Mulder as early as 1840, though by other methods.

Then, in 1898, Hopkins and S. N. Pinkus described an improved method of crystallizing proteins. When egg-white was treated with ammonium sulphate according to Hofmeister's method, the ammonia produced apparently rendered crystallization difficult by making the mixture alkaline. If, however, the mixture were filtered and treated with slight quantities of ammonium sulphate and acetic acid, the albumen crystallizes within four or five hours, giving a large yield of needle-shaped crystals. It was possible to secure large quantities of crystallized egg-proteins within twenty-four hours of the beginning of the process.

It had generally been considered that perfectly fresh eggs were necessary. 'We have found those ordinarily to be bought at the shops as new-laid give perfectly reliable results, and, although the degree of freshness undoubtedly affects the ease with which crystals are to be obtained, we have secured quite satisfactory preparations from the cheapest eggs in the market, rejecting, of course, such as are obviously decomposed.'

This paper closed the London period of Hopkins' life. His next papers, continuing the same subject, were communicated from Cambridge. He showed that the crystals obtained by Hofmeister's method, and the improved method that he had himself recently described, were probably mixed. He now described a further improvement of the acetic acid method which gave a large yield of pure albumen crystals with extraordinary ease and rapidity.

As compared with the older crystalline albumens, there could be no doubt of their true 'chemical individuality', as their optical properties and content of sulphur remained absolutely constant after repeated crystallizations.

Hopkins was finding proofs for his profound belief in the variety of proteins, each with its specific character and chemical individuality. He strove with all his strength against the conception that there was one substance or one molecule, such as 'protoplasm' or 'biogen', which was characteristic of living matter. He regarded such a notion as obscurantist, and an 'inhibitor of thought'.

Hopkins' first big discovery at Cambridge arose, as has already been mentioned, out of his teaching of physiological chemistry and his technical skill in elucidating an accidental observation. When it had been noticed that the Adamkiewicz reaction for the detection of proteins would not always work, he sought for the cause, with S. W. Cole's collaboration.

They showed that the mechanism of the reaction had been 'wholly misunderstood'. The change of colour was not due to acetic acid but to an impurity usually present in the acid. This impurity could be separated from the acid by distillation.

They tried to imagine how the impurity might have been produced from acetic acid, and thought of possible products of oxidation of the acid, such as glycollic acid. They secured a pure specimen of this, but it did not give the violet colour in the presence of proteins. So they oxidized the glycollic acid into glyoxylic acid. This, they found, gave the violet colour beautifully.

It had been supposed that the Adamkiewicz reaction depended on carbohydrate groups in the protein molecule. The discovery that it was due to glyoxylic acid showed that there was in fact no evidence for this view. So Hopkins and Cole searched for the part of the protein molecule which was actually responsible. They found that when protein was broken down into amino-acids by digestion or proteolysis, a constituent could be separated, and prepared in the form of beautiful crystals, 'fine white glistening plates of highly characteristic appearance', which gave the violet colour with glyoxylic acid perfectly.

Claude Bernard had noticed that a substance was produced by the putrefaction of protein whose presence may be detected by the production of a red colour in bromine water. This hypothetical substance, which could be obtained by tryptic digestion from protein, was named 'tryptophane'.

Hopkins and Cole found that their fine crystalline plates gave the 'fine red-rose colour' of the tryptophane reaction, as well as the violet colour of the Adamkiewicz reaction. So there was 'indeed not the smallest doubt that our substance is the much-sought tryptophane.'

They concluded that tryptophane is 'skatol-amido-acetic acid', a result which required correction when Ellinger and Flamand synthesized the substance in 1907 and proved that it was a related indol-amino-propionic acid.

After the completion in 1903 of his original work on tryptophane, Hopkins did not publish again for three years. In 1906, however, came two classical papers. He had been pursuing the elucidation of the role of proteins, and their constituent amino-acids, of which tryptophane was one in the building-up and nutrition of the tissues.

He was opposed to the old view that the glands of the body had a special power of producing particular substances, such as adrenaline, out of indifferent precursors. He believed that the assumption of such powers to an animal cell, when there were simpler chemical possibilities, was a resort to vitalism and contrary to what was known of the processes of metabolism. French investigators had suggested that tryptophane was a precursor of adrenaline. With one of his colleagues, Miss E. G. Willcock, he had found experimental evidence that 'an individual amino-acid may be essential for special purposes in the body, other than tissue-repair.'

Hopkins and Miss Willcock described a feeding experiment which showed that a diet in which nitrogen was provided in the form of zein, a protein obtained from maize, was incapable of maintaining growth in young mice. The addition of tryptophane, which is not one of the constituent amino-acids of zein, did not enable the diet to maintain growth, but it greatly prolonged the survival of the animals and improved their well-being. The addition of tyrosine, which is one of the constituents of zein, had no such effect. 'It is suggested that the tryptophane is directly utilized as the normal precursor of some specific "hormone" or other substance essential to the processes of the body.'

They remarked that a deficiency in the diet may not necessarily be one of quantity. The form of the nitrogen, the kind of molecule in which it is available, may determine its efficiency.

Hopkins pursued the implications of this work in his remarkable address to the Society of Public Analysts, in November, 1906. He chose this occasion to make the first clear and exact enunciation of the existence of what are now called vitamins. It was an infinitely graceful tribute to the profession through which he had entered the scientific life, and he opened his address with one of the finest examples of his courtesy: 'In venturing to address you this evening, I am deeply impressed with the sense that the excuse for my intrusion is a small one. Seldom, I imagine, has this learned Society welcomed a visitor less obviously entitled to occupy its time. So fully do I realize this that I must ask you at the outset to listen to a few words of explanation, the personal aspects of which you will kindly excuse.'

He explained that he had been told that his views on the modification of the Institute of Chemistry's examination in pharmacology and therapeutics might be of interest to a wider audience, and the flattery of this invitation had committed him to the effort of making the address. As, however, the discussion of the examination could not take more than a few minutes, he would have to expand this narrow text into something wider.

He spoke of the relations of public analysts and medical officers of health, championing the rights of his old profession against those of the doctors.

'The care of the body in sickness, with all the delicacies of human relationship which it involves, must remain always an entire and carefully guarded prerogative of the physician', but the scientific material with which the doctor has to deal should not be a medical monopoly. He illustrated his point by reference to recent studies in dietetics. In this subject the doctor is the recognized authority, but he is dependent on the chemical physiologist and analyst for scientific knowledge. He ventured to foretell that the latter would in the near future find 'an efficient source of income' if they were prepared to work in this field.

'While on the business of prophecy, I am tempted to put another series of prognostications before you . . . the influence of minimal qualitative variations in dietaries will one day become recognized as of great practical importance.'

He described how his mice had failed to maintain growth on a diet of zein, with abundant fat, carbohydrate and salts. When, however, the zein was replaced by casein, they grew at once. When tryptophane was added to the zein, the animals did not grow, but were in much better health than on zein alone. It was evident 'that a group in the protein molecule may serve some purpose in the body other than that of forming tissue or supplying energy.' Food was usually supposed only to repair the tissue and supply energy. 'But the body has other and more subtle needs equally urgent. Here, there or elsewhere in the organs must appear special, indispensable, active substances which the tissues can only make from special precursors in the diet.'

The indol grouping in the protein molecule had some such function, in addition to supplying material for the repair of tissue. These qualitative differences in protein had no small significance in diet. He believed that it might account for the fact that wheat-eating peoples require a much wider dietary than rice-eating peoples. It may explain 'variations in nutritive values which at present we feel and recognize only vaguely.' In the future the analyst would not only have to determine the quantities of proteins, but also their qualities.

No animal can live on a mixture of pure protein, fat, carbohydrate and salts. Countless other substances were to be found in the natural food obtained from plant and animal tissues, and he believed that 'physiological evolution' had made some of these essential constituents of diet.

'In diseases such as rickets, and particularly in scurvy, we have had for long years knowledge of a dietetic factor; but though we know how to benefit these conditions empirically, the real errors in the diet are to this day quite obscure.' But they were certainly of the kind of 'minimal qualitative factors' which he had been considering. Other less striking conditions, but with a most important effect on the health, were due to 'unsuspected dietetic factors'.

He could not do more than hint at these matters, but he asserted that in the future 'the science of dietetics will deal with factors highly complex and at present unknown.'

He drew the attention of the analysts to these questions because it was clear that further progress depended on their efforts. It was necessary to develop an interest for identifying unknown substances with unknown properties in complex mixtures. The pure chemist had lost this instinct, as the physiological explorer found when he turned to him for help.

He was not afraid to assert that progress in dietetics was about to react on professional chemical practice through the inspiration of 'fresh problems and new ideas'.

Hopkins strove for the next six years to identify one of these 'minimal qualitative factors'. He delayed and delayed the publication of his experiments, but he did not succeed. He was often depressed by the apparently inexplicable difficulties of the work, and on a day in 1910 he came in to the laboratory tea and murmured despondently 'It is not creatine.' In the same year he was overcome by his nervous breakdown owing to overwork.

Finally, in 1912, he published results, which seemed to prove conclusively the existence of what he now called 'accessory food factors', though they had not yet been chemically identified. He showed that while animals would not grow on diets of pure proteins, fats, carbohydrates and salts, the addition of, for example, milk, 'in astonishingly small amount' led to normal growth and well-being. He expressed his results in the form of graphs, which presented the effect in a most striking way, and had a large part in carrying conviction of the existence of 'accessory food factors'.

He had two sets of rats, one of which was fed on the ration of purified food only, and the other with an addition of milk, whose content of solids did not exceed from 1 to 4 per cent. of the total weight of the ration. The rats without milk soon stopped growing, and began to lose weight, while those with milk grew steadily and normally.

After eighteen days he reversed the diets. The second set of rats stopped growing and declined, quite in parallel with the first set of rats in the first experiment. Similarly, the first set of rats in the second experiment, now getting milk, grew in close parallel to the second set of rats in the first experiment.

These results could be obtained with remarkable regularity on each repetition and reversal. Several investigators had previously obtained rather similar results, but when their data were plotted, they did not give such tidy, regular and convincing curves.

Hopkins refined his control experiments in considerable detail. The two sets of rats were not chosen at random. No. 1 rat in each set was of the same breed, similar ancestry, weight, and sex, and the rats were kept in pairs so that they should not pine away in solitude.

In his early experiments Hopkins had flavoured the pure synthetic diets with traces of extracts from meat or yeast, in order to make them appetizing. He found, however, that the rats would eat them perfectly well without. Nevertheless, they lost weight and condition. But with the traces of extracts, they flourished on the pure diet. Evidently the extracts had a powerful effect on nutrition. This observation led him to use small additions of milk.

Though Hopkins' paper of 1912, owing to its perfection of technique and scientific quality, had a decisive effect on the acceptance of the existence of vitamins, it became clear years afterwards that the quantities of milk added to the diet could not have contained sufficient vitamins to have the effect on growth and well-being which he observed. It was generally believed that the effect was due to the phenomenon of refection in the intestines of the rats. Some of the starch eaten by the rat remained undigested and was attacked by bacteria, which produced the vitamins needed for the growth and well-being.

In 1945 Hopkins published with Miss V. R. Leader repetitions of his experiments. He was convinced that the effect was not due to refection, though he found that his results could be repeated only with potato starch as the carbohydrate.

Hopkins published only one paper specifically devoted to the demonstration of the existence of vitamins, his paper of 1912, and even that, in spite of its apparently convincing nature, subsequently proved to be inconclusive.

When he received the Nobel Prize, jointly with Eykman, for his work on vitamins, he disclaimed being their discoverer. He referred to the feeding experiments of Lunin in 1881, who had shown that milk contained besides its principal ingredients small quantities of unknown substances essential to life. Eykman had proved in 1890 that beri-beri, a disease described by the Chinese 1300 years ago, was due to eating polished rice. He showed that the bran contained a constituent, soluble in water and alcohol, which was capable of preventing beri-beri. He supposed the disease was due to a poisonous protein or salt in the rice, which was normally rendered innocuous by this substance in the bran.

Eykman's pupil, Grijns, showed in 1901 that this substance was used by the body and was necessary for health, and Eykman subsequently adopted this view.

Another Dutchman, Pekelharing, following Eykman's tradition, published in a Dutch journal in 1905 a remarkable statement, which remained little known because it was overlooked by the abstracting journals.

Pekelharing reported that 'till now my efforts, constantly repeated during the last few years, to separate this substance and get to know more about it have not led to a satisfactory result, so I shall not say any more about them. My intention is only to point out that there is a still unknown substance in milk which even in very small quantities is of paramount importance to nutrition. If this substance is absent, the organism loses the power properly to assimilate the well-known principal parts of the food, the appetite is lost and with apparent abundance the animals die of want. Undoubtedly this substance occurs not only in milk but in all sorts of foodstuffs both of vegetable and animal origin.'

The discovery of vitamins was not clear-cut, and was inevitably a subject of sharp difference of opinion. Funk protested in 1925 against the ascription of the discovery of vitamins to Hopkins, though he agreed that he had shown prophetic insight.

The merit of Hopkins' work on vitamins lay less in priority and decisive experimental results than in his intellectual grasp and philosophic vision of the subject. He combined uniquely an insight into the function of vitamins, their chemical origin and role in the body, their connection with normal health, rather than with disease, critical experimental method in their investigation, and, finally, exquisite exactitude of expression, exemplified by the superiority of 'accessory food factor' over 'vitamin' as a scientific term.

His combination of all these qualities caused him to be described by F. Ropmann as 'the spiritual father of vitamin research'. Wieland wrote that he was 'the first to show that carbohydrate, fat, protein, and inorganic salts, even when applied in adequate amounts, do not suffice to maintain metabolic equilibrium. His experiments, planned on an extensive scale, were the first to be carried out with rats as experimental animals, and showed that other substances besides the above categories must be present in the food; these, the accessory food factors, were to be named vitamins, as the result of a less happy suggestion by C. Funk.'

While Hopkins was grappling with the fundamental chemical problems of nutrition, he collaborated with Morley Fletcher in a classical investigation on the chemistry of muscular contraction. Again, a single great paper, published in 1907, sufficed for the description of the results.

Morley Fletcher had been working on the physiology of muscle since 1898; Hopkins collaborated with him on the chemical side of the work.

The presence of lactic acid is a prominent chemical feature of muscle, and since 1865 there had been conflicting accounts of its formation and role. It was evident that an exact quantitative study of the lactic acid changes in muscle was necessary. Fletcher and Hopkins found that the former conflicting results were due to defects in technique in handling the muscle-tissue. When the tissue was disintegrated and extracted, its chemical equilibrium was destroyed. The old observations which seemed to show that fresh muscle contained as much lactic acid as one in rigor were due to the effects of disintegration produced by prior treatment. It was essential in measuring the amount of lactic acid not to irritate or stimulate the muscle in such a way as to make it produce more lactic acid. Owing to the interference caused by the water and alcohol previously used in extracting the muscle, no exact measures of the amounts of lactic acid in muscle at rest and in action had yet been made. They found, however, if the muscle were extracted with ice-cold alcohol, the surprisingly large stimulation of lactic acid production by alcohol at room temperature was avoided.

They concluded that when muscle is exercised, lactic acid is formed in it. During rest and recovery after exercise, oxygen comes into the muscle, leading to the disappearance of the lactic acid. For if the muscle is prevented from having access to oxygen, it is found that the lactic acid accumulates, whereas if it has proper access to oxygen the lactic acid fails to accumulate at all.

The investigation was facilitated by the use of a new colour test for lactic acid which Hopkins had discovered two years previously.

These fundamental experiments of Fletcher and Hopkins showed that it was possible to follow, by chemical methods, the building-up and breaking-down of tissue as it proceeds in the cycles of normal living processes. They were the starting-point of the modern theory of the contraction of muscle, based on the fermentation of the muscle itself.

Hopkins' next major research arose jointly out of his work on vitamins and on the role of lactic acid in muscular contraction. His early searches for vitamins had led him to look for them among substances containing sulphur groups.

Then, subsequently, he had been prompted by the muscle researches to see whether lactic acid could be replaced by aceto-acetic acid derived from fat in the process of muscular contraction. Through these experiments he had been led to apply the nitroprusside test for detecting sulphur groups to many tissues. Thus he returned to his earlier unsuccessful search for sulphur-containing vitamins.

The existence of an important sulphur-containing substance in tissues had long been known, and had been named *Philothion* by de Ray Pailhade. Hopkins' new searches with the sulphur-test enabled him to confirm that such a substance did indeed exist, and he succeeded in identifying it. He concluded that it was a dipeptide containing glutomic acid and cystein. He named the substance glutathione.

This notable discovery was published in 1921, when Hopkins was sixty years old. He was particularly proud of this intellectual child of his old age.

He indicated that glutathione had an important role in the chemical dynamics of the cell, having a part in the process of oxidation and reduction of materials. It transported hydrogen atoms to oxygen atoms. Hopkins thus gave the first clear formulation of the notion of a respiratory carrier in the cell.

The sulphydryl group in glutathione appeared to be associated with the oxidation of materials utilized in the cell-structure. He published a series of papers on the investigation of the oxidizing and reducing role of glutathione.

Then, in 1927, came a shock from America. Hunter and Eagle showed that Hopkins' constitution of glutathione was incorrect. It was a tripeptide, containing a third amino-acid besides glycine and cystein.

This stimulated a further brilliant effort, at the age of sixty-eight, from Hopkins. He published in 1929 a new, beautiful, and easy method for preparing pure crystalline glutathione. 'The grave discomfort involved in making an admission of previous error is mitigated by the circumstance that I am now able to describe a method, not without special interest in itself, which with ease and rapidity separates from yeast and from red blood corpuscles a pure crystalline' compound with a peptide structure.

In 1931 Hopkins and K. A. C. Elliott announced that pure crystalline glutathione, unlike the crude substance he had first obtained, did not stimulate oxidation when added to washed tissues. However, investigation showed that in the living tissue, glutathione undertook a certain respiratory function by carrying hydrogen to oxygen.

In 1936, at the age of seventy-five, with E. J. Morgan, he showed that glutathione could protect the vitamin ascorbic acid in the tissues from oxidation.

The discovery of glutathione concentrated Hopkins' attention, from 1921, on the mechanism of oxidation in the cell. He announced in that year that 'oxidation in the living tissues offers to biochemistry what is perhaps its chief problem.'

So, for his remaining years, the main work of his Institute was on this most difficult and most profound of biochemical problems. His last paper, published in 1948, the year after his death, was on glyoxalase, the enzyme of which, as Lohmann had shown, glutathione is the specific coenzyme. He reported in this paper that the glyoxalase system contained a protein activator, apparently of a new type.

Thus ended his unremitting life of experimental elucidation of the processes of living tissues. He published, by himself and with collaborators, more than 132 papers, one half of these after he had reached the age of sixty-two, when he received for the first time an adequate laboratory.

He had a singular gift of choosing problems of profound significance, and deriving from them, by unique technical methods of his own, profoundly important results, opening new lines of investigation. Yet in virtually every case his results were subsequently found to require correction.

No doubt this contributed to his deep feeling of being an intellectual amateur and his antagonism to pure organic chemists. Why did his experimental results remain significant even after correction? Because his experiments, for all their originality and brilliance, were not the main part of his genius. This was in the realm of creative biochemical theory, by which he saw into the way in which the mechanism of living nature works.

VI. BIOCHEMICAL PHILOSOPHER

When Hopkins was young, before the end of the nineteenth century, chemistry in England was not in a flourishing condition. Organic chemistry was influenced by the needs of the not-very-large synthetic dyestuffs industry, and physical chemistry was meeting the demand for the more efficient management of chemical manufacturing processes. Under these circumstances, both organic and physical chemistry were concerned almost entirely with inanimate processes. It was not surprising, then, that British organic and physical chemists had little feeling and insight for the application of their sciences to the elucidation of the chemistry of living processes. Medicine provided a medium for the investigation of living processes, but it subordinated chemistry to its own limited field and ends. General biology subordinated chemistry to itself in an even cruder way. It propounded a hypothetical chemistry of its own. Living organisms were supposed to be made of 'protoplasm', which did not obey the laws of ordinary chemistry because it was living. Some contended that living tissues were built of a peculiar 'biogen' molecule characteristic of life. Any chemical process in a living organism could be dismissed as a characteristic property of 'protoplasm' or 'biogen', and therefore unanalysable.

The organic chemists were absorbed in the development of their intricate dead geometry of molecular structures. The physical chemists were concerned with the transformations of energy in systems of uniform phases. Even the colloid chemists, who at first sight seemed much more promising, were concerned mainly with the properties of featureless surfaces.

In England the divorce between biology and chemistry seemed profound. Nowhere were the two sciences being combined on equal terms. There were virtually no posts and no literature.

On the Continent the situation was better. In Germany the study of chemistry in every branch was more advanced. Great organic chemists like Emil Fischer had applied their technique to the elucidation of the constitution of living tissues, with magnificent success. Hofmeister, who had been appointed to his chair of experimental pharmacology in 1885, suggested in 1901 that the orderly sequences of chemical processes in the living cell were the result of the chemical organization of substances by individual chemical agents in the cell.

By 1900 thirty-one volumes of the German *Journal of Physiological Chemistry* had appeared, while the British *Biochemical Journal* was not founded until 1906.

Hopkins entered this scene in the 1890s with the conception of a chemical science, with its own principles, methods, and autonomy which would deal with organic chemistry, medicine, and biology on equal terms. It would explain the chemistry of living processes as well as the chemical constitution of living materials.

He developed his ideas for many years in his Cambridge lectures and discussions. He expounded them in a famous address to the Physiology Section of the British Association in 1913, on the 'Dynamic Side of Biochemistry', and on many occasions in his later years, especially in an address on 'Organicism' in 1927, and on the 'Influence of Chemical Thought on Biology' in 1936.

He described how Liebig, the founder both of organic chemistry and of modern animal chemistry at the beginning of the nineteenth century, had enthusiastically anticipated the great aid that organic chemistry would give to biology, but had been sorely disappointed. As a matter of fact, even in Germany the combination of chemistry with biology never reached, within the nineteenth century, the degree for which Liebig had hoped, and in England it happened hardly at all.

He thought that the fault lay partly in Liebig himself, who, in spite of his brilliance as a chemist and his intense desire to aid animal physiology and agriculture, did not succeed as he had wished because he lacked a biologist's training and a biologist's instincts. The other great mind of the time, Pasteur, was also not altogether fortunate in his influence, for when he entered biology from chemistry he 'became almost too much of a biologist' and tended to believe that nothing useful could be discovered about organisms except by studying them as wholes. This strengthened the biologist's belief that protoplasm contained elusive properties which were not susceptible of analysis, and that the chemist's investigations could not touch on the realities with which biology, as the science of life, was concerned.

Hopkins fought the notion of 'protoplasm' and 'proteid', which carried its mental associations, as Pasteur fought the notion of 'spontaneous generation'. He drove the word 'proteid' out of circulation, and when Michael Foster's picture, with this word 'proteid', had to be dated, somebody said at once that it must be 'pre-Hopkins'.

When the journal *Protoplasma* was founded in 1929, his colleagues waited with amusement to see whether he would allow it to be taken in his Institute's library.

Chemical investigation had revealed that the living tissues contained complicated substances such as proteins whose constituent molecules are large. It was difficult to elucidate the structure and properties of these big molecules, so the organic chemist left them severely alone.

The animal tissues yielded sticky, intractable substances, which prompted the organic chemist to the cynical comment that 'Thierchemie ist Schmierchemie', that 'animal chemistry is the chemistry of mess'. So the organic chemist concentrated on his own nice, clean chemistry of the simpler molecules, on which he practised his 'magical synthetic art' that 'provided him with substances made not by nature

6. Sir Frederick Gowland Hopkins in his laboratory, about 1941. He is making a remark in the vein of his characteristic story: 'A very distinguished organic chemist, long since dead, said to me in the late 'eighties, "The chemistry of the living? That is the chemistry of protoplasm; that is super-chemistry, seek, my young friend, other ambitions."'

face p. 234

but by himself', and which 'were in the highest degree suitable for the development of clear ideas concerning molecular structure.' Thus clear thinking on molecular structure became divorced from biological material.

The biologists seized on the big molecules and revelled in their vagueness and undefined qualities, attributing to them any properties that they pleased.

Hopkins' own patron and admirer, Michael Foster, had in 1895 supposed that when atoms of oxygen passed through the blood into the tissues, they disappeared 'into some protoplasmic complex'. The steps in which the oxygen passed through the processes in the cell before it reappeared in molecules of carbon dioxide had not as yet been traced. 'The whole mystery of life lies hidden in that process and for the present we must be content with knowing the beginning and the end.' With such ideas was associated the notion of a special chemistry of living matter. Hopkins said that in the 1880s a very distinguished organic chemist, on being questioned by him concerning the chemistry of living things and his desire to pursue it, replied: 'The chemistry of the living? That is the chemistry of protoplasm; that is superchemistry; seek, my young friend, for other ambitions.' And so, as he tells us, his youth was fretted by the antagonism between organic chemistry and animal chemistry, and the belief of the organic chemists that the latter was an inferior discipline. Even in 1913 there was still a widespread distrust of methods of Biochemistry, which were supposed to be amateurish and inexact. He did not, by the way, find the new name 'biochemistry' very attractive. It seemed to contain a hint of biologism, with its theory of a special chemistry of life.

At first sight, physical chemistry seemed more sympathetic than organic chemistry to the elucidation of living processes. It at least was concerned with movement rather than the static geometrical shapes of molecules and structure, with function rather than morphology.

But in the 1890s a movement, led by Wilhelm Ostwald and Ernst Mach, arose out of the triumphs of the generalized theory of energy, which in turn had been born of the steam-engine and the universal impersonality of the industrial revolution, which set out to describe the world in mathematical formulae, without reference to its particular characteristics. They emphasized the current absence of rigid proofs of the existence of individual atoms and molecules, and therefore proposed that these hypothetical entities should be banished from science, together with the visualizations and pictures of them which scientists used to aid them in their researches.

This movement greatly hindered the application of physical chemistry to biology, because it denied the biochemist the use of visualizations of atomic processes which would assist him to imagine how the chemical processes of living matter might operate.

The proof by Jean Perrin of the reality of atoms, and the rise of atomic physics, with its vivid pictures of the atom, the revelation of the ordering of atoms in crystals by X-ray analysis, the extension of this discovery by Langmuir to the elucidation of the ordering of molecules on surfaces ended the non-atomistic, non-visual interlude of Ostwald and Mach.

In fact, the progress of physics has imposed on the physicist and on the organic chemist, at least for a period, 'a mode of thought which is essential for the biologist—the visualization of a mechanism.'

Hopkins told his audience in 1913 that 'in the study of the intermediate processes of metabolism we have to deal, not with complex substances which elude ordinary chemical methods, but with simple substances undergoing comprehensible reactions.' By simple substances he meant those with which organic chemists are accustomed to deal. The biochemist was not concerned only with the separation and identification of the products from the animal, 'but with their reactions in the body; with the dynamic side of biochemical phenomena'.

Thus Hopkins indicated that living processes did not consist of simple combinations of many different kinds of unanalysable, fundamental units, but of complicated combinations of simple units. Life was an enormously complicated variation on a few simple themes, so it might be successfully analysed, in its chemical aspects, by the unremitting pursuit of these few simple and therefore intelligible themes. On the other hand, if life consisted of comparatively simple variations on many unanalysable themes, there was little prospect of progress in its analysis.

The view of the living unit as 'very large and very labile' was 'as inhibitory to productive thought as it is lacking in basis'. There was an obsession that the molecules in tissues were necessarily of high molecular weight, and that simple molecules were of no importance. But, in fact, it was from the study of simple molecules that progress was to be expected. The extraordinarily wide influence of adrenaline in the body was a striking illustration of the importance of a simple molecule in its processes. The subtlety of living processes arises from the complication of their dynamics, and not from complexity of their units.

The life of the living cell 'is the expression of a particular dynamic equilibrium which obtains in a polyphasic system.' It 'is a property of the cell as a whole because it depends upon the organization of processes, upon the equilibrium displayed by the totality of the co-existing phases.'

Thus the presence of substances must not only be proved, and their nature identified. It was necessary to elucidate their role in the dynamics of the cell. Their quantity and prominence were not necessary signs of their importance—indeed, they might be there because they were not working in the cell; the important working molecules might, on the other hand, be too busy to be seen.

So he was certain that the search for simple tissue products would in the future receive important rewards, 'so long as physiologists are alive to the dynamical significance to all of them.'

One can deduce from the response of the living cell to foreign molecules of simple constitution results of profound biological significance. Before it had been observed that benzoic acid, when eaten by an animal, is converted in the body into hippuric acid, it had been believed that the animal body was incapable of performing chemical syntheses.

If one conceived the living cell as a system in dynamical equilibrium, one might be able to elucidate that equilibrium by studying the way in which it was disturbed by the intrusion of a new molecule.

The object of the process of digestion was to present the body with simple molecules. When one viewed the cell as a dynamical system, one gained the prospect of being able to follow these simple molecules through all the processes of metabolism, of their breakdown and oxidation.

Certain molecules may be of 'great importance, not to the structure, but to the dynamics of the body'. The hormones are instances. The structure of the living cell, which forms the most obvious parts, is made of highly complex, stable molecules. This character, and the colloidal condition in which they exist, give it its definite though mobile shape and its mechanical properties. The colloidal complexes form within the cell a milieu in which the dynamic chemical processes can proceed, providing, as it were, special apparatus and an organized laboratory. Within the cell, simple molecules undergo reactions promoted or catalysed by ferments or enzymes.

During the last fifteen years a doctrine had grown up which, while it had difficulties, had 'the supreme merit of possessing an experimental basis', and hence of providing a direction for further work. This was the conception that each chemical reaction within the cell was 'directed and controlled by a specific catalyst'. The evidence for the existence and activity of enzymes within the cell was extensive and increasing rapidly. One need not therefore pause at the 'myriad nature of the army of enzymes that seems called for' by the great multiplicity of chemical reactions which occur in the body and the narrow specificity with which each enzyme acts.

With the huge array of examples before us, we may logically conclude that 'all metabolic tissue reactions are catalysed by enzymes'.

The very complexity of the cell, with its myriad reacting enzymes and multitudes of exchanging atoms, all proceeding in a dynamic equilibrium, suggested that, underlying it, 'we may discover a simplicity which now escapes us'. Hopkins expressed the opinion that the investigation of the cell with the guidance of this dynamic conception was suitable to 'our national tastes and talents'.

He invited organic chemists to join in the work, though he advised them to spend a year or two acquiring a biological discipline. The organic chemist could do useful work by determining the constitution of substances obtained from tissues, but he will not so become a biochemist. In order to do that, he must learn how the chemical reactions run in the organism. He must get in touch with the animal, which usually does the unexpected. The difficulties of his material and technique may cause the professor of organic chemistry to call his work amateurish, and he will find that, unlike the physical chemist, he can rarely put his results into mathematical form. He may occasionally meet the spectre of Vitalism and wonder, quite unjustifiably, whether his work may be to no purpose.

He should be content to describe the living animal in the sense that the morphologist has described the dead. If 'life' contains some residuum which will always remain elusive, then he can take comfort in the truth expressed by Robert Louis Stevenson:

> To travel hopefully is better than to arrive,
> And the true success is labour.

Sherrington, apparently, heard Hopkins deliver this address in 1913. He wrote in an obituary reminiscence in 1947 of how 'In an uninviting gaunt lecture room he descanted for nearly an hour on the cell as a theatre of chemical processes. Without any deliberate attempt at eloquence, and in a voice that, as he became more interested, fell to such a purely conversational level as to be a little difficult to hear, he conjured up for his audience—some thirty persons all told—a picture of the cell as a tiny sponge-work containing perhaps a thousand foci of different actions co-operatively confined within a unitary whole. An organized factory manifoldly hydrolizing, pulling to pieces, and contemporaneously constructing and reconstructing. And this unity bounded outwardly by a mosaic of countless poles and leaking like a sieve. As I listened I felt I was being privileged for the time to see something of the micro-cosmic world in which my friend's scientific thoughts took shape and did his bidding. One mental factor which, it seemed to me, such thinking must demand, was a peculiar intensity of visual imagination, continuously checked in factual knowledge.'

Through that address, Hopkins became the biochemical philosopher of his generation.

In addition to his onslaught on Vitalism, in the form of the theory of 'protoplasm' endowed with unanalysable qualities peculiar to life, he attacked the Organicism of J. S. Haldane and the Holism of Smuts. Haldane had asserted in 1885 that the problem of physiology is not to obtain piecemeal physico-chemical explanations of physiological processes, but to 'discover the details of structure and activity in each organism as expressions of its nature as one organism.' Haldane held that a living organism was not a problem for analysis. Its essential attributes are axiomatic. Heredity, for example, is not a problem, but an axiom. He believed that biology and physical science would one day meet, but he confidently predicted that if 'one of the sciences is swallowed up, that one will not be biology.'

Latterly he had said that 'the attempt to analyse living organisms into physical and chemical mechanism is probably the most colossal failure in the whole history of modern science.'

Hopkins said that there was some subtle truth in what Haldane had been trying to say, but that unfortunately he never seemed to have succeeded in finding the right words in which to express his meaning. 'After some years of respectful effort' he could not pretend to have understood Haldane completely.

If the biochemist accepted and acted on Haldane's views, then he must retire gracefully from scientific endeavour. There is no doubt that, as a 'whole' the organism has properties not deducible from its parts, and that these properties are dominant. But the biochemist cannot proceed unless he believes that he is able to reconstruct from his data some significant aspect of the whole organism.

Biology started from the visual description of organisms, their forms, and evolution of their forms. It was concerned with 'what is visible in the phenomena of life', and it produced a kind of biological language which was 'apt to the description of the visible'. On the other hand, the chemist worked for a long time without concern for visible structure, and saw only in his mind arrangements of invisible atoms.

The biochemist must always picture in his mind the molecular events which underlie the changes of form which interest the biologist. If he is to obtain a picture which will be of any use to him, 'it must be frankly mechanistic'. But he recognizes that the organism is not strictly a machine, for a machine loses its meaning unless it operates for some exterior purpose. An organism is 'adjusted as a mechanism in order that it may be itself'.

Haldane's teaching implies that if the results of experimental biology are to be of value, they must be made on the intact organism. This is contrary to experience. For instance, such very important agents as the hormones, which harmonize the actions of the body, could never have been discovered by experiments on the intact organism.

Where is the mistake of the organicist and the holist? It lies, in part at least, in exaggerating the difference between the whole and the sum of the parts; that the separation is so wide that the acquisition of knowledge of the parts goes only a little way towards the understanding of the whole.

No biochemist will forget that the materials with which he works are part of or have come from 'a sentient organism endowed with attributes which, having neither spatial extension nor numerical values, are beyond his scope.' The organicist, however, goes beyond this. He implies that the parts themselves are different when studied individually, than as in the whole. The biochemist does not desire to be assured of more than the fact that there is no such fundamental difference in category. He will not exaggerate the importance of his picture of living processes, but he claims for it a certain degree of validity.

He was far from maintaining that such fundamental properties as inheritance, growth, and differentiation might not depend upon organization above the chemical level. If there be a hierarchy of levels of organization, then the chemical level will have its place, which will be that of 'self-maintenance'.

He boldly claimed that the living cell is already able to maintain itself as an individual entity in equilibrium with its environment by its physical and chemical attributes, without calling on others at a possibly higher level. So, already, through these attributes, operating through the agency of highly specific catalysts, the cell is able to display one of the characteristic properties of a living organism. He believed that biochemists implicitly held this, or a similar, faith.

It would not be necessary to discover unknown principles before proceeding to describe in physical and chemical terms the cell's capacity for maintaining its identity.

Hopkins combated vitalism, biologism, organicism, and holism. He courteously but firmly fought the organic chemists who scorned the difficult molecules found in living tissues. He explained the error of a physical chemistry preoccupied with the thermodynamics of homogeneous systems, whose philosophers were trying to abolish the visualization of processes in their science, so that its power of dealing with complex living processes was 'strangely lessened'.

He described how, by dynamic equilibrium, the cell could preserve stability of form through incessant change. He marshalled the evidence which showed that the living cell is an interplay of simple molecules which in themselves might easily be known. This enormously increased the biochemist's confidence in the face of the prodigious complexity of living processes.

He produced conceptions which illuminated the biochemist's path, and revealed to the young the possibilities of research and discovery. He protected his colleagues from the miasmas of false philosophies. He provided the biochemist with a picture of the running internal mechanism of the cell, as a guide to thought and experiment. Sherrington compared his role in biochemistry with that of Kekulé in organic chemistry.

He objected to the impression sometimes given by mathematicians and cosmologists that biology and life were unimportant. Bertrand Russell had once ventured to suggest that in passing from physics to biology one is conscious of a transition from the cosmic to the parochial, because from the cosmic point of view life is a very unimportant affair.

But 'those who know that supposed parish well are convinced that it is rather a metropolis entitled to much more attention than it sometimes obtains from authors of guide-books to the universe. It may be small in extent, but it is the seat of all the most significant events.'

Hopkins had begun his scientific life as a naturalist. He was well acquainted with the 'inhabitants of the metropolis', and remained so always. His instinctive biological feeling and insight made his dynamic chemical picture of the living cell all the more fertile and remarkable. It was the product, not of an engineering, but of a biological mind.

Hopkins conducted all his teaching in the most engaging language and exquisite style. Marjory Stephenson rightly said that 'he alone amongst his contemporaries succeeded in formulating the subject.' This was his unique achievement, which, together with his experimental investigations, and his inspiration of others, made him, as E. Mellanby generously and justly said, 'in his sphere the greatest scientist of his time'.

VII. STATESMAN OF SCIENCE

Hopkins' small figure, quiet voice, and tentative manner were not obviously those of a statesman. His manner did not convey any immediate compulsion and he had no taste for public affairs. Yet he had an exceptional influence on science and its social relations, and thereby became one of the chief statesmen of science of his generation. This was due to several causes. The first of these was his interest in socialism. Sherrington thought that, after science, this was his chief interest. He had had a difficult youth. He became sensitive to social injustice. As an analytical chemist, he disliked the patronizing social superiority of the medical profession. Intellectually he opposed the conceit of the narrow-minded organic chemists of his youth. His natural mental gifts enabled him to express his criticism of these defects in society and in science in a clear, logical manner. He became a rationalist, and Ray Lankester, who also was a rationalist, communicated his first memoir to the Royal Society. Hopkins fought for his ideas and for biochemistry all his life.

As an analyst in Stevenson's laboratory, he became acquainted with crime and the circumstances and conditions in which it arises. As a medical student walking the wards of Guy's, he saw the dependence of health on social conditions. He became aware of the unsolved problems of human nutrition.

At Cambridge he laid the foundations of the modern science of nutrition, contributed towards the discovery of vitamins, and was the first to explain and secure an understanding of their full significance. He shared a Nobel Prize for this work in 1929.

As an original member of the Medical Research Committee, he proposed the appointment of his collaborator Morley Fletcher as the first Secretary of the Medical Research Council, whose work had profound influence on the development of medical research in Britain, and which was continued by E. Mellanby, one of his pupils.

Hopkins' authority was the greatest single contribution to the new British national awareness and understanding of nutrition, and the formation of a policy to improve it.

He was elected President of the Royal Society in 1930, and he occupied this premier position during the economic and governmental crisis of 1931. A nutritionist of unrivalled authority, democratic in spirit and sympathetic in understanding occupied the chief scientific position in the country during five of the worst years of unemployment and social misery. He never allowed any consideration of political convenience to deflect him in the slightest from the steadfast explanation of human nutritional needs, not only in quantity, but in quality.

In his first four annual Presidential Addresses to the Royal Society, in 1931–34, he adopted an uncompromisingly progressive view of the advance of science, and in his fifth and last, in 1935, he spoke directly on Science in Modern Life.

In 1931 he said that progress in experimental biology was 'not less revolutionary than the progress of modern physics'. He instanced the advances in the knowledge of genetics and the physiology and biochemistry of nerve and muscle. He saw in Pavlov's research on conditioned reflexes 'a supreme proof of the success of the experimental method in analysing even such apparently transcendental phenomena as those which underlie the higher functions of the brain'. He thought that research on viruses 'may make it necessary to modify some fundamental biological concepts and indeed be as revolutionary as their effects on the breaking-up of the atom.'

In 1932 he had the courage to recommend a cautious, but unprejudiced study of Gurvitsch's claim for his mitogenetic rays.

In 1933, he made a survey of 'progressive science'.

In 1934, he spoke on the relations of chemical medicine and science, and how the advance of the former depended on the latter, and that consequently, expenditure on clinical research should not be increased at the expense of laboratory research. In this address he remarked that 'no outstanding advance in the theory or practice of medicine occurred in the eighteenth century.'

He began his discourse in 1935 on science and modern life by observing that the scientific enquiry which began after the Renaissance performed the ethical task of releasing human thought from 'the shackles of authority and dogma'. Darwin and Lyell had 'reorientated all thought, not only biological thought', and thereby had 'subtly encouraged the growth of intellectual honesty', for which Huxley and others had fought so hard. Science had established the idea of progress. From 'this faith, hope is fed, human efforts towards betterment are encouraged and ethical values emerge.' It dispersed that depressing belief that led to other-worldliness, and the contemptuous indifference to social wrongs.

Such were the words that Hopkins uttered from the Presidential Chair of the Royal Society.

The social lessons to be drawn from nutritional science included the appreciation of the subtlety of nutritional needs, 'a failure in quality may be just as deleterious as a failure in quantity.' Then it was necessary to realize that 'certain diet habits are cultural', and are not to be departed from without harm to the individual. In Britain, particularly, was there a need for good cooking. 'In no country is so much good food ruined by bad cooking.'

He supported the League of Nations' work on nutrition, one of the more successful of that institution's activities.

As president of the British Association in 1933, he attacked the pessimism that had followed the economic depression and his predecessor Ewing's disillusion with the benefits to humanity of his own science of engineering. He repudiated the implications of C. D. Broad's remark that 'physics and death have a long start over biology and life.'

The study of biology would help man to understand himself and escape from his inhibitions. 'Inadequate as at first sight it may seem when standing alone in support of so large a thesis', he would be content by referring to a single biological example, that of nutrition. Since the adoption of the method of controlled experiment twenty years ago, it had been shown that the needs of the body were far more complex than had been previously believed. The argument that such knowledge was unnecessary because the human race had survived until now without it was 'very old and never right'. It had been learned that 'nurture can assist nature' to an extent not previously admitted. Biology has many, and will have more, lessons to teach.

During the last twenty years, the number of scientists in the world had increased by a factor of eight to ten. Science is making discoveries at ever-increasing speed. Sir A. Ewing had feared that mankind was gaining command of nature before he could command himself. Hopkins commented that, of course, the indictment was not of science but of mankind. He thought that, apart from war, science had done little to aid the display of 'man's irrational impulses'. The worst that it did, perhaps, was 'to give to clever and predatory souls that keep within the law, the whole world for their depredations, instead of a parish or a country as of yore.' He thought there was more danger to be expected from 'Money versus Man'.

The proper use of machinery and science would require planning, and 'will impose a new structure upon society.' The form that would emerge 'may well differ in some essentials from any of those alternative social forms of which the very names now raise antagonisms.'

He believed that the progress of science would release mankind from some of its labour and fatigue and provide it with leisure. Mankind in recent years had shown itself to be educable.

How was science to be put at the service of the people? He thought that Bacon's suggestion of Salomon's House, though conceived too much in the 'mood of a Lord Chancellor', was a great one. It would be 'devoid of politics' and conceived with appraising and synthesizing knowledge and the social adjustments needed for mankind to benefit from its ever-continuing advance. It would contain authorities of all kinds, not those of science alone.

He did not 'fear the gift of leisure to the majority.' If the political and economic problems of mankind could be sanely solved, the progress of science would provide mankind with the opportunity for every individual to 'be free to develop his own innate powers.' For those who cultivate it 'in a right and humble spirit, science is one of the humanities; no less.'

As President of the British Association, Hopkins supported the movement for the serious study of the Social and International Relations of Science, a division for which was subsequently added to the Association's organization. He supported, too, the international movement in this direction, started by the International Council of Scientific Unions in 1937.

In 1935 he was President of the Association of Scientific Workers. In his presidential address he said that, in spite of his years, he remained an optimist. The Association would be of real value to scientists as long as its enthusiasm was maintained. Its fundamental function was to raise the economic status and security of scientific workers. In order to achieve this, it was necessary for scientists to acquire public sympathy and understanding, by explaining the needs of scientists and the possibilities of science.

He thought that the three National Research Councils were doing well, but he sympathized with the Association's desire for a wider and more planned policy of national research.

He agreed with the Association's intention to strive for a 'really generous national attitude' towards expenditure on research. He even thought that a Ministry of Science ought seriously to be contemplated.

L. J. Harris has justly referred to Hopkins' 'deep social progressive conscience'.

In his presidential address to the Children's Nutrition Council, he described how in his youth *laissez faire* reigned and the ravages of individualism were mitigated only by sporadic, wholly inadequate charitable activities. A decade or two earlier, even charitable enterprises of any kind had been condemned by many.

The industrial revolution had brought wealth and power to the middle-classes, who had a code of morality 'which appears the more remarkable the more one thinks about it'. Highly-respected and self-respecting men practised for six days of the week a code of behaviour flatly in contradiction with the tenets of the religion they professed on the seventh.

It was natural that these men should have been actuated by the profit motive, but 'as professors of a merciful religion' they might well have believed that they ought not to have counted profits before social costs, and have realized that insecurity was inherent in their industrial system, and that therefore workers should be protected by a system of social insurance. If that had been instituted, 'there would have been less suffering in the past, and a better balanced society today.'

The reforms slowly acquired by a century of effort had improved this situation. Today, the comfortable classes understood the need for unemployment insurance, but not how indispensable a part 'adequate and right nutrition plays in the maintenance of health and happiness.' Unfortunately, the education of the governing classes, until quite lately, had left them ignorant of the elementary facts of human physiology.

He explained how malnutrition may not be evident, and yet leave serious lasting effects, especially if suffered in infancy. He thought that a national food policy should be instituted and the essential foodstuffs should be removed from the 'crude operation of the laws of supply and demand'. Its correct distribution would lead to the disappearance of 'innumerable surplus middlemen and retailers, with their heavy costs.'

In his Birkbeck College Foundation Oration of 1936 he gave his remarkable reminiscences of his rebellious old uncle who found in himself 'a rebel in embryo'. His uncle told him of Dr. Birkbeck's personality, and the 'feverish and even fierce' enthusiasm of those who had started the movement for the education of mechanics and the working class. It was opposed by many who regarded the movement as 'revolutionary, dangerous, and even evil'. The class that had reaped the chief benefits from the industrial revolution bitterly opposed the education of the class that had suffered most from it; 'one of those incidents in our social history which we would fain forget.'

Years after he had heard of Birkbeck College from his uncle, Hopkins had to attend its classes in order to study practical biology for an examination. It was the only place where he could obtain this instruction at a critical moment in his career.

Hopkins then described how, after presiding at a hundred meetings of the Royal Society, he had been profoundly impressed by the variety and rapidity of the advance of science. Would science continue to increase at an ever greater pace? And if so, what would the world be like a century hence?

He certainly would not ask his listeners to visualize a future on the lines of Aldous Huxley's *Brave New World* or even of H. G. Wells'

Shape of Things to Come. He was concerned on this occasion only with the future of science itself. Would it increase at such a rate that, as Bertrand Russell had suggested, the frontier of knowledge would move so fast that the would-be investigator would never catch up with it? Or would the human race get tired of investigating Nature? 'We should, I think, have a robust faith that science will never commit suicide or suffer defeat.' He believed that the growth of science would increase mankind's curiosity, not diminish it, and 'he will acquire more and more curiosity concerning every scientific aspect of his own complex self.'

He believed that literature appealed to what he wished to be permitted to call 'high intelligence', while science was concerned with a more active process of the mind, which he would call 'hard thinking'. It is not in the nature of literature and art to reveal anything fundamentally new. Their materials, in spite of their infinite variety, remain the same today 'as when the Hexameters of Homer were fresh in men's ears or when the Mycenian artists wrought their works of beauty.' All the stories of emotional adventure may have been told, but those of intellectual adventure are only just beginning.

Wordsworth, Tolstoy, Carlyle, Ruskin, and many other writers had suggested that the data of science are trivial compared with those of the world of values. They supposed that natural happiness had been known by mankind and that science and progress had taken it away. He could not accept that view. 'Nature unconquered' had often been 'very cruel to man.' The old simple happiness of ancient times was a myth.

He thought it unfortunate that the Greek philosophers had come to despise the experimental approach to Nature. If they had not, philosophy and science might have been more mutually helpful.

He had studied the writings of Dr. Joad, the Head of their Philosophical Faculty, in which he had dealt with the philosophical implication of science. He agreed with his criticism of 'the metaphysical structures which certain modern scientists have erected, in which the universe becomes wholly an externalization of the scientist's own mind.' He welcomed Joad's view that science is, after all, a new thing, and it is not reasonable to suppose that mankind should at once discover the best way of managing it.

How was the effect of increasing specialization to be dealt with? How were scientists to be made aware, even amongst themselves, of the general progress of science? How were the people to find out about the possibilities of science, so that science itself should be encouraged, and its results utilized for the benefit of man?

At the age of seventy-five, with characteristically enduring enthusiasm and prophetic vision, Hopkins foretold the creation of a new profession to help to solve these problems. 'I do not know whether other specialists will agree with me in my belief that a certain form of help for us may be forthcoming now. An increasing number of very able writers are today producing books in which the most significant aspects of the progress in this or that branch of science are assembled and appraised with accuracy and literary skill. I am, of course, not thinking of efforts meant merely to popularize science. Even such books as those I have in mind are perhaps at present mostly written for the intelligent lay public, but the best of them are of real value to the specialist. I venture to think that every encouragement should be given to this form of literature, which at the high standard of which I am thinking is something new and of an importance which will steadily increase. It can be produced at its best only by men with special gifts who are enabled to give all their time to it. . . . There are, however, some gifted individuals who find in themselves an inborn sympathy with the nature and aims of science together with an urge for literary expression rather than for experiment. If such elect to devote their lives to wide surveys of scientific progress and the interpretation of its significance for their colleagues who are immersed in specialized researches, I for one think that science should honour them as being among those who really advance it. . . . I am thinking, of course, of the development of a literature in which the essential and significant items of current progress are selected with highly educated judgment and presented with literary skill, while what is speculative is avoided and no special weight is given to the sensational. . . . My very real desire has been to suggest that literary efforts of this kind should be encouraged by us all; they are likely to be a great help in the future.'

Finally, though the advance of science, and its increasing acceleration, may produce difficulties, it is capable of 'benefiting both the minds and the bodies of mankind in ways beyond our present dreams. The pace is great, and the pace is growing greater.'

BIBLIOGRAPHY

Hopkins and Biochemistry, 1861–1947. Edited by Joseph Needham and Ernest Baldwin (1949).

Les Prix Nobel en 1929, Stockholm (1930).

'The Pigments of the *Pieridae*,' F. G. Hopkins, *Phil. Trans. Roy. Soc.*, B. **186,** 661 (1895).

'The Chemistry of the Urine,' F. G. Hopkins: *Schäfer's Textbook of Physiology*, 571–638 (1898).

'Note on a Yellow Pigment in Butterflies,' F. G. Hopkins, *Proc. Chem. Soc.*, 117 (1889).

'Pigment in Yellow Butterflies,' F. G. Hopkins, *Nature*, **45**, 197 (1892).

Obituary Notices: *Lancet*, **1**, 728 (1947); *Roy. Soc. Obit.*, **6**, 15 (1948); *Biochemical J.*, **42**, 161 (1948); *Brit. Med. J.*, **1**, 742 (1947); *J. Chem. Soc.*, 713 (1948).

WILLIAM BATESON

1861–1926

I. A GENETICIST'S ANCESTRY

THE most serious scientific controversy that has so far occurred in the twentieth century has been concerned with the science of heredity which was named Genetics by William Bateson in 1905. Why is this controversy so important and why is it so remarkably fierce?

The answer to the first of these questions may be given at once. Biological heredity is one of the fundamental properties of living things. Owing to it, living organisms have ancestors, relatives, and descendants, which form species that, as such, endure very much longer than themselves. Besides giving life to the species, heredity operates in such a way as to assist in providing it with variety. Duration and variety are two of the conditions necessary for the possibility of evolution.

Genetics, the science of biological heredity, provides the facts and theories which are essential for an understanding of certain of the most important properties of living things. These underlie not only agriculture and medicine, and all applications of biology, but the properties of man himself. Any misunderstanding or mistake about them may lead to actions that cut several of the main roots of life. Genetics is therefore much more immediately important than many other branches of science.

The vehemence of the genetics controversy arises in part from the extreme and immediate importance of this science. Other contributory causes will become clearer in following Bateson's life and work.

He was born at Whitby on August 8th, 1861, a son of the classical scholar and divine, Dr. William Henry Bateson, who was Master of St. John's College, Cambridge, from 1857–1881.

William Henry Bateson was the son of a prosperous Liverpool merchant. He was educated at Shrewsbury School under Dr. Samuel Butler, the famous headmaster who became Bishop of Lichfield and grandfather of Samuel Butler, the satirist. He married at the age of forty-five, in the same year as he became Master of his college. His wife Anna, a beautiful and vigorous woman, was the daughter of James Aikin, a Scotsman who had settled in Liverpool when a boy, and had by his own efforts made a very successful shipping business. He owned a fleet of clipper ships, which traded mainly with the West Indies. He was active in municipal politics as a Liberal. Aikin was noted for his practical ability and his terse and epigrammatic speech and his command of repartee. He was a formidable man, striking in appearance, mentally and physically strong, and energetic. He died at the age of eighty-six through catching a cold after his habitual sea-bathe before breakfast. He retained a freshness of spirit throughout his long life. As a self-made man, he had been very determined and severe. His own children held him in awe, but his grandchildren adored him. He had a simple directness that gave confidence, and a lively conversation and zest in life which made his presence always interesting.

Miss Anna Bateson, the sister of William Bateson, who recorded these observations, said that her brother resembled his grandfather in many ways, especially in physical energy and capacity for making anything in which he was interested interesting to others.

Bateson's social ancestry was not unlike that of W. E. Gladstone, the statesman.

William Henry Bateson was already forty-nine years old when his eldest son was born. He was a man of clear logical intellect and acute judgment with strong critical power, particularly good at detecting the weak points in any proposal. He was an excellent chairman, who instinctively sensed the feeling of a committee, and a very capable and industrious man of affairs, 'beneath a slight external coldness of manner lay a heart remarkably kind.' He was the leader in the struggle for liberal reforms in the University.

The distance between a father who had married rather late and was a master of a college and had already been Vice-Chancellor of the University, and a babe-in-arms, can be very great. William Bateson's infancy, however, was happy. His interest in nature was manifested early, and was encouraged. His mother gave him one of Lubbock's books on the natural history of insects, and he acquired a copy of Gray's *Anatomy*, in which he was deeply interested. He had gifted sisters, one of whom, Mary Bateson, became a historian of distinction. She discovered that the laws of many British towns were based on the little town of Breteuil in Normandy, and not on Bristol, as had previously been believed. His sister Anna became a botanist.

He was sent to a preparatory school, one of the type which specially prepared boys for entry into the 'public' schools. At the usual age of about fourteen he sat in 1875 for a scholarship examination for Rugby, of which his father was a governor, and was successful. The head-master, Dr. Jex-Blake, wrote to his father saying that his English Essay, in the Socratic style, was 'one of the best things done in the Examination'.

It consisted of a dialogue between the Captain of the Eleven and Socrates as to whether it was worth while winning the match. In reply to Socrates' question, the Captain replied that winning would not be of much benefit to him, but would redound to the honour of the school. Socrates observes that he seems to think that honour can be gained by one set of eleven boys hitting a ball better than another. The Captain is doubtful of this. Socrates suggests that perhaps he thinks that honour may be gained by the Master and boys of his school being richer than the other, but the Captain is again doubtful. Perhaps, then, says Socrates, honour is gained by one set of boys hitting a ball for four or five hours, while others are improving their minds. The Captain is sure that honour is not to be gained in this way. Socrates presses the point, and asks whether it is honourable to anybody to assemble and bet and drink on the performance of a game of ball.

The Captain is still more sure that this is not honourable. So Socrates asks, what is honour? and the Captain says he does not know exactly. Socrates says that he does not, either, and concludes that exercise is permissible for the benefit of the health, but not to think only of amusement.

It is not surprising that William Bateson was an unqualified failure at Rugby. One should not start, at the age of fourteen, by applying the Socratic method to the examination of the principles on which the new imperialist governing class of Britain were being trained. The special educational system that had been evolved for this purpose had been worked out by the famous Headmaster, Dr. Arnold, at Rugby itself, and had been thoroughly established in a generation of practice when Bateson entered the school in 1875.

In spite of his scholarship and his obvious ability, he was near the bottom of all his classes during the four years he was at the school. He was unpopular with the boys and masters, and the Headmaster disliked him.

After he had been there nearly three years, he wrote in 1878 to his mother the joyous news that one of the masters, whom he had met casually on a walk, had been kind to him: '. . . he was not like a master at all.' He tells his mother that he doesn't think he will ever be happy; the only moments of happiness he can remember are when he won his scholarship, and when one of the masters said he was doing better. 'I never get on with anybody for long; at home even I am always in some scrape except when I am alone.' She might say that this was all morbid nonsense, but it was true.

He enclosed a letter in the same envelope for his father. It consisted mainly of excuses on having spent so much on exercise books, and ends: 'I do so hope you won't despise me for what I have said to Mamma.

I know I have no pluck. Fellows have kicked that out of me long ago. . . . I could not tell you before, but now it would come.'

Shortly afterwards he heard of the death of his remarkable grandfather, James Aikin. He felt so glad that he had seen him once more after all his bad school reports.

The young Bateson pursued his naturalist interests through all this period, with a dogged endurance which was described as 'self-satisfaction' and was offensive to the Headmaster. He used to sit dreaming during the services in the School Chapel, wondering what he would do if some rare butterfly fluttered in through an open window. Ought he, or ought he not, to catch it? And if he tried, would he be allowed, and what would the others say?

His father, the Master of St. John's, now sixty-six years old, congratulated a clerical friend on having a son who at least knew what he wanted to do. He could detect in his own eldest boy no sign whatever of wishing to follow any particular pursuit in life.

At the end of 1878 Dr. Jex-Blake wrote to the Master, that his son's schoolwork was 'scandalous'. His divinity, which was taken under himself, was 'very poor'. And he had to say that unless the boy did very much better in the next two terms, he could not advise him to send him back after the summer holidays; and it was very doubtful whether 'so vague and aimless a boy will profit by University life.'

The boy was certainly capable and 'very much to blame'. Dr. Jex-Blake was exceedingly disappointed in Dr. Bateson's son; he had, however, seen the boy's sister, and had been delighted with her cheerfulness, and had found her 'so nice in all the real points'.

He regretted that he could not think well of his son 'while so self-satisfied, indolent, and useless'.

During the next year, Dr. Jex-Blake reported to Dr. Bateson that his son was doing a little better, but he was wise to remove him from the school early. He is 'very self-satisfied and desultory—even indolent in most things . . . but he has vigour and character, though a little abnormal.' He did not despair of his making a mark in the world yet.

Thirty-six years later, in an address to the Salt Schools in 1915, Bateson said that he looked back on his Rugby education as 'a time of scarcely relieved weariness, mental starvation, and despair'. His sole moment of illumination occurred near the end of his school period, when he entered the chemical laboratory and came in contact with real facts, and not mere pedantry. The staple of the school work was Latin and Greek, of which he made nothing, though he gained some emotional pleasure from the Greek tragedies. Now, in after life, he, like thousands of others, had found that Greek contained the materials which, more than any others, might have inspired him and stimulated him to development if they had been presented properly.

The Bateson family had had a governess, Miss Lakin, who, though not learned in natural history, had encouraged the boy's interest when home from school for the holidays. She retired to Reading, and in later years, when he frequently visited the gardens of Messrs. Sutton, the seed-merchants, in search of variations, he never failed to call on Miss Lakin, returning home with excited and grateful memories of her kindness in his boyhood.

Bateson left Rugby in 1879, and entered St. John's, his father's college. He failed in mathematics in his first attempt to pass the elementary matriculation examination for the University.

Bateson in 1905 described how he had been coached at school in mathematics and was again coached at Cambridge, but failed. He was coached yet once more, and passed. Now he had forgotten all his mathematics, and the hundreds of hours he had spent on it were wasted.

He recounted this in his arguments in favour of the retention of a compulsory paper in Greek in the university matriculation examination. He contended that mathematics was compulsory because it was useful 'in trade and professions for the making of money'. He supported the compulsory study of Greek because the Greek language held the key to art, and the entry to a world where 'utility' did not count. This world contained many things that have no value in the market, yet are without price in their value for humanity.

At Cambridge he was a brilliantly successful student. He secured a first-class degree in Natural Sciences, specializing in zoology, but unfortunately his father was unable to witness and take pride in these achievements, for he died in 1881, in the year before his son achieved the first of his academic successes.

Bateson subsequently learned that his father was one of those who had done most in his day to introduce progressive changes in the University. He referred to him as 'most unworldly', 'thoughtful and gentle', and had since realized that he was 'of the old régime, courtly and delicate', yet essentially a strong, clever man of action.

In the weeks before he died, the Master tirelessly supported the progressive policy in the Commission on University reforms, and steadily wore down the opposition on point after point. Bateson said that, at the time, he knew nothing of his father's activities in these matters.

Bateson found Cambridge a very free institution after Rugby, and pursued many interests with intense vigour. He began to develop his appreciation of the art of painting, which became the second mental passion of his life, and he read widely. He thoroughly mastered French and read German. His favourite French authors were Balzac and Voltaire, and he often carried a copy of *Candide* about with him.

In 1882 he canvassed very hard in the Cambridge parliamentary election, as a Radical supporter of the Liberal candidate. He never showed much interest in parliamentary politics afterwards.

He attended church and chapel assiduously, and was deeply interested in the ritual of the sects, of which he had learned much.

He subsequently described law, political history, geography and economics as 'ephemeral', and believed that natural science was the only path to profound and permanent knowledge. He had persuaded his sister Anna to study botany, and when it appeared that she might have done better at history, he consoled himself for having persuaded her, possibly, to have chosen an unsuitable subject by the reflection that if she had done well at history, she would only have 'despised such a success.' Bateson's prejudice against history was made all the more striking by the achievement of his sister Mary, who became a distinguished historian.

When Bateson was a student, Francis Maitland Balfour's genius was shining in its full power. Balfour was a younger brother of A. J. Balfour, the statesman, and through his mother a grandson of the second Marquis of Salisbury, of the Cecil family. His sisters presently married, respectively, the great physicist Lord Rayleigh and Henry Sidgwick the philosopher. He had been elected a fellow of the Royal Society at the age of twenty-seven, and had succeeded in publishing his classic treatise on *Comparative Embryology*, in 1881, in the year before he was killed in an Alpine climbing accident at the age of thirty-one. He had profound influence on biological research at Cambridge. His wonderful work extended the scope of systematic evolutionary biology, and laid the foundation of the modern study of embryology, a subject to which he had been originally directed by Michael Foster. It was based on the biogenetic ideas of Baer and Haeckel, that the history of the individual recapitulates the history of the species.

Balfour emphasized and demonstrated that evolutionary relationships are often more illuminatingly revealed by comparing organisms at the various stages of their embryological development than in their adult stages.

It was natural that Bateson should have begun research under the influence of Balfour and his demonstrator W. F. R. Weldon. The latter was Bateson's closest friend and a fellow-scholar of St. John's College.

Balfour's studies had led him to elucidate relationships of groups of animals especially among those which verged from the invertebrates into the vertebrates. There were certain kinds of worm-like creatures in these border cases whose classification was difficult. It was not certain whether they were the last of the invertebrates or the first of the vertebrates. Among these was *Balanoglossus*, which had been classed as an invertebrate on the study of its adult form. Its embryology had not been thoroughly investigated, and it was possible that this might throw further light on its relationships.

Such an idea arose naturally in the Balfour school, and Bateson decided to study this creature. He wrote later that Weldon had enabled him to make his 'first start on *Balanoglossus*'. His wife has recorded in her *Memoirs* that Bateson noticed this animal during his studies for his degree, and the possibility that it might be allied to the vertebrates occurred to him then.

The creature is found in the beaches of Chesapeake Bay, so Bateson decided to go there in the summer of 1883 to study its embryological development. Dr. W. K. Brooks of Johns Hopkins University kindly found him a place in the Chesapeake Zoological Laboratory, which was then at Hampton, Virginia.

The worms, which are eight to twelve inches long, existed in profusion, buried in the sand at half-tide. Bateson soon showed that previous descriptions of their morphology were very incomplete. He secured many of the creature's eggs, and succeeded in hatching them under observation. He wrote home to his mother in the greatest excitement, telling her that he was already observing things that had not been recorded before.

He found the conversations with Brooks even more stimulating. They lived in a very mean boarding-house, whose menu consisted almost entirely of shore-crabs and small fish, against which Brooks and he were daily deciding to strike in protest. Brooks used to lie on his bed in his shirtsleeves, discussing for hours the fundamental ideas of biology, with Bateson sitting by. Brooks was writing a book on heredity, which he believed had a physiological basis capable of independent study. His discussions may have helped towards turning Bateson's attention in that direction.

Bateson returned to England and began the publication of his observations in his first paper, which appeared in 1884. At the end of this paper he thanked Sedgwick and Weldon for rendering him 'much valuable assistance and advice in connection' with the research.

In the following summer, in 1884, he visited North America again as Brooks' guest, working first at Beaufort, North Carolina, and then at Hampton. He fell dangerously ill, and was devotedly nursed back to health by Brooks and his students.

He proceeded with the description of his work in a second paper published in 1885. He had devised a way of extracting, in an hour or two, all the *Balanoglossus* larvae living in several hundredweight of mud, and collecting them in a pint of water, where they could be sorted by the aid of a simple microscope.

He published two papers in 1886. In the first he completed the description of his observations, and in the second he discussed their significance. The second paper, with the title *The Ancestry of the Chordata*, is classical, and quite different from his earlier papers. It is written in a strong capacious style, revealing his power of biological discussion. It suggested that his feeling for argument and expression was greater than his feeling for experiment, in spite of the success of his practical work.

He starts the paper with a discussion of the morphological meaning of segmentation, and shows that there was 'reason for not accepting the view that the vertebrate segmentation was of such a kind as to necessitate the existence of a series of segmented ancestors to account for it.'

Then he says that: 'The decision that it would be profitable to analyse the bearing of the new fact in the light of modern methods of morphological criticism, does not in any way prejudge the question as to the possible or even probable error in these methods.

'Of late the attempt to arrange genealogical trees involving hypothetical groups has come to be the subject of some ridicule, perhaps deserved. But since this is what modern morphological criticism in great measure aims at doing, it cannot be altogether profitless to follow this method to its logical conclusions. That the results of such criticism must be highly speculative and often liable to grave error, is evident.'

These magisterial remarks in the sacred atmosphere of Cambridge comparative embryology sounded boldly irreverent. However, he concluded the paper with the proposal that Balfour's order of *Chordata*, to which *Balanoglossus* had been supposed to belong, should be subdivided into four, instead of three, classes; and that the first of these should be a new class named *Hemichorda*, containing *Balanoglossus* and its allies.

Bateson's *Balanoglossus* thus became the first of the vertebrates in the evolutionary scheme. At the age of twenty-five he had established one of the important features of evolutionary classification, mentioned henceforth in every textbook of the subject.

He was elected a fellow of his college in 1885, on this work, and his future as an orthodox evolutionary comparative embryologist appeared brilliant.

But the bold critical rumble in the opening of his classical paper on the ancestry of the vertebrates broke with a flash of lightning and a clap of thunder. He threw over comparative morphology, saying that the morphologists were freely hypothesizing biological variations in all directions, without any exact information of whether such variations had in fact occurred, and if so, how. His discussions with Brooks had interested him in the process of evolution. What variations of living organisms actually occurred in nature, and how? He began, in 1885, a systematic search for variations.

He had made one of the most triumphant applications of the Balfour comparative embryological method before he was twenty-five years old, and he exploited the status that this gave him to secure support for research in this entirely different field, in which, as he was told by an influential colleague, Darwin had already 'swept the board'.

The brilliant Cambridge school, supported by the prestige not only of its intellectual achievements in refreshing orthodox evolutionary zoology by the development of the comparative embryological method, but also by the social connections of Balfour and Trinity, began to regard him as a deserter and a renegade.

Bateson's contest was not purely scientific. He belonged to the Liberal *rentiers*, descended from the trading and industrial classes, whereas Balfour belonged to the Conservative landed aristocracy. The spirit of the Liberal *rentiers* was centred in St. John's College, whereas that of the Conservative landed aristocracy was centred in Trinity. The rivalry between the two colleges reflected the struggle between these two social classes. Bateson's ideas and work were influenced by and were in some degree a part of this class struggle.

II. HUNTING FOR VARIATIONS

Bateson sought a way of looking for variations which might help to reveal their cause. Attempts to identify variations in animals and then to look for features in their environment which might have produced them had been unsuccessful. He therefore proposed to examine the problem from the perspective of the environment. He would find an environment which existed in many slightly different forms, whose chief features could be exactly measured and defined. He would then study a species of animal which was common to all of the different degrees of the environment, and find how it varied from degree to degree, and see whether there was any connection between the degrees of the environment and the variation in the animal.

The innumerable salty lakes in Kazakstan seemed to offer such an environment. The salinity and other properties of the water varied from lake to lake, and could be exactly measured. He would examine how common shell-fish varied from lake to lake, and whether their variations could be related to the present and past properties of the lake waters.

He proposed to make an expedition to Kazakstan for this purpose. He proposed also to seek vestiges of a former Asiatic-Mediterranean sea, which, it had been suggested, once included the Caspian and Aral seas and extended to the Arctic Ocean.

On the strength of his achievement in comparative embryology, he secured strong support from the Royal Society, and other quarters. He made very complete preparations and left for Russia in the spring of 1886. The journey was a considerable one, for at that time the railways to Siberia and Tashkent had not yet been built. He provided himself with weapons, clothing, food, medicines, instruments, presents, and various stores. He obtained letters of introduction from T. H. Huxley, Michael Foster, Bryce, and others.

On arriving in St. Petersburg, he called on the British Ambassador, Sir Robert Morier, and was invited to dine with him. He found that no one in the Ambassador's party spoke a word of Russian, or had even seen the pictures in the Hermitage. Dinner was served by eight footmen. Morier promised to introduce him to several ministers, who might facilitate his journey in Kazakstan.

He called on the Grand Duke Nicholas with an introduction, arriving first at the servants' entrance, and then being requested to present himself at the front entrance, which he did. After being received by a secretary, he was told that the Duke was 'not at home'.

He was helped in his arrangements by M. Semenov, the Vice-President of the Imperial Geographical Society, who intimated to him that he regarded Siberian prisons as 'pretty comfortable'.

The Governor of Omsk, General Kolpakovsky, wrote that all measures would be taken to apprise the local authorities of his arrival in the steppe.

He wrote home that he was 'altogether being fêted'. He was rather coldly shocked by the backwardness of Czarist Russia. Almost his first note on crossing the frontier was that everyone wore high boots and sheepskins, and stank. He warned his relatives not to put censorable political comments into their letters. He noticed that the comic papers especially were censored. One had a longing to know what the joke could have been.

He was struck by the crowd of beggars outside St. Isaacs Cathedral, and their incessant crossing of themselves, which reminded him of a St. Vitus' dance, and the relation between religious fervour and hypochondria. He noted the magnificence of the priests' vestments and their habits of spitting through the service. He found no cafés in St. Petersburg, and saw no ladies even in restaurants. Telegraphists, bakers, and chemists were generally Germans, and many were Jews. He found the Kremlin churches small, but a 'blaze of gloom and splendour'.

He arrived at Orenburg and was received by Prince Dolgorouki in a lonely maisonnette, an oasis in the steppe where he even grew trees six feet high, and entertained Bateson with splendid wines and cigars.

He wrote home that he was astonished by the 'amount of physical and material civilization that the Russians carry with them', but that 'of course it is only skin deep'.

He found that the information given to him by Russians was unreliable. It was not that they meant to lie, but they did not know what truth is. It was the same with their work—it was all shoddy.

As for the Russian peasant, he was 'beastly beyond words'. He did nothing and could do nothing. He was filthy, grew no vegetables, and was too lazy to make butter. Yet educated persons 'never tired of singing the praises of the "Moujik". It is just like America in that respect, that though everything is beastly, everybody insists on your praising it, and saying that it is *very European.* That is the untiring duty of the foreigner.'

He had to proceed to Irghiz, which was north of the Aral Sea, but the people in Orenburg could not tell them anything about the town. He was struck by the poverty of the steppe. It was hard to realize—there was no wood, no metal, no clay. Pieces of string and paper were precious. One did not realize how useful stones were until one had to do without them. The sheep, however, were splendid.

He left Irghiz with a train of four camels, four horses, two Kirghiz policemen provided by the authorities, and two servants. His equipment included a boat, as well as various instruments.

He found the Kirghiz people very peaceable and pleasant. They had wonderful manners, and great physical strength and endurance. He saw a girl of eighteen catch a runaway bullock by its horns, and stop it. They never washed, and suffered from much skin disease. They were perfectly helpless, for they had no arms, no ideas or knowledge of either men or things, and no resources, internal or external. They were curiously devoid of any ambition or love of comfort. They lived in the most abject poverty, and nothing could be expected of a people that lived on such very poor rations. Yet mutton was a penny a pound and eggs twenty-five for fourpence.

Bateson had a severe struggle to become accustomed to their food. The mutton was sometimes good, but often putrid. Tea was drunk with butter in it. Bread was a luxury, made in flat cakes, cut in small squares and fried in fat.

The absence of hygiene, the use of eating utensils without being cleansed, tried him severely. The cooking was very bad, but when he attempted to do his own cooking, he found that he had committed a grave social solecism. He found that if he was to preserve his prestige, he must do no manual work. The conditions tried his health, and he had considerable difficulties with his men.

He met a mullah who told him that he had heard that he hated the Russians.

Bateson behaved very much as a young English gentleman from Cambridge, as indeed he was. He showed considerable power of external description but very little social insight.

His search for vestiges of 'Asiatic Mediterranean' was a complete failure. The Aral Sea was 212 feet above the Caspian, and there was no evidence that they had been joined in recent geological times. Much less had they been joined to the Arctic. He found that there were no deposits of shells at a height of more than fifteen feet above the level of the Aral Sea, so that its waters could not have spread very far from their present limits.

The hope that the vestiges of the 'Asiatic Mediterranean' might contain shells which would throw light on variations in existing animals in the Aral Sea, and the other lakes, from Tschalkar to Balkhash, was entirely disappointed.

He collected specimens of cockles from about 500 lakes and isolated waters. His study of them led him to conclude that while one animal may be profoundly and uniformly modified by certain conditions, these same conditions may have no palpable effect on an allied animal of a different sort.

He made a short visit to Egypt in order to examine the shells in the brackish lakes near Alexandria.

He concluded that the diversities of environment did not, in general, produce variations in shell-fish, though there were exceptions. One of these occurred, apparently, in cockles, for these showed variations which could best be explained by this cause.

He had spent eighteen months on the Russian expedition, thirteen of them in the steppe and the remainder in various towns. He regarded it as a failure. He wrote in a letter from Omsk to his friend S. F. Harmer, that he had 'played for a big stake, and—well, I lost.'

In after life he regretted that he had concentrated so completely on the cockles, and had missed opportunities of investigating many other things of wider interest.

He spent the winter of 1886 at Kazalinsk, near the north-east shore of the Aral Sea. While there, he received a letter from which he learned that he had been passed over for the Balfour Studentship for 1886. He wrote home that it was probably a good thing for him. His success over *Balanoglossus* had been too easy. The work was not really valuable, and in five years' time would be forgotten and despised. It came at a 'lucky moment and was sold at the top of the market.' But it had no bearing on the things that biologists really want to know. His work was of a kind which did not appeal to the Cambridge biologists. But he had gone into the business with open eyes, knowing that chance of promotion was indefinitely deferred, and he was prepared to go on, on that understanding. In one of his last letters from the steppe, he wrote that if it was expedient for him to teach in Cambridge, he would pocket his pride and do so.

However, in spite of the modest results of the Russian expedition and the Egyptian visit, and the lack of sympathy of the Cambridge school for his ideas, he was awarded the Balfour Research Studentship in 1887. He planned a comprehensive collection of materials for the study of variation, which was to be summarized in a book.

He undertook this task with great energy. He searched the scientific journals and literature, watched the gardeners' and agricultural periodicals, and went to flower and animal shows, where he became a familiar figure to breeders and fanciers. He ransacked the gardens of amateurs and of the seed-merchants. He talked to gardeners and drovers, and everyone who might give him information. He wrote innumerable letters to explore any promising hint, and made visits wherever possible, in order to see the material for himself. His approach was eclectic: he gathered data of every kind which appeared to have any significance for biological variation.

He continued busily with this programme. But he learned in 1890, from his teacher and good friend Alfred Newton, that his Balfour Research Studentship would not be renewed. Newton advised him to increase his income by writing magazine articles on his travels in Russia and elsewhere. Adam Sedgwick told him bluntly that he would not wish to have him as successor to Weldon, who was leaving for University College, London.

Bateson's connection with St. John's College kept him going during the next years. His fellowship continued, and in 1892 he was appointed steward of the College, for which he received an additional income. His duties included the supervision of the College farm and kitchen gardens in which he was keenly interested. The combined incomes from the College provided him with enough to live on, and a position of independence in the University with regard both to its policy in biological research and in academic affairs.

The independence of college and university, at that time such a prominent feature of the Cambridge and Oxford system, enabled a number of men with peculiar and unpopular views to survive. Bateson owed much to his college, which was loyal to him, not only as one of its fellows, but also as the son of a former Master. The traditional rivalry between the aristocratic conservatism of Trinity and the bourgeois liberalism of John's contributed to the Johnian support for Bateson in the face of the Trinity men who had so much influence in making appointments in the biological departments.

The contemporary social situation and traditions in Cambridge had an important influence on Bateson's development. In 1890 he applied for the post of deputy to Moseley, the Linacre professor of comparative anatomy at Oxford. Moseley (whose son H. G. J. Moseley, the great physicist, was then two years old) was in failing health. Lankester, after T. H. Huxley the most prominent British biologist of the day, was appointed, and no doubt would not have accepted the position unless it had been understood that he would be Moseley's successor.

Bateson's application was therefore of no practical importance. But it gave him the occasion, in 1890, at the age of twenty-nine, to make a considered statement of his scientific aims, and how he would carry them out if he had the opportunity.

He described how, in his work on *Balanoglossus*, he had become dissatisfied with the current morphological method of biological research, and had started on another method. Five years of experience had confirmed his belief that it was the correct one, and the most hopeful for the fertile development of zoology.

When the theory of evolution was proposed, it seemed necessary to see whether organisms had in fact evolved from predecessors. This was substantially proved by Darwin. The next task was to determine the ancestry and relationships of each organism and the forms through which it had evolved. Von Baer's law, that the history of the individual is the history of the species, had suggested that the methods of comparative embryology might provide the best way of accomplishing the second task. If two organisms were found to have similar stages of embryological development, then this was strong evidence that they had had a common ancestor. This had been the chief line of zoological research since Darwin, and it was now tolerably complete in its main features for the most important organisms. It was, therefore, now possible to estimate its value.

One difficulty was that the method rested on 'an error in formal logic', which lay in the assumption that von Baer's law was true. This had been inferred as a general truth from the observation of particular cases. Then, with this uncertain basis as a general truth, it had been immediately applied to particular cases.

His experience in applying the method to *Balanoglossus* had impressed him with its dangers. The larva of one of its species crept in the sand, and of another swam on the surface of the sea. How can one determine from two such very different forms which was the ancestral type? Morphological studies had provided a magnificent body of facts, but they were riddled with such difficulties. There was no rule by which they could be solved, and every investigator postulated such solutions as he thought best. The study of anatomy and embryology revealed the relations between successive forms of organisms, but it could never show 'their genetic relations nor can it forecast their future.'

All biologists, both Lamarckians and Darwinians, were agreed that evolution had proceeded through the occurrence of variations in organisms. Until the modes of variations were known, no fundamental advance in zoology could be made; and every zoologist was at liberty to postulate any kind of variation that he thought fit.

For these reasons, then, it was his ambition to devote himself to the study of variations in animals and plants, using the methods of Darwin himself. He would concentrate on familiar organisms, and especially domestic animals, by 'promoting experiments in breeding and cultivation', by the 'comparison of local varieties', and the 'amassing' of data on the nature and 'modes of variation'.

He would, 'above all', encourage the study of animals '*as they are*, in the field and on the sea', looking on them as 'living things and not as diagrams or types.'

The division between scientific zoology and natural history had been most unfortunate. 'When the history of the study comes to be reviewed, it will seem a strange thing that the first result of the work of Darwin, who was above all men a field-naturalist, has been to bring the study of Natural History into contempt.'

He thought, then, 'that the proper way to develop Zoological study is to pursue Darwin's problems and to employ Darwin's methods.'

With this object he had made his collections of shells from brackish waters in Western Siberia and elsewhere, to see whether diversities of environment produced variations in structure, and he had specially endeavoured to familiarize himself with the breeds and varieties of domestic animals.

He was classifying the material that he was amassing on variations under four headings:

1. The variation of multiple parts and of symmetry;
2. Variations in the size, number, and fertility of animals;
3. Variations apparently due to conditions;
4. Local varieties.

He proposed to publish such collections of facts in order to encourage others. They were urgently needed as a condition for the progress of zoology. No such fresh collections of facts had been made since Darwin, and all new biological theories were still being based on the facts that Darwin and Wallace had collected.

He concluded his application with statements that he had had little teaching experience, and would aim at creating a small group of students of high attainment, rather than organizing the smooth teaching of zoology to large classes.

Finally, if Professor Lankester were a candidate, he would withdraw, for though he stood by the scientific opinions that he had expressed, he fully recognized Professor Lankester's pre-eminence, and held that in the interests of science and the University, he should be chosen before all others.

Bateson's application was able and undiplomatic. It gave a remarkably well-expressed statement of his views at an early date in his career. One of the most interesting technical features of his statement was the absence of any reference to the statistical analysis of variations. He spoke of the 'magnitude of the integral steps', with its hint of the discontinuity, and difference in size, of individual variations, but not of the quantitative investigation of their frequency.

Bateson was unable to improve on his Cambridge situation, and continued with the collection of materials for the study of variation. He obtained a contract for the publication of a book on the subject, with that title, with the stipulation that he and the publisher should share the profits equally. The book, *Materials for the Study of Variation*, treated with special regard to Discontinuity in the Origin of Species, was published in 1894. It is a closely printed work of nearly 600 pages, packed with details of 886 variations. The state of the subject, and the implications of his researches, are examined in a preface and a formidable introduction of eighty pages. It was published when he was thirty-three, and whatever other qualities it might possess, it was clearly the product of enormous labour.

In the preface he explained that the work had arisen out of his research on *Balanoglossus*. When he had tried to interpret his results, he was confronted with difficulties, which raised the suspicion that the current belief that biological variations are continuous must be wrong.

He then repeated, with almost unchanged wording, the critique of the comparative embryological method which he had made in his application for the Linacre Chair at Oxford. The aim of the naturalist was to explain how living things had become what they are, and discover the laws which govern their forms. Though thirty years had passed since the *Origin of Species* was published, these problems had not yet been solved. This was no reflection on Darwin. It was through his work that mankind had realized that there was a reasonable hope that they might be solved. He had converted natural history from a vain riddle into a problem. Those who have taken this achievement for granted since their childhood can scarcely imagine how great it was. But Darwin and his successors had not solved the problem of the origin of species. It was necessary to look at the problem, and at the material, afresh.

The laboratory study of the forms of animals had led to an enormous accumulation of facts. But the explanation of the change from one form to another was based on theoretical assumptions that variations of such and such a kind, favourable to the survival of the animal, had occurred. Zoologists, who would have poured scornful ridicule on Paley or Butler for using comparable assumptions, argued habitually in this manner without a qualm.

Clearly, nothing more could be learned from the immense store of morphological facts until a well-founded theory of variation was established; and if a new set of facts was required to accomplish this, then the first aim of contemporary biology must be to collect them.

The first duty of the naturalist was, therefore, to collect and codify the facts of variation. Why had he not done this? Because in the past he had collected the facts of Nature as 'beautiful in themselves.'

He thought that one of the reasons why variations had been neglected by naturalists was because natural history had been made a medium of elementary education, and it was not suited to this purpose. Teachers tended to ignore exceptions and abnormalities in order to simplify their work. Consequently, one found even patent facts missed out in textbooks. For instance, the cockroach, which is much used as a type for elementary dissection, has notable variations in its anatomy, but these are ignored in the textbooks. Thus teaching, in its tendency to uniformity, undermined the 'habit of spontaneous observation' which natural history was 'best fitted to teach.'

What are the most striking characteristics of living things? Undoubtedly, one of them is diversity. The innumerable different kinds of species are one example of diversity. The variations of the individuals within each species are another.

Darwin succeeded in giving an ordered account of this diversity, and out of it arose the method of comparative embryology and the study of adaptation. The importance of the first was as a test of the theory of evolution, and of the second as a test of the theory of natural selection.

The fundamental problem in adaptation was the accurate measurement of the survival-value of a character to an organism. As Darwin had pointed out, characters which visibly differentiate species are not usually the most important parts in vital organs. Any organ or instinct might be of use; the real question is 'how much?' It was impossible to determine the value of a character without a complete 'profit-and-loss' account of its use in the organism's life.

There had been no organized effort to study variation since Darwin's first treatment of the subject in *Animals and Plants under Domestication*. How well we knew 'the offspring of Lord Morton's mare, the bitch "Sappho", the Sebright bantams, the Himalaya rabbit with pink eyes, the white cats with blue eyes, and the rest!' Surely their use was to point the way along which further facts might be found.

It was almost universally assumed that variation occurred continuously, by very small increments. This theory presented very serious difficulties. The study of plants and animals living in nature showed that a character is useful to them only when it is perfect. But if a character has arisen through an accumulation of minute variations it must have passed through various imperfect stages before it reached the perfect form. But the principle of natural selection cannot protect imperfect forms. How then could they survive, and so gain the opportunity of evolving into perfect forms? Continuous variation presented a serious obstacle to the operation of natural selection.

The study of variation may be expected to reveal the origin of the discontinuity of which species are an objective expression. As the inorganic environment in general changes continuously, this biological discontinuity would not seem to be due to the effect of the environment. May it not then arise in the living thing itself?

It was now time to try to see the processes of evolution actually taking place, and might be done by directly observing variations and how they occur.

Species are discontinuous. May not the variations from which they arise also be discontinuous?

One of the most universal characteristics of living organisms is symmetry. This is subject to variations. It is probably related to the phenomena of cell-division.

The pattern of organisms is an 'outward and visible expression of that order and completeness which inseparably belongs to the phenomenon of life.'

Apart from variations in the symmetry and pattern in the tissue of organisms, there are variations in the constitution of the tissue itself. He called such variations substantive. The two kinds of variation might be quite distinct. For instance, the *Narcissus* flower normally has six parts, but variants have seven or four. Though the number depended 'upon the characters which were present in the original mass and upon the modes by which the parts were divided out from it.'

Then he gave some examples of his hunting for variations. In the Farne Islands off Northumberland, he and his sister had found enormous numbers of earwigs. 'They lay in almost continuous sheets under every stone and tussock, both among the sea-birds' nests and by the lightkeepers' cottages.' The males were of two sizes, one very much larger than the other. The percentage of the large ones was much higher on the Farne Islands than in Cambridge gardens. The two sizes of males appeared to be living in exactly the same environments 'in close communion', so it did not appear that their remarkable difference could be due to environmental causes.

He gave examples of mutual exclusiveness in the coloration of organisms; for instance of black and tan in dogs, green and yellow in birds, grey and cream in mice and cygnets, red and blue in eggs, scarlet and blue in the pimpernel flower.

He gave the pedigree of swans kept by St. John's College, which illustrated the alternative character of their inheritance.

The frequency of colour variations suggested that they might depend on a range of chemical derivatives of certain pigments. Their discontinuity was connected with chemical changes, which might be compared with the red and blue colours produced in litmus by acid and alkali treatments. It was much more reasonable to attribute colour variations to chemical discontinuity than to 'hypothetical fitnesses', of which 'we know nothing'.

His discussion of colour variations led him to conclude that the change from one to another was not gradual, but occurred in one complete jump. For instance, in a butterfly whose coloration changed from red to purple, the original change occurred suddenly, so that the process of natural selection was offered red or purple, but not a succession of intermediate shades. The suddenness of the colour change was due to a change in the chemical composition of the wing pigments.

'This case is a particularly interesting one in the light of the fact that, as Mr. F. G. Hopkins has lately shown me, this purple colour, dissolved in hot water', gives a substance which cannot be distinguished from the orange or yellow colouring matters generally found in the *Pieridae* butterflies. Mr. Hopkins had shown that they were related to uric acid, an excretory product. Hence colour variation might be the sign of a change 'in the chemical economy of the body, and that when an animal is said to be selected because it is red or because it is purple, the real source of its superiority may be not its red colour, or its purple colour, but other bodily conditions of which these colours are merely symptoms.'

Bateson's recognition of Hopkins' work on butterflies, his going to see him to have the facts personally demonstrated to him, and then drawing important conclusions from them, is one of the most striking examples of his method of collecting facts on variation, and the energy, enterprise, and intelligence with which he carried it out.

Hopkins was still a little-known demonstrator at Guy's Hospital, of the same age as himself but academically much his junior. Bateson sought the facts of variation where he could find them, and had a remarkably direct and unprejudiced vision in assessing them.

Besides the discontinuous colour variations in butterflies, he discussed also those in coloured cats, dog-whelks, black mice, ladybirds, and many other instances.

The variations of reversal, where left-hand shapes are transformed into right-hand, and *vice versa*, are noticed in slugs, flat-fishes, etc.

Discontinuous variations in animal covering, feathers which become silky through absence of a normal part of their structure; nakedness in man, horses, mice and other animals, is illustrated.

Cases of bulldog-ism, due to a variation in the foreparts of the skull, are noted not only in dogs but in codfish, carp, trout, chub, and minnows. We hear that the skeleton of a bulldog had been found in an Inca grave in Peru.

He quoted Darwin's celebrated passage on how nectarines can grow from peach stones and *vice versa*, and hence that the nectarine was a variety of peach.

There were cockroaches with four-jointed legs, which were a variety of the insect with five-jointed legs. Both the four-jointed and five-jointed kinds were normal creatures. The sport was as definite and as perfect as the normal. Hence the perfection of the more numerous five-jointed kind would not be due to natural selection.

Sex was a striking discontinuity which strongly resembled a variation. Not only were there variations which affected an organism as a whole, there were variations which occurred in particular parts or organs, as if they possessed an 'individuality of their own. If ever it should be possible to form a conception of the physical processes at work in the division and reproduction of organisms, account must be taken of both of these phenomena.'

He compared the discontinuities caused by disease with those of variation, and quoted Virchow's remark that every deviation from the type of the parent animal must have its origin in a pathological accident.

The terms heredity and inheritance, the use of which he had tried to avoid, were metaphors taken over from the transmission of property. They suggested that 'the actual body and constitution of the parent are thus in some way handed on', and Darwin had developed this material view of descent into his Theory of Pangenesis. This idea that the body of the parent is somehow 'remodelled into that of the offspring' led to many errors, the chief of which was that 'variation must necessarily be a continuous process', for the remodelling could hardly proceed in jumps.

Weismann's theory of the continuity of the germ-cell suggested that it was more likely that if the relation of parent and offspring had any analogy with the succession of property, it was rather as 'that of trustee than of testator'.

He hoped that 'Hereafter, perhaps, it may be found possible to replace this false metaphor' of inheritance 'by some more correct expression.' He did this himself, eleven years later, when he introduced the term *Genetics*.

He believed that further progress in the study of variation required 'systematic experiments in breeding', and attention by the naturalists to unusual as well as normal forms.

He concluded his Introduction with a quotation from Sedgwick or, rather, from the translation of Claus' textbook of zoology by Sedgwick and Heathcote, that 'it is only *natural selection which accumulates those alterations, so that they become appreciable to us* and constitute a variation which is evident to our senses.' Then he quoted Claus' own words in German, and added that the crude belief that the forms of organisms are impressed on 'plastic conglomerates of miscellaneous attributes' was a fancy that received no support from the study of variation.

He believed that the study of variation bore the same kind of relation to evolution as that of the study of affinities and reactions to chemistry. 'We might almost as well seek for the origin of chemical bodies by the comparative study of crystallography, as for the origin of living bodies by a comparative study of normal forms.'

He concluded his book with the opinion that variation was not caused by the environment, nor by adaptation, but had its origin in 'the intrinsic nature of organisms themselves', as manifested in the discontinuity of variation.

This discontinuity was the final proof that the accepted theory was inadequate. In some unknown way, the discontinuity of characters was one of their fundamental properties, and not directly due to natural selection.

The only way of probing farther into this fundamental property was by systematic breeding experiments, which called for more patience and resources than any other kind of biological research. But 'sooner or later such investigations will be undertaken, and then we shall begin to know.'

The materials of variation were particularly suited to experimental analysis by simple means. Anyone could collect a few hundred shells or butterflies, and then examine, for example, their colour variations statistically.

For historical reasons, two approaches to the study of living organisms were then in vogue, which appealed to different and exclusive tastes and temperaments. One approach appealed to the intellectual and analytical mind, which accepted the 'challenge of Nature' and attempted to discover how she works. This type was disgusted with the superficiality of the naturalists, and sat down in the laboratory to probe deeper and deeper into the structure of the organism, 'hoping that the closer they look, the more truly they will see'. They cared little for life out of doors.

The other school was that of the naturalist, who was attracted by the living organism, and its ways in its environment. He regarded the outlook of the laboratory worker as 'frigid and narrow'. He was ignorant of the depth and accuracy of the laboratory method, and the weight of the theories advanced on its authority. With his 'senses quickened by the range and fresh air' of his own acquaintance with the natural world, he was conscious only of their inadequacy, and the smallness and conventionality of the world for which they were devised. So the naturalists tended to become antagonistic to theory, and degenerate into collectors. The laboratory workers tended to ignore the real world of living things, and the naturalists to ignore scientific analysis.

Both schools must combine in order to attack the problem of evolution, and if the origin of species is to be discovered, 'it must be by the study of variation.'

Bateson's book was published in 1894. It did not sell, and was soon remaindered. However, Bateson was elected a fellow of the Royal Society in the same year.

Why was this remarkable book not an immediate success? It was outspokenly unconventional. Its author was not only a renegade of great talent from morphology, but had carried the war directly into the enemy's camp. Considering the heinousness of his behaviour from the conventional biologist's point of view, Bateson could have been treated far worse than he was. He certainly could not complain at being elected to the Royal Society at the age of thirty-three, in the year of the publication of his highly controversial book.

Why did it have such a poor immediate sale? It was quite unsuited to any existing university course. Both in content and manner of presentation it was of no use as a textbook. As a philosophical work to be read, it ran flatly counter to accepted ideas.

Its early failure was, however, mainly due to a serious fundamental limitation. Bateson had assembled a very interesting collection of materials on variation, which amply demonstrated the discontinuity of their nature, and he had commented on the subject with great power and interest. But he had not discovered any theory which provided the key to the riches that he had collected.

His book did not become generally interesting and necessary reading until that key appeared to be provided by another. After Mendel's laws had been recognized, Bateson's book acquired a new interest, because it directed biologists to mines of material by which Mendel's laws might be illustrated and investigated. The new Mendelians found that Bateson had not only done much preliminary collection of materials for them; he had submitted to critical examination both the material data and the theoretical ideas with which they would have to work. But six years passed before Mendel's work was rediscovered, and even after that, the appreciation of Bateson's book, and of his experimental researches, did not increase swiftly.

III. MENDEL RECOGNIZED

In July 1899 Bateson spoke to the Royal Horticultural Society on the 'study of Hybridization and Cross-breeding as a Method of Scientific Investigation', several months before he learned of the recognition of Mendel's work. His address gave a very well-expressed statement of the views at which he had arrived. He said that the pursuit of the experimental investigation of cross-breeding could create a new Natural History within twenty-five years.

The chief garden flowers, Narcissus, Begonia, Gladiolus, Orchids, Rhododendrons, had been produced by hybridization. May not new species have arisen in the same way? Experiments showed, however, that this was not so.

By integral steps of what size did a new species arise, and how did it persist? The answers to these questions contained the keys to fundamental progress in biology.

The horticulturalist saw, even in his own life-time, a period of negligible brevity in the time-scale of evolution, a development in all the best horticultural plants so rapid that it must be called 'not Evolution but Revolution'.

No further *general* ideas on evolution were needed at this time, but more *quantitative particular facts*. These could be collected by cross-breeding, which was to biology what the study of experimental reactions was to chemistry. A few experiments could reveal more than ages of theorizing by systematists.

Bateson was asked to address the Royal Horticultural Society again in 1900. He prepared his address, which was to be delivered on May

8th. He took with him in the train a copy of Mendel's original paper on peas, and saw at once the bearing of his method and results on his own work. He added an account of them to his already written lecture.

He had opened this with an inspiring call to action: 'An exact determination of the laws of heredity will probably work more change in man's outlook on the world, and in his power over nature, than any other advance in natural knowledge that can be foreseen.'

He spoke with the clarity, confidence and power of a leader. He was completely harnessed and prepared to rush forward into the exploration of the process of evolution by the means of experimental breeding.

We had 'no glimmering of an idea as to what constitutes the essential process by which the likeness of the parent is transmitted to the offspring.'

He had already incorporated in his address a reference to Galton's Law of Ancestral Heredity, a conception which had first been proposed by William Wollaston in 1722. Following Galton, he had supposed that laws of heredity must be statistical, and hence give no information on particular instances; like the laws of rainfall, which give a probability as to whether rain will fall, but no certain information as to whether it will rain at any particular moment.

But since writing this, he had read Mendel's paper, and those of Correns, de Vries and Tschermak. Mendel had started from the conception that the characters of the peas with which he had experimented were separable, and that some characters took precedence of others, which he called *dominant* and *recessive* respectively. He had shown that in the first generation of cross-breds, the ratio of the number of dominants to recessives was as *three* to *one*. These, and other results of a related kind, had a remarkable arithmetical precision, fundamentally different from the statistical imprecision given by Galton's law. It appeared that Mendel had made a major contribution to biological science, and that Galton's theory of heredity was seriously defective.

Bateson now came in violent conflict with his former most intimate friend, Weldon, by whom he had been inspired to devote his life to science. Their common interests had not been confined to science: they were both passionately interested in pictorial art. Weldon had been a year senior to Bateson, and had never lost the sense of being his mentor, if not patron. Bateson used to say that Weldon had tended in their early days to treat him as his 'bottle washer'. Apparently, he could never become accustomed to the emancipation of Bateson, whose deep frustration he had done so much to relieve.

They had diverged since Bateson had accomplished his first work in comparative embryology. Both were convinced that the subject had been worked out, and a new concept was required. Weldon had adopted the statistical method of Galton, who believed in the orthodox theory of continuous variation, while Bateson had been influenced by W. K.

Brooks. Weldon accepted Bateson's *Materials for the Study of Variation* as a challenge to continuous variation.

The first public dispute between Bateson and Weldon occurred in 1895 over the interpretation of the variations of *Cineraria.* Bateson was supported by Lynch, and Weldon by the botanists Botting-Hemsley and Thistleton-Dyer. This spurred Bateson's breeding experiments, and he secured the voluntary assistance of his first colleague, Miss E. R. Saunders, a botany lecturer of Newnham College, who joined him in 1897.

Weldon was the son of the inventor of the Weldon process for regenerating the manganese dioxide used in the manufacture of chlorine. His father had been a journalist and Swedenborgian who believed that he had made this invention, from which he derived a large fortune, not through science, of which he claimed at the time to be ignorant, but through mystical inspiration.

Weldon had inherited his father's fortune, and was free from the lack of means which hindered Bateson. After having been devoted to the Cambridge tradition of morphology, he began, about 1887, to doubt whether this method of laboratory research could solve the problems of variation, heredity, and natural selection. Studies of marine animals in their natural environment and at the Biological Laboratory at Plymouth strengthened this belief.

Bateson sought an advance through the acceptance of discontinuity in variation; Weldon continued to accept the doctrine of the continuity of variation as firmly as the morphologists, but sought to advance by applying statistical analysis to the measurements of the characters of organisms. He applied Galton's statistical methods to the characters of shrimps, and calculated the degree of correlation between the same individual characters in four local races. He was elected a fellow of the Royal Society in 1890, in recognition of two papers that he had published on this subject. These were the starting point of the science subsequently called biometry. In the same year, he was appointed Professor of Zoology at University College, London, where the mathematician Karl Pearson joined with him in the application of statistics to biology.

Weldon expressed the view in 1893 that variation, heredity, and selection 'are all questions of arithmetic, and when we know the numerical answers to these questions for a number of species, we shall know the deviation and the rate of change in those species at the present day—a knowledge which is the only legitimate basis for speculation as to their past history, and future fate.'

This led Pearson to publish his first paper on the *Mathematical Theory of Evolution.* In Pearson's words, Weldon 'first formulated the view that the method of the Registrar-General is the method by which the fundamental problems of natural selection must be attacked, and that is the essential feature of biometry.'

In 1894 Weldon wrote that 'the questions raised by the Darwinian hypothesis are purely statistical, and the statistical method is the only one at present obvious by which that hypothesis can be experimentally checked.'

Weldon set himself to learn the necessary mathematics and read the French masters of the theory of probability, making himself competent to follow them.

In 1894 Galton had persuaded the Royal Society to set up a committee, with Weldon as a member, on the application of statistics to biology. Bateson was invited to join it in 1897, but he declined at first, because it was dominated by the biometricians. The committee was later reconstituted as the Evolution Committee. Bateson joined it, and presently became secretary, and its dominating personality. Its reports were virtually the only free medium for the publication of the results of his researches in genetics.

Weldon resigned, and in 1900 founded, jointly with Pearson, *Biometrika*. With his wealth he could support any project in which he was interested. He wrote to Pearson that the contention among biologists that 'numbers mean nothing' was a very serious thing and must be fought. A journal would be of assistance to that end. Pearson agreed, and proposed the name 'biometry' for the subject.

Weldon's influence was further increased in biological circles by again succeeding Lankester. In 1899 he was appointed Linacre Professor at Oxford.

The antagonism between Bateson and Weldon, which had been tense, but relatively private, now burst into the open. Bateson was a great naturalist, a notable morphologist, and a competent experimental breeder, but he had not learned any mathematics in his youth, nor in later years.

Weldon was a fine morphologist who also entered the naturalist's field. He was a magnificent lecturer. He had acquired mathematical technique and the collaboration of an eminent statistician. But he had not spent long years on breeding experiments.

The biometricians had nicely got their new science launched, in which they had staked their claim to the application of mathematics to biology.

And then, just as they had done this, Mendel's papers were rediscovered in 1900. Here was an interloper who had been applying statistical methods to biological material in 1865. He could not know the new refined statistical methods which had been elaborated since his time.

Weldon submitted Mendel's paper to a patronizing biometrical critique. This was published in 1902, in the first volume of *Biometrika*, under the title: *Mendel's Laws of Alternate Inheritance in Peas*. Weldon started by defining heredity in his own way. He laid down that if two organisms, each with a differing corresponding character, were allowed to breed together, then the offspring might be affected by the parallel characters in any one of three ways. The characters of the parents may appear intimately blended in the offspring, which was the most usual case. In the second case regions on the offspring's body resembled the character of one of the parents while the rest of the body resembled the character of the other. In the third case the offspring might have the complete character of one of the parents, and none of the other. He thought it probable that the difference between these three kinds of inheritance was only one of degree, but it was convenient to speak of them separately.

Galton and Pearson had given a mathematical expression for calculating the effects of blended inheritance. Knowledge of '*particulate* or mosaic inheritance, and of alternative inheritance is, however, rudimentary.'

He remarked that Mendel did 'not waste time in discussing the question whether all his races belong to one species' or not. He merely described a cross of any two of his races as a 'hybrid'.

Mendel's first general result could be stated as follows: 'If peas of two races be crossed, the hybrid offspring will exhibit only the dominant characters of the parents; and it will exhibit these without (or almost without) alteration, the recessive characters being altogether absent; or present in so slight a degree that they escape notice.'

Weldon would call this the law of Dominance and it at once explained the terms dominant and recessive.

Mendel's second result, rewritten by Weldon, was that 'if the hybrids' generation, produced by crossing the races of peas which differ in certain characters, be allowed to fertilize themselves, all possible combinations of the ancestral race-characters will appear in the second generation with equal frequency, and these combinations will obey the Law of Dominance, so that characters intermediate between those of the ancestral type will not occur.' Weldon would call this the Law of Segregation.

He then applied the new statistics of deviation and probable error to Mendel's figures of the distribution of characters in the offspring of his crosses. He found that his results accorded 'remarkably well with Mendel's.' In another of Mendel's experiments, the statistical analysis showed that his statement was 'admirably in accord with his experiments.'

He commented that 'Mendel's results are based upon work extending over eight years. The remarkable results obtained are well worth even the great amount of labour they must have cost, and the question at once arises, how far the laws deduced from them are of a general application. It is almost a matter of common knowledge that they do not hold for all characters, even in Peas, and Mendel does not suggest that they do. At the same time, I see no escape from the conclusion that they do not hold universally for the characters of Peas which Mendel so carefully describes. In trying to summarize the evidence on which my opinion rests, I have no wish to belittle the importance of Mendel's achievement. I wish simply to call attention to a series of facts which seem to me to suggest fruitful lines of enquiry.'

Many of the inconsistencies between the results obtained by various breeders were due to the 'neglect of ancestry, the tendency to regard offspring as resembling their parents rather than their race.'

Weldon concluded his examination with the judgment: 'The fundamental mistake which vitiates all work on Mendel's method is the neglect of ancestry and the attempt to regard the whole effect on offspring produced by a particular parent, as due to the existence in the parent of particular structural characters; while the contradictory results obtained by those who have observed the offspring of parents apparently identical in certain characters show clearly enough that not only the parents themselves but their race, that is their ancestry, must be taken into account before the result of pairing them can be predicted.'

Bateson was very angry after reading this paper. He hurriedly wrote a book of 208 pages, on *Mendel's Principles of Heredity*, which was published within a few months, in 1902. Exactly one half of the book was devoted to a reply to Weldon's critique.

Bateson wrote that it had become clear to him, as to many others, that Mendel's work was destined to play a part in the study of evolution comparable only with that of Darwin. 'After the weary halt of forty years we have at last begun to march.'

He recognized that in the post-Darwin period one notable advance had been made by Galton and Pearson, with their law of Ancestral Heredity. But there were great groups of facts to which it did not apply. This indicated that these facts had a different physiological basis, and this is just what Mendel's work had enabled us to conceive.

Weldon's restatement had left out perhaps the most significant of Mendel's experiments. 'The perfect simplicity and precision of Mendel's own account' was destroyed. He had not mentioned Mendel's conception of the hybrid as a distinct entity with characters proper to itself, which was 'the most novel thing in the whole paper'.

Weldon had suggested that Mendel's work was on an insecure basis because of his neglect of the Law of Ancestral Heredity. He nowhere directly questions Mendel's facts, and 'even expresses a mild enthusiasm.' But there was no mistaking the import of the paper. It was to produce the impression that the importance of Mendel's work had been greatly exaggerated, and those who accepted the Law of Ancestral Heredity need not feel disturbed.

Weldon's title of his paper appeared to restrict Mendel's discovery to peas. But Mendel himself had described his experiments as on Plant-Hybridization. He was aware that he had discovered a law that applied far beyond peas. His aim was to discover the law of inheritance in hybrids: he chose peas only as means to the end.

Weldon had introduced 'alternative inheritance' into his restatement of Mendel's results, in order to try to make them fit in with Galton and Pearson's law of Ancestral Inheritance, which based the calculation of the probable structure of each descendant on the structures of each individual ancestor.

Pearson had claimed 'if Darwinian evolution be natural selection combined with *heredity*', then the Law of Ancestral Heredity 'must prove almost as epoch-making as the law of gravitation.'

Already, within two years, Pearson had become more cautious in claims for the law. Indeed, 'the law of Ancestral Heredity after the glorious launch in 1898 has been home for a complete refit. The top-hamper is cut down and the vessel altogether more manageable; indeed she looks trimmed for all weathers.'

There is no more comparison with the law of gravitation. Tycho Brahe and Kepler, not to mention Newton, have receded into the distance. These high hopes had in fact been disposed of long ago by the notable experimental breeders whose works had been ignored by the prospective Newtons of biology.

Weldon had stated that Mendel had discovered a 'Law of Dominance'. But Mendel never stated any such law. Weldon had made an 'eager gathering in an unfamiliar literature' and had made a 'scrutiny of samples' in order to try to find discrepancies between Mendel's results and those of other observers. He did not refer to any of his own breeding experiments, and presumably had not made any. If he had had actual experience of breeding peas, he would have been more cautious in his criticisms. Though it might seem to be incredible, he appeared not to be aware that seed-coats in pure strains of peas are a maternal character.

He did not appear to distinguish between transmission of character and variation. Mendel was concerned with the former, not the latter. He was perfectly aware that a hybrid might show something seen before in either parent, or it might show something intermediate or something new.

Weldon had offered the sententious opinion that these things could not be understood without the Law of Ancestry. 'This is the idle gloss of the scribe, which if we erase it not thoroughly, may pass into the text.'

Mendel did not deny that characters might be other than sharply defined. He had carefully chosen material in which they were, and had rejected material in which they were not. It was owing to this choice that he had secured clear results and hence been able to make his great discoveries.

Weldon had made a wide search for discrepancies, but in the spirit of collecting missiles; he gathered them as 'smooth stones from the brook'. He had in particular, examined the papers of Knight and Laxton, noted pea-breeders, and had drawn attention to some interesting new points with regard to these, and to other matters. He had concluded from Knight and Laxton that certain of their strains of peas showed departures from what he himself called 'Mendel's laws of dominance and segregation'. But it was clear that these peas contained from the beginning characters which were not sufficiently sharply defined to follow the Mendelian rules in a simple way. Other factors had to be taken into consideration, such as stability.

Weldon had quoted a long passage from Laxton, published in 1866, on the inheritance of colour in peas. He had written that 'the fourth generation produces numerous and wider variations, the seed often reverting partly to the colour and characters of its ancestors of the first generation, partly partaking of the various intermediate colours and characteristics, and partly sporting quite away from any of its ancestry.'

There, with the word 'ancestry', said Bateson, Weldon's quotation ceased. It was unfortunate that he had not included the next sentence, which read: 'These sports appear to have become fixed and permanent in the next and succeeding generations; and the tendency to revert and sport thenceforth seems to become checked if not absolutely stopped.'

Bateson suggested that if Weldon had appreciated this passage, his article on Mendel might never have been written.

He concluded that Weldon's statement that he simply desired to suggest fruitful lines of enquiry and stimulate interest in Mendel's work was, to borrow a phrase of Horace Walpole's, 'about as likely' as 'to light a fire with a wet dish-clout.'

But the truth was that a continent of discovery had opened before us, and 'in ten years' time' we should be looking back to the present as 'days of our captivity'. The breeder will be 'second only to the chemist in resource and foresight'.

Weldon retaliated in a second article in *Biometrika* in 1903. Bateson sent a reply to *Nature*, whose editor informed him that they were 'not prepared to continue the discussion on Mendel's Principles', and returned his papers.

Bateson could not send his reply to *Biometrika*, because it had become closed to him on account of a paper he had published in the *Proceedings of the Royal Society*, criticizing a memoir by Pearson on 'Homotyposis'. Pearson replied to Bateson's paper in *Biometrika*, but when Bateson sent a rejoinder, *Biometrika* refused to publish it. So Bateson had his reply privately printed in the same type and format as *Biometrika* and circulated it to his correspondents.

Bateson was elected President of the Zoological Section of the British Association at its Cambridge meeting in 1904. He realized that it would be a great opportunity to present the case for Mendelian heredity, and he spared no pains in preparing his presidential address. Punnett in his *Early History of Genetics in England* has given a valuable recollection of the events. Bateson worked on the address for weeks, emerging from his table in the garden to discuss difficult points thoroughly. He carefully planned with his collaborators the topics on which they would speak, so that the whole school would deliver a balanced account of their work.

The meeting opened on a morning in the large lecture hall of the new Adam Sedgwick Geology Building. Bateson said that it was customary for presidents to say that all was well with their science. He could not do that. An eminent foreign professor had recently told him there were not half-a-dozen persons in his country who were interested in Mendel's discoveries. They appeared to be of more interest to his students than to his colleagues. Yet, said Bateson, if discoveries of comparable importance had been made in physics or chemistry, they would at once have been repeated and extended everywhere. He thought that a cause of this situation was the enormous success that had attended Darwin's work. It had given a first general solution of evolution, heredity, variation, and natural selection, and had opened up many new lines of highly important research. In pursuing these, the actual kind of work which Darwin himself had done, his problems and methods, was respectfully put on one side as a finished masterpiece.

The general idea of variation concealed a prodigious confusion of particular ideas and facts. The struggle of the evolutionists to 'exorcise the demon of immutability' had led to the banning of the idea of a definiteness of species. The belief that they were continuously variable became a dogma.

The evidence of definiteness of species provided by the gardener, the breeder, and the fancier was neglected, and even treated with contempt. This wide field from which Darwin had drawn his facts had lain untouched for forty years.

Meanwhile, the biological philosophers speculated how natural selection might impress bodily symmetry on formless protoplasm—'that monstrous figure of the mind'. The professional zoologists regarded the breeders as unscientific, and yet they themselves substituted evolution in the abstract for evolution in the concrete, which was what the breeders worked with.

As early as 1719 Marchant had suggested that new species might arise by the process which de Vries had named mutation. The decision as to whether a mutation should survive must rest with natural selection. But the business of selection was to *select*, not to *create*.

'The creature is beheld to be very good after, not before its creation.' Prophecy had little part in the breeder's work. He noticed a desirable novelty and then tried to work up a stock of it.

The popular belief that evolution is 'effected by mass-selection of impalpable differences' arose from imperfect analysis. The statistical indications of continuity were derived from the 'fluctuations to which all normal populations are subject'.

The 'gross statistical method' was a 'misleading instrument' when applied to this intricate material; 'the imposing Correlation Table into which the biometrical Procrustes fits his arrays of unanalysed data is no substitute for the common sieve of a trained judgment.'

Bateson then explained that winning canaries in the 'Clear Yellow Norwich' class were not got by breeding from previous winners. The Brown Leghorn cock and hen champions came from different strains.

The observer contemplating the magnificent ranges of fine shades of modern sweet peas might imagine that the extreme types had resulted from selection from a more or less continuous series of previous intermediates. 'That is the reverse of the truth. The multiple series of fine shades have been produced by most laborious breeding from a few extreme shades.'

The prospects from the application of Mendelian methods were enormous. 'Breeding is the greatest industry to which science has never yet been applied.'

Bateson concluded his address with a few general remarks. Could the science of heredity help to produce 'healthier, wiser, or more worthy' people? A competent breeder endowed with full powers could already rid the human race of certain diseases. Science already suggests a way, but 'no one, however, proposes to take it'. So long as 'superstition remains the guide of nations', there was 'nothing to hope or to fear' from this science.

The student of heredity must answer to the philanthropist that he had no evidence that transmission of characters could be affected. From the moment of fertilization, no exhortation or teaching hygiene can pick out from the cell the particles of evil, or put in one particle of good. From seeds in the same pod of sweet peas may come some plants capable of growing five feet high, while others are dwarfs that grow along the ground even when provided with sticks. Education and sanitation merely provide opportunity to qualities already there.

After he had delivered his address, Miss Saunders spoke on Stocks, C. C. Hurst on poultry, and Punnett on fowls' combs.

The discussion was postponed until the afternoon. The audience retired for lunch. When the meeting was resumed, the hall was packed, people sitting on the window-sills.

A. D. Darbishire, a protégé of Weldon, described breeding experiments with mice, the results of which, he thought, could be explained partly by Mendel's laws and partly by the Law of Ancestral Heredity.

Weldon made a speech of 'impassioned eloquence'. Beads of perspiration dropped from his face. Punnett records that he heard an

Oxford man sitting in front of him comment: 'Clever beggar that—he hasn't got to stop and think.' He recited a number of results of breeding which appeared to be best described by the Law of Ancestral Heredity.

Karl Pearson proposed there should be a three years' truce to controversy, and allow the experiments to give the final decision. The discussion was concluded by the Reverend T. R. Stebbing, a great authority on crabs and crustacea, a little, mild man. He began to speak in a quiet voice, as if he were about to counsel compromise. 'We have heard what Professor Pearson has suggested,' he began diffidently, and paused. The hearts of the Batesonians sank. Then with a sudden rise of voice he continued: 'But what I say is let them fight it out.'

Punnett said that Bateson's generalship had won all along the line.

But Alfred Russell Wallace, E. B. Poulton and J. Arthur Thomson continued to support Weldon, and *Nature* continued to refuse to publish the communications of the Batesonians.

Punnett has described how Bateson was persuaded by C. C. Hurst to recommend a paper on the genetics of horses to the Royal Society. Hurst had collected data from the Stud Book which suggested that chestnut was recessive to bay in the coats of horses. Weldon had heard that the paper would be read, so he diligently examined the data in the Stud Book. After Hurst had spoken, he pointed out that there were recorded cases in which chestnuts mated together had given bay or brown offspring. Bateson was exceedingly angry when he saw, as he thought, that he had been let down by Hurst, and his fury was increased when 'Hurst blandly assured Weldon that he was mistaken, and that these alleged exceptions were mere errors of entry' in the Stud Book.

Bateson withdrew the paper, and went off with Miss Durham and Punnett to dinner in a thundering temper.

The Stud Book was thoroughly studied during the ensuing Christmas Vacation. The most serious problem was presented by a horse named 'Ben Battle'. It had been registered as a chestnut, but had sired bays from chestnut mares. One day a telegram was received from Hurst: 'Ben Battle never *ran* as a chestnut.' He had looked at the files of the betting men's paper *Form at a Glance*, a journal little known in learned circles, and found this evidence that the entry in the Stud Book was wrong. Thus, after all, Hurst had unconsciously been right. Bateson recommunicated his paper to the Royal Society and it was published.

The debate in the Zoology Section at the Cambridge meeting of the British Association in 1904 secured a place for Mendelian principles in Britain, but not an easy one. Bateson and his followers had many difficult years before them.

In 1906, however, Weldon died, at the age of forty-six. He had insisted on leaving his bed before recovering from influenza, and developed pneumonia. His widow survived until 1936. She had shared her husband's interests in mathematical statistics, and she became a munificent donor of paintings to the Ashmolean Museum in Oxford. These included works by Blake, Claude, Veronese, Corot and Sisley, and a gallery of the Museum was named after her. She bequeathed the greater part of her residual fortune of £68,453 for the endowment of a Chair of Biometry at University College, London. J. B. S. Haldane was appointed the first occupant in 1937.

When Bateson heard in 1906 of Weldon's death, he wrote to his wife that he had been more intimate with him than with anyone except herself, yet 'if any man ever set himself to destroy another man's work, that did he do to me.'

Yet to Weldon he owed 'the chief awakening' of his life. Before he had received his inspiration 'failure and uselessness' had been his 'accepted destiny'. 'Such a debt is perhaps the greatest that one man can feel towards another; nor have I been backward in owning it. But this is the personal, private obligation of my own soul.'

Weldon had begun to embitter himself towards Bateson about the year 1890, and had used all his influence to hinder him.

IV. THE FIGHT FOR GENETICS

For long years Bateson was unable to obtain a proper appointment in Cambridge. As Punnett puts it, he 'kept himself by looking after the kitchen at John's.'

He married Beatrice Durham, a daughter of a senior surgeon of Guy's Hospital, in 1896. She has written that then 'she began to learn what life may be.' She found her husband to be a man of untiring energy, deliberate and critical, accomplishing much work, yet having time for social pleasures. He was impatient, an excellent companion, very absent-minded, and careless of clothes.

In 1897 Bateson received a small grant from the Royal Society's Evolution Committee, which enabled him to hire an allotment, on which he began poultry and plant breeding.

His first son was born in 1898, and he and his wife looked for a larger house, with a garden in which further plant and animal breeding could be pursued. They found Merton House at Grantchester, which had a good paddock and outhouses and a very good garden and orchard. They moved there in 1899 after the birth of their second son. The gardens of Merton House became a world-centre of genetic research, attracting scientific visitors from every land.

Their early years were very happy, though strenuous. Bateson, with his wife's help, began breeding poultry, peas, poppies, and other plants. They had one gardener, and shared in the manual labour. They sorted, sowed, dug, hoed, weeded, staked, and watered the plants. They ran incubators of 100 egg-power, cleaning them out, trimming wicks, and filling with paraffin. They fed chicks and larvae with their own hands. They opened thousands of half-hatched eggs, a messy job. Mrs. Bateson kept the detailed records of all the experiments. Correspondence was dealt with at night.

When the experiments had encroached too far on his kitchen garden, Bateson secured an allotment on the University farm, which was four miles away. He cycled there, towing his wife in a trailer, with log-book and microscope, to record the progress of their breeding experiments. He was assisted by voluntary experimenters, the first of whom had been Miss E. R. Saunders, who worked especially on stocks, Miss Sollas bred guinea-pigs, Miss H. Killby goats, Miss M. Wheldale *Antirrhinum*, and Miss F. M. Durham (Bateson's sister-in-law and a physiology lecturer at Newnham) mice. R. Staples-Browne bred pigeons, C. C. Hurst poultry and rabbits, L. Doncaster moths, and Miss Darwin *Oxalis*.

In 1899 a suggestion was made to the Royal Society that Darwin's old home, Down House, should be converted into an institute for plant and animal breeding. Bateson did not favour the proposal, because of the isolation of the house from other scientific institutions and libraries. The study of evolution should not, he said, be detached from science as a whole.

He sought persistently for years to secure endowments. He found it even more difficult after the recognition of Mendel's work than before. He had in his files a large envelope marked 'Begging Letters', which he had written at various times trying to raise funds.

In 1900 Andrew Carnegie, following a suggestion of Herbert Spencer, wrote to F. Darwin, asking whether it would be worth while endowing experimental research on the inheritance of acquired characters. His enquiry was referred to Bateson, who bluntly replied that it would not. But he ventured to suggest that he might endow plant and animal breeding, etc. Bateson heard no more. He had not handled the opportunity with diplomatic skill.

The magnitude of his financial arrangements was strikingly described in 1903, when he wrote to Punnett inviting him to join in the breeding experiments. He had just received the promise of £150 a year from Mrs. Herringham for the assistance of his researches.

He said: 'My expenses this year were about £90, and it was a cheap year. I made about £30 by sales and produce consumed, and I had a Royal Society grant of £36, so I only lost about £24.

'For next year in addition to the £150 I have a British Association grant of £38. As I shall not be able to spend anything substantial on the work in future, I propose to use £70 of the £150 for expenses. This leaves £80 which would be offered to a partner.'

So Bateson offered Punnett the £80 a year. He told him he would have to share in the menial work, but he would raise no difficulties on questions of priority on future discoveries. These might have arisen with a young student, but Punnett was already a senior man, a fellow of a college with an independent reputation. Punnett accepted the invitation to collaborate, but refused the payment.

Bateson was appointed deputy to Professor Newton in 1899. This gave him his first opportunity for regular university teaching. From that date, he recruited colleagues from among his pupils. His first group of collaborators were interested amateurs, the second came to him as a result of his university lectures. It included R. H. Lock, L. Doncaster, R. P. Gregory, Miss Wheldale, Miss Sollas and Miss Marryat.

Both groups of collaborators in the main obtained their livelihood by other means. Bateson did not have a research school in the modern sense, with research workers enjoying fellowships.

In 1902 he received an invitation to a post at Brooklyn, in the United States. He and his wife spent days in anxious consideration as to whether he should accept it, but he finally decided against it, as he hoped that ultimately a research station for genetics would be founded in Britain.

Bateson enjoyed much more popular success in the United States than in Britain. All of his visits there were very successful. His forceful, hard and clear style, without nuance, and his individualist ideology appealed to the Americans.

In 1903 the Wisley garden was bequeathed to the Royal Horticultural Society, and Bateson was invited to outline a plan for converting it into a plant research institute. He did so, but the Society did not adopt the plan, as they preferred to spend their available money on their new Exhibition Hall in Vincent Square, London.

He was sounded as to whether he would accept the Secretaryship of the Zoological Society. He was unable to allow his name to go forward, because he feared that the post would absorb most of his energy in administrative work.

In Cambridge he still had no adequate academic position. In 1903 the large Quick Bequest was made to the University for the study of plant and animal biology. Bateson submitted in writing a proposal that part of the bequest should be used to endow a small department for plant and animal breeding. His letter was returned to him.

Two years later, the Vice-Chancellor of the University invited him to submit a plan for a 'Quick Institute for the Study of Heredity and Variation'. In a letter to Adam Sedgwick, dated April 18th, 1905, concerning this scheme, Bateson said that no word in common use sufficed to describe the study of heredity, and that one was badly needed: 'If it were desirable to coin one, GENETICS might do.' The origin of the name of a branch of science is rarely so precisely and definitely known.

In spite of the Vice-Chancellor's enquiry, the Quick Bequest was not devoted to genetics, but to protozoology.

Bateson was bitterly disappointed. He wrote to a friend that he must bear the responsibility of failing to bring about 'the revolution in biological study that sooner or later must certainly come to pass'. He had failed with his contemporaries, but hoped to be more successful with posterity.

In 1907 Newton died, and Bateson was a candidate for the vacant chair of Zoology. He had been Newton's deputy since 1899. Newton had been an eminent ornithologist, whose lectures attracted only small classes. The teaching of morphology, which was attracting very many more students, had for years been done by Adam Sedgwick, the Reader in Zoology.

Sedgwick was elected to the chair, and Bateson himself agreed that it was a correct appointment.

In the meanwhile, Bateson was invited to give the course of Silliman Lectures at Yale University for 1907, one of the most distinguished lecturing appointments in the world.

He left for the United States in the summer of 1907, and received a magnificent reception not only at Yale but also at the International Zoological Congress at Boston. The subjects of heredity, cytology, and experimental biology dominated the Congress.

He wrote home that he had 'never before felt what an exhilarating thing it is to speak to a really large and enthusiastic audience. It is dreadfully intoxicating.'

On his return home, after this international triumph, the University of Cambridge offered him a Readership in Zoology with a stipend of £100 a year. He felt deeply humiliated, and at first was inclined to refuse the appointment, but was presently persuaded to accept it. He wished to leave Cambridge.

In 1908 the University decided to found a chair of Biology, on the occasion of the centenary of Darwin's birth in 1809, at a minimum salary of £500 a year. The chair was offered to Bateson and accepted. So at last, at the age of forty-seven, he had an appropriate academic position. But he felt compelled to resign from his college stewardship, so he was little better off financially.

Then, in 1910, the trustees of John Innes decided to use the funds for the creation of a Horticultural Institution for research on plants. The income from the funds was about £18,000 a year, so there were magnificent possibilities for development.

Bateson was invited to become the Director of the new Institution, plan its construction, and create its gardens and programme of research. He accepted the invitation, and resigned his chair of Biology.

In 1912 the Cambridge chair of Biology was changed into one for Genetics, and Bateson was invited to return to Cambridge. He felt, however, that his first duty was with the John Innes Horticultural Institution. He regarded the Cambridge invitation as a great honour, but the Institution provided a great opportunity. He remained at the Institution, which was established at Merton, in the suburbs of London.

When he arrived at Merton, there was bare land and no laboratories. He started on the laborious task of creating the gardens for experimental plant breeding, which cannot be done very quickly.

Then, in 1913, he had a slight heart attack, and knew that he would not be able to be as physically active in the future as he had been in the past.

His problems were not only practical. Many were of the opinion that John Innes had not intended his fortune to be devoted to horticultural research, but to charitable objects, and they resented Bateson's programme.

In 1914 the First World War started. He was interrupted in his work by loss and shortage of staff. The breaking of international scientific relations affected him severely, for he held the old liberal view that science was an international intellectual activity which was above war and politics. The world hegemony of middle-class culture was not yet shattered.

His first son was killed in 1918 only fourteen days before the Armistice, and in 1919 he had to resume, in bitter spirits, the development of the Institution. His second son committed suicide in 1922. He was now sixty-one years old, and deeply shaken.

He visited the Soviet Union in 1925. In the same year he told his Council that he was approaching sixty-five, and was due to retire. He wrote that he had failed to find a successor. All those talented men who formerly 'would have been zoologists or botanists' are today 'biochemists, almost every one'. This remark suggests that Bateson had in mind the influence of Hopkins, the acquaintance of his youth, who had had such great success in creating a new scientific school.

Why had Bateson failed, whereas Hopkins succeeded? He lacked Hopkins' sociological insight, and had an excessively individualist psychology. During his most active period he was unable to secure research fellowships for his pupils. He depended on voluntary helpers personally attracted to him. They were a devoted group but not a research team. When he went to Merton at fifty it was probably too late for him to develop the team system. To add to his other misfortunes he was hindered by the unsuitability of the site, which had poor soil and was being swiftly encircled by suburban houses and smoke. Several of his ablest collaborators, L. Doncaster, R. H. Lock and R. P. Gregory died young, and finally, he was hindered by his basic scientific ideas.

He did not accept the gene theory of chromosome inheritance until 1921, when a visit to Morgan's laboratory in New York and a thorough inspection and study of his material convinced him that it contained important truths. But, as he showed in his last major paper, on *Segregation*, published in 1926, he was cautious and critical of the chromosome theory to the end.

Punnett has said that Bateson's belief in Boveri's views on the individuality of the chromosomes deflected his attention from considering the possibilities of their structure.

Bateson was unable to radiate from Merton an inspiration for the development of genetics in Britain. He had more effect on visiting students from abroad. One of these was N. I. Vavilov, who worked with him in 1913, and subsequently organized Batesonian genetic research in the Soviet Union on an immense scale. After thirty years of experience, many persons in Soviet agricultural, scientific, and governmental realms concluded that this programme had been mistaken, and that there were fundamental errors in the scientific principles on which it had been based.

In this connection, Bateson's philosophical and scientific preconceptions have much significance. He launched into experimental breeding for philosophical reasons, because he had concluded that this was the key to the discovery of the process of evolution. He was primarily a philosopher of natural history. He pursued his breeding researches with the utmost enthusiasm and determination, and, with his collaborators, obtained many important results. But these are impressive in quantity and variety, rather than for comprehensively fundamental importance in any particular case. He was not a great natural experimenter, like Rutherford or Hopkins.

The most important of his experimental discoveries was that of linkage of hereditary characters. In their Report to the Evolution Committee of the Royal Society in 1906, Bateson, Punnett and Miss Saunders described the phenomenon. If a sweet pea with purple flowers and long pollen grains was crossed with one with red flowers and round pollen grains, then in the offspring these respective pairs of characters were found together from generation to generation. They ascribed the effect to pairs of hereditary factors being associated in the gametes, or germ-cell, of the pea.

Subsequently, linkage was explained as being due to the two hereditary factors being in the same chromosome.

They mentioned also the case of the Cupid Sweet Pea, in which the entirely different characters of procumbent habit, or growing along the ground, and dwarf stature were always associated, whereas in culinary peas this was not so.

They showed, too, in the same report, that a particular colour was not necessarily due to one unit, but might be due to two acting together. When white sweet peas with long pollen grains were crossed with white with round grains, purple flowers were obtained. Each white flower had a single unit, which was unable to produce colour by itself. But when the single unit from each of the two white flowers came together, the pair of units produced the purple colour.

In 1908 Bateson and Punnett described a cross between a White

Silkie fowl and a White Dorking. All the offspring were coloured. They interpreted the case in terms of the presence or absence of two factors, but the appearance of colour was due to the interaction of two factors neither of which could produce colour by itself.

In 1906 his talented collaborator L. Doncaster, who died prematurely in 1920 at the age of forty-two, produced the first clear experimental breeding analysis of sex-linked inheritance. He had learned from the Rev. G. H. Raynor, a fancier of the Currant Moth *Abraxas*, that the *lacticolor* variety, which bears much less black coloration than the *grossulariata*, was known only in the female form. Doncaster saw that the association of coloration and sex in this moth might be analysed by experimental breeding. Hitherto, the best-known examples of characters associated with sex were such as colour-blindness, which occurred in man, and therefore were not suitable for analysis by experimental breeding.

Doncaster arranged for Raynor to make appropriate breeding experiments with the moth, and analysed the results. They produced the missing *lacticolor* male, and Doncaster elucidated the scheme of linkage of the coloration character with sex. He showed that the *grossulariata* female produced two kinds of gametes, those which bear this character and may produce males and those which do not and may produce females.

Punnett worked out sex-linkage in poultry. If a black cock is crossed with a barred hen, black chickens are produced, which remain apparently indistinguishable for some time. These grow up into barred cocks and black hens. The black male chickens have a whitish spot on the top of their head, by which they can be identified. Thus a commercial poultry breeder can pick out all the males from his day-old chicks, and save much of his hutching space and feeding.

Bateson's collaborator C. C. Hurst produced the first definite proof of Mendelian inheritance in man. He showed that eye-colour was inherited according to the Mendelian rules.

Bateson himself produced two improved crop plants by Mendelian breeding. Ordinary sugar-beet is liable to 'bolt', or put up flowering shoots, in its first year. This occurs in about 5 per cent. of the crop. He planted beets very early, so that about half of them bolted. Then he selected beets for further cultivation from those that did not bolt. After several years of selection, he obtained a strain that did not bolt under ordinary conditions. In another series of experiments, he fixed a giant strain of flax containing fibres of the maximum length.

Under Bateson's influence, R. Biffen began the researches on wheat-breeding which led to his discovery of wheats resistant to 'rust' disease, a demonstration of the possibility of the general principle of breeding strains of organisms resistant to particular diseases, a result of great biological, medical, and economic significance.

Bateson investigated the heredity of chimaeras produced by graft-hybrids in *Pelargonium* and of 'rogues' in culinary peas. He was interested in the chemistry of the inheritance of colour. His collaborator Miss Wheldale (Mrs. Onslow) started the researches on the inheritance of colour in flowers, which led to the remarkable combination of the genetics of anthocyanins with their chemistry, as worked out by Sir R. Robinson and his school.

He studied the inheritance of polydactylism, or abnormal numbers of toes, in cattle; of alkoptonuria, a rare disease of the human urine; of red-legged partridges; and albinism in mice and men.

The magnitude of Bateson's experimental work is illustrated by the researches carried out with R. P. Gregory and D. de Winton on the Chinese primrose *Primula sinensis.* They bred 200,000 plants and examined them for mutations, and found that no mutation appeared more than once, and on the average one mutation appeared in every 20,000 plants.

He made innovations in terminology, the most important of which were the term *genetics* itself in 1905, and *homozygous* and *allelomorph* in 1902. As late as 1925 he introduced *anisogeny,* to describe the phenomenon of the appearance of different characters in reciprocal crosses, or, in his own words, the contributions of the two sexes are unequal.

Bateson was aware of the limitations of his experimental work. In his paper on 'Segregation', published in 1926, he remarked that there had been a long spell of dullness in the early years of Mendelian research. He attributed it to a lack of curiosity and 'a too facile acceptance of impermanent doctrines'.

He criticized the conception of 'unit-characters'. For him they had always been merely ingredients in the collective stock that makes up the organism.

The work of the Morgan school had shown that the transferable characters must in some way be associated with the chromosomes. But 'we have no idea what the nature of that association might be.' It would be necessary to provide the chromosome theory with 'much elasticity and emendation', in order to bring it into accord with the facts of plant genetics.

He devoted much of his last paper to the discussion of the genetics of chimaeras, graft-hybrids, and rogues. He believed that the analysis of these complicated phenomena would broaden the conceptions of chromosome genetics. He suggested that the study of bud-sporting in plants would throw more light on the nature of segregation, and he commented on the vague and multiple meanings given to the term 'mutation'.

He concluded that no disservice would be done to genetical science if acceptance of the many extensions and implications of the chromosome theory were postponed. It was necessary to distinguish fact from hypothesis. It was true that in certain animals the inheritance of some characters was associated with particular chromosomes. This was a very extraordinary advance. But the hope that it might 'be safely extended into a comprehensive theory of heredity seems to me ill-founded.' He believed that a wide survey of the genetics of plants would scarcely confirm it.

During the last fifteen years, many genetical facts had been discovered which could be explained by the simple multiplication of subordinate hypotheses. He considered that it became increasingly certain that 'the problem of heredity and variation is intimately connected with that of somatic differentiation', and that research along that line offered the best prospect of success. 'Pending that analysis, the chromosome theory, though providing much that is certainly true and of immense value, has fallen short of the essential discovery.'

This was the last scientific judgment of Bateson. The great naturalist, intimately acquainted with a vast range of living creatures, could not believe that the chromosome theory of inheritance contained the essence of the principle of heredity.

The facts of inheritance, as observed in the vast range of nature, rested on the operation of a much deeper and wider principle, yet to be discovered.

He was a naturalist who wished to draw conclusions from looking at the whole of Nature, and was not a biological specialist who tried to derive general principles from intensive and narrow investigation of one particular organism.

His first collaborators were all persons who secured their living by other means than the study of genetics. Punnett has written that Bateson was a most delightful man to work with. He was dominant, but not domineering. He has described how Bateson once lost his temper with him for introducing the breeding of silky fowls without his knowledge. The experiments had an important bearing on the development of the doctrine of sex-linkage.

Punnett has described, too, how Bateson was revolted by the sight of the 'cretin' sweet pea. 'Do throw away that monstrous thing', he said. But the crossing of this pea helped them to discover the principle of linkage.

Bateson inspired independent and mature workers, but he did not create a research team. This was partly due to his early lack of contact with students, as he did not then have a teaching position in the University, and also to the absence of adequate research grants. He could help those who could help themselves, but he was not a teacher for the immature.

He was notably unsuccessful in persuading the English authorities in universities and government of the importance of genetics. This was

7. William Bateson: drawing by W. Arnold Forster

face p. 288

not due to lack of contact with practical men. He was a most assiduous and well-known attendant at shows and visitor to breeding stations. His personal relations with plant and animal breeders were sympathetic and extensive. He took infinite pains in organizing shows of poultry, canaries, and flowers, and bringing scientists, fanciers, and commercial men together. But he had little influence on their attitude towards the application of science to plant and animal breeding.

His difficulty in securing funds for research was due in part to his social outlook. He was of comfortable middle-class Liberal origin, descended from well-to-do North of England trading families. He was not well off himself, but he had the ideology of a class which regarded automatic provision of living and education for its members as one of their natural rights. He drew the sharpest distinction between science, and trade and industry, regarding science, and even more art, as highly spiritual activities. He looked down on trade and industry with an admonitory and patronizing eye. In 1905 he wrote that 'we recognize in the feudal system a nearer approach to the natural plan' for the organization of human society. Owing to his origin, he was out of sympathy with the landed aristocracy of his day, in spite of his admiration for the feudalism of the past. He was passionately interested in plants and animals, but not as the materials of agriculture. He was a naturalist, not an agriculturalist.

In 1904, in his presidential address to the zoological section of the British Association, he posed the question whether the new knowledge of heredity is of any use. 'That is a question,' he replied, 'with which we here have, fortunately, no direct concern. Our business in life is to find things out, and we do not look beyond.' Mendel had been 'untroubled by any itch to make potatoes larger or bread cheaper.' He had set himself the task, in the quiet of his cloister garden, of finding the laws of hybridity, and 'so struck a mine of truth, inexhaustible in brilliance and profit.'

It was true that 'breeding is the greatest industry to which science has never yet been applied' and that the prospects were of the most striking character. But in Bateson's philosophy there was no necessary connection.

The old land-owning classes contained elements that loved the land and the processes of the cultivation of plants and animals. These elements in Russia, for example, under the leadership of Dokuchaiev and others, founded modern soil science.

The modern landowners in the Soviet Union, the Soviet people as a whole, have produced, under the leadership of Lysenko, a new biology, out of their love of the natural and cultural life of their country and their desire for its improvement and permanence.

Bateson, in spite of his great natural gifts, was prevented by his social ideology and the social situation in England from finding a proper combination of science and agriculture. As a consequence, his science was stunted, and the benefit of genetics to British agriculture was limited and developed slowly.

V. THE CAUSE OF EVOLUTION UNFOUND

In 1904 Bateson said that he was convinced that the investigation of heredity by experimental methods offered the only way by which progress could be made in investigating the fundamental problems of evolution.

Professional zoologists regarded breeders and fanciers as unscientific persons whose concerns were beneath notice. They spurned the concrete facts produced by the breeders in favour of abstract speculations on the mechanism of evolution, of how, for example, 'natural selection could impose bodily symmetry on formless protoplasm.' But the function of selection is to *select*, not to create. The breeding-pen is to the zoologist what the test-tube is to the chemist, and Bateson suspected that 'variation is a novel cell division.'

In 1906 he remarked that the 'application of Mendelian rules to mankind has not made the progress that was to have been expected.' In 1909 he reviewed the state of the science of heredity and variation. He quoted Darwin's statement that he was 'convinced that natural selection has been the main but not the exclusive means of modification', and he expressed the opinion that Darwin gained his first illumination on the process of evolution not from Malthus, but from the study of his own first notebook on the transmutation of species.

In the thirty years since Darwin's death, Darwinian genetics had 'suffered attrition', but in every other direction biology had enjoyed a golden age.

He considered that Weismann's demands for evidence for the inheritance of the effects of use or disuse had given a negative result, but had served a useful purpose. His 'elaborate speculations on the genetic meaning of cytological appearances' stimulated the minute examination of the division processes in germ cells. Bateson pointed out that Samuel Butler had put forward similar ideas in 1878.

But 'what we know of the behaviour and variability of the chromosomes seems in my opinion quite incompatible with the belief that they alone govern form, and are the sole agents responsible for heredity.'

He regarded Mendel's demonstration of the phenomenon of segregation as of the 'utmost novelty'. It implied that the characters of organisms depended on 'units or factors capable of combination in various ways', and this suggested that the order of heredity arose from the physical and chemical properties of things.

The demonstration of segregation and the existence of stable unit characters removed the difficulty of 'swamping' of characters. Darwin had originally believed that 'sports' were an important source of variation, but later came to believe that these abnormal characters would be swamped in the inter-crossing with the very numerous normal forms. He was almost thankful that Mendel's laws had not become known to Darwin. They might have prevented him from forming the theory of evolution with the bold simplicity which was so necessary in his time.

Natural selection determines the branch along which evolution shall proceed, 'but it does not decide what novelties that branch shall bring forth.'

The greatest advance that could be foreseen would arise from the discovery of the connections between 'the geometrical phenomena of development with the chemical.' He agreed with Butler that 'the origin of variation, whatever it is, is the only true "Origin of Species".'

He admitted that the widespread existence of the phenomena of adaptation is an almost 'unsurpassable difficulty in all non-Lamarckian views of evolution.'

In his Presidential Address to the Agricultural Section of the British Association in 1911, he reviewed some of the possibilities of Mendelian research in agriculture, and remarked that 'in the mysterious properties of the living bodies of plants and animals there is an engine capable of wonders scarcely yet suspected, waiting only for the constructive engine of the human mind.'

In 1914 he was President of the British Association during its meeting in Australia. He delivered two addresses, one in Melbourne and one in Sydney, the first dealing with pure genetics and the second with its implications for sociology.

He was still of the opinion that the failure of cytology and the chromosome theory to throw light on the mechanism of heredity was complete. He thought that genetic factors could not be material particles, but phenomena of order and arrangement. He believed that systems of inheritance quite different from the Mendelian might be discovered.

The appearance of contemporary variability, of variation taking place before us, had proved to be an illusion. No case had been conclusively proved.

Some of his colleagues had spoken of genetic factors as 'permanent and indestructible'. It was true that they have a 'relative permanence', because they commonly passed through the process of segregation unchanged. If genetic factors were indestructible, their evolution might be conceived as 'an unpacking of an original complex which contained within itself the whole range of diversity which living things present.' This theory of evolution as a process of unfolding of the complex into the simple is similar to Lemaître's theory of the evolution of the universe as the disintegration of a single complex primeval atom into all of the particles and radiations of the universe.

Variation was probably not a change in material particles, but of order.

Bateson's theoretical views were expressed again in his Sidgwick Memorial Lecture in 1917. He repeated that hereditary characters were like chemical molecules, but he was still unconvinced that genes had a linear existence, as rows of particles in chromosomes.

He considered that there was no evidence at all of the existence of life apart from matter. Physical and mental qualities were decided, without distinction, by physiological determination. He quoted as confirmation that two feeble-minded parents produced feeble-minded children, and that the offspring of a 'pacing' horse and a 'trotting' mare were all 'pacers'.

He was of the opinion that, except in certain special cases, the most careful research had still failed to reveal any transmission of the effects of experience, and not even any deflexion attributable to them. 'The doctrine of transmitted experience as the origin of modification asks us to turn away from the whole course of genetic experiment.'

He quoted with approval the definition proposed by Cuvier, and adopted by Michael Foster, that living creatures are vortices of chemical and molecular change, and anticipated that the explanation of the process of heredity would be found in terms of physics and chemistry. The hereditary factors would be found to be forces rather than materials, and their heterogeneity was due to differences between forces rather than to differences in material composition.

The war of 1914–18 increased Bateson's isolation. The John Innes Horticultural Institution, founded in 1910, had hardly had time to settle down to an established pattern of research when its staff was dispersed. The rhythm of his scientific work was interrupted.

Like some other leading scientists of the day, such as Sir Oliver Lodge, who had publicly proposed that the British Association in Australia should ignore the outbreak of war, on the ground that science was above the battle, Bateson also questioned it. He believed that the war was 'almost purely a commercial matter'. He despised chauvinism and resented the disruption of international scientific relations. For him the international world of art, science, and learning was above commerce and politics, and he was genuinely amazed to find it split by national prejudices.

He wrote to Denmark in order to get German scientific journals, and remarked in one of his letters that 'British hypocrisy' is a 'real and astonishing phenomenon.'

He was horrified by the destruction of young men and the disruption of the European community of learning and wrote: 'I greatly fear that a more or less permanent state of war, for a generation at least, is the most probable outcome of it all.'

In 1916 he felt he must do some active work in connection with the war. In spite of a weak heart, which had become manifest in 1913, he went to France to lecture to the troops. After one of his popular talks on the determination of heredity, a Scottish soldier said to him: 'Sir, what you're telling us is nothing but *scientific Calvinism.*'

In 1918 the Ministry of Information invited him to write propaganda on genetics. He replied that such an activity was objectionable and incompatible with the spirit of science. So far as his own branch was concerned, he was happy to say that it was free from chauvinism. The British Government had done far more harm to science by closing museums than good could possibly be done by such propaganda. He regarded the writing of literature on national scientific achievements as 'not good scientific manners'. If he had to write on these matters, he would have to point out that while the British Government had been closing museums, the Prussian Government had, in 1916, opened a new Institute of Genetics at Dahlem in Berlin.

His bitterness was increased by the death of his son John, who was killed on active service fourteen days before the Armistice of 1918. He was invited in 1919 to be president of the new International Biological Union, but when he discovered that the German scientists were to be excluded, he succeeded in getting the project dropped.

In the same year he threatened to resign from the Linnean Society when he heard that it was proposed that they should not re-subscribe to the German scientific journals.

He expressed the opinion in 1918 that people such as himself existed only on sufferance, and to think for oneself was, in most societies, a crime. All that they could do was to make the best of the situation and find a niche where they could be left alone.

In the winter of 1921–22 he visited North America, where he was warmly received, and lectured very successfully. He gave an address on 'Evolutionary Faith and Modern Doubts' at Toronto in 1922, in which he recorded the state of his thoughts after the war of 1914–18. He opened it with an expression of joy at the reestablishment of international scientific relations, after the 'dangerous isolation' of scientists that had been caused by the war.

He related how W. K. Brooks had introduced to him in 1883 the idea that there was a special physiology of heredity capable of independent study, but that this idea had not been explored experimentally.

The first discussions of evolution started by Darwin gradually faded because no real progress was made.

The study of the shapes of organisms, morphology, was minutely investigated, but it did not provide the key. So attention was concentrated on variation and heredity. Many striking facts on the discontinuity of variation, its proceeding by jumps, were accumulated, and the dogmatic doctrine of evolution by blending inheritance, of the accumulation of small variations was disproved. Then Mendel's work was recognized. It offered an explanation of the origin of varieties and of how the phenomenon of segregation maintained their identity. The application of Mendelian analysis to more and more plants and animals gave a coherent explanation of a vast range of variations, but less and less was heard of evolution, until mention of it was virtually dropped. The new genetics had thrown no light on the problem of the origin of species.

Why was this? Bateson thought that it was because biological evolution conceived in the eighteenth and nineteenth centuries had been derived from the study of the whole organism, or zygote, as distinguished from its germ, or gametes.

Bateson now announced that after visiting T. H. Morgan's laboratory in New York, he had been converted to the chromosome theory of heredity. Biology had returned to its proper task of investigating 'the structure and properties of the concrete and visible world.' Variation was daily witnessed, but not origin of species.

In dim outline, evolution was evident enough, 'but that particular and essential bit of the theory of evolution which is concerned with the origin and nature of *species* remains utterly mysterious.' The event for which he waited was 'the production of an indubitably sterile hybrid from completely fertile parents which have arisen under critical observation from a single common origin.' J. B. S. Haldane has remarked that such an example has subsequently been found.

Julian Huxley and other authorities are not convinced that infertility on crossing is an absolute criterion of species.

The validity of natural selection as the chief directive agency of evolution also failed with respect to specific differences. It was evident that every species must have the fitness to survive, but no one could tell how the specific differences enabled one organism to survive better than another. Consequently, 'natural selection as the chief factor in the determination of species' has 'been discredited'.

Bateson said that he had frankly explained the reasons why he considered that the 'actual mode and processes of evolution' still remained unknown. 'When such confessions are made the enemies of science see their chance.' It was therefore necessary to state that faith in evolution is unshaken. 'The obscurantist has nothing to suggest which is worth a moment's attention.' The reality of evolution was certain and he did not doubt that further research would ultimately solve the problem of the origin of species.

In spite of these categorical statements at the end of his address, Bateson's lecture produced uproar in the United States. The press reported 'Darwin downed.' A new wave of anti-Darwinism swept through the United States. W. J. Bryan introduced the topic into his Presidential Election campaign, and in Kentucky the House of Representatives passed a Bill suppressing the teaching of evolution.

Shortly after Bateson returned from his exciting visit to America, he was stricken by the death of his second son Martin, who committed suicide in April, 1922. He was profoundly depressed by this disaster. In the following month, May, 1922, he was offered a knighthood, but did not accept it.

In 1924 he made a review of biology during the last hundred years, in a Birkbeck Centenary Address. He remarked that there was nothing more curious in the history of science than that man should not have learned until the middle of the nineteenth century that organisms grow by division, a fact that could have been observed at any time. The process could not be investigated accurately until the microscope had been perfected and the conception of the cell as unit was introduced. But this 'was not essential, and may not improbably be replaced hereafter.'

The Mendelian analysis of transferable characters had not led to the discovery of the origin of species, but it had closed a wrong road of investigation.

He described the discovery of Morgan and his colleagues that the transferable characters are determined by elements in the chromosomes as of 'romantic brilliancy'.

Little was known as to what the rest of the cell was doing. 'Perhaps the true specific characters belong to the cytoplasm', though this was only an idle speculation.

He remarked that in 1844 Darwin had spoken with contempt of Lamarck's theory of the transmitted effects of the conditions of life, but by 1876 he appeared to accept Lamarckism in an obscured form. This was due, he thought, to Darwin's conviction that the fact of evolution was certain, whether or not its mechanism could be found. He was anxious to find a plausible explanation of this mechanism, which would make the fact of evolution, which was much more important than its actual mechanism, more acceptable.

He considered that Weismann's demand for proof of the transmission of adaptive responses had not been met, though it had been shown that violent treatment of the germ cells could produce permanent consequences in their posterity.

Bateson devoted his life to the search for the cause of variation, the key to evolution. He discovered many things, but the completeness of his failure to achieve his main aim became clearer and clearer, until it was absolute. To the end of his life the origin of species appeared to him to be 'utterly mysterious'.

VI. OBSCURANTISM

Bateson had been a frustrated and unsuccessful schoolboy. He became strongly prejudiced against law, politics, commerce, history, and sociology, and remained ignorant of mathematics.

Natural history, biology, was for him more than a study, it was the medium through which he recovered the control of his personality and life. It was the standard by which he measured most of the other realms of theory and practice, though not the realm of art.

In 1905 he asked the question: 'What will happen when civilized society thoroughly grasps what heredity means?' The knowledge of the physiology of inheritance, of the laws of genetics provided man with the instrument with which he could 'really shape his own destiny.'

He agreed with Galton that nature contributed much and nurture 'incomparably little' to the qualities of organisms. He regarded 'the alarming increase in the relative numbers of the insane, the utility of teaching the minds of starving children, the relation of the state to the unemployed, and all questions of grave national anxiety' as problems of 'national physiology'.

The political economists had regarded man as uniform units. Social enactments had been made for this 'purely ideal conception', which ignored differences in qualities and did not 'reflect the facts of nature'. The feudal system, with its 'ladder of lords,' more closely fitted the natural plan. The presence of the king on his throne and of God above was an 'unconscious recognition of the struggle for existence as the scheme of social order.'

The advanced social equality, which had led to the collapse of the feudal order, was a temporary and transitional phase, due to the supply of necessities having increased more rapidly than the population. He believed that some nation eager for 'national efficiency', probably not the English, would begin to apply biological breeding principles to the human population. This could accomplish more 'in three generations than all the criminal and sanitary enactments that the centuries have devised.'

When people understood that semi-insane persons were being allowed to breed, and their insane progeny had to be supported by the state, they would 'not stand it'. The ruin of any farmer who proceeded with the same plan would be assured. Unlike the farmer, we applied every 'resource of science to the reduction of infant mortality, and to the preservation of the unfit', and we drew on the public rates for the purpose.

When educated people had come to understand the situation, something like a new era in civilization would begin. 'Contemporary socialism strives for the elevation of the unfit; that of the future will probably aim at their extinction.'

The science of heredity was providing a powerful instrument for changing the composition of nations, and 'in some country, at some time, not, perhaps, far distant, that power will be applied'.

Bateson expressed these opinions in a review of Archdall Reid's *Principles of Heredity*. This book, which recommended the application of selection to human populations, was the subject of a symposium in the *Morning Post*, a journal which has itself subsequently disappeared through the exercise of selection by the people.

Bateson elaborated his views in his Herbert Spencer Lecture at Oxford in 1912 on *Biological Fact and the Structure of Society*. He said that the 'right direction of social progress' must be deduced from 'biological observation and experiment'. The present social structure in civilized states is extremely unstable.

Science had provided mankind with unexampled power of control over human life, and its application had led to changes in the composition of the population, through, for example, the prolongation of the lives of those defective in mind or body.

Social and economic measures modifying the distribution of wealth and opportunity had direct social effects, but their indirect biological effects, through altering the breeding rate of social classes, were far more profound. 'Every legislative encouragement given to one class and every repression of another has an effect on the future of the race.'

It was not yet known how far Mendelian analysis would go, but it was certain that it extended far, and would cover most of the features of 'mind or body' in a mixed population like the human. However, it had to be explicitly stated that, at this date of 1912, 'unimpeachable evidence' for the transmission of an ordinary attribute of normal men according to Mendelian rules existed in only one case, that of the colour of the eyes.

Bateson described how he viewed the splendid animals exhibited in the pens of a cattle-show, and then the farmers and sightseers who were looking at them. He saw big men, little men, 'men who fill out and "mature" early, as they say in the meat market', spare men or bad 'doers', as the farmers would say. He saw tame men, vicious men, sharp and dull, dark and fair; shepherds, grooms, salesmen. Could they be arranged to breed? Yes, a pen of shepherds might be made, and grooms would not be put in it by mistake.

The conditions of life provided opportunity for development, but they did not increase the original endowment.

Bateson said he was entirely opposed to the interference of the state in human breeding. He did not believe that British marriages should be made at Westminster, but preferred, as heretofore, that they should be made in Heaven. Present knowledge of human genetics was far too slender for such action. He disagreed with Galton's proposals that the families of unexceptionable parents should be subsidized. The examples of 'civic worth' which would be infallibly chosen are not very attractive.

How are unexceptionable types to be chosen? 'The monogamous pigeons sitting on the barn roof perhaps are scandalized at the polygamy of the fowls in the yard. Such decadence, they hold, is disgusting and should be stopped. The fowls no doubt would reply that they may be polygamous and even polyandrous, but as for decadence, they at least do not limit their families to two. Such degeneracy is race-suicide, and they think it should be punished.'

Genetic science could enable men to take 'more natural views of themselves and of each other'. He was profoundly impressed with the variety, the polymorphism of man. He agreed with the acute Cobbett, who wrote long ago that he was satisfied 'that there are as many *sorts* of men as there are of dogs.'

The political philosopher had to devise a coordinated community in which these different elements could, on the whole, enjoy the maximum of contentment with their lot. Bateson said that he had neither the desire nor the qualification to deal with this problem himself.

He had noticed that publicists rarely cared for natural knowledge. Canning did not learn until late in life that tadpoles change into frogs, and Gladstone believed that the normal number of human teeth was twenty-eight.

He found the law courts sickeningly cruel, and considered that the administration of criminal justice should be mainly conceived as a medical problem.

He questioned the conventional view that infant mortality was deplorable. He was in favour of the limitation of families. He agreed with Malthus' distinction between a 'redundant population and one actually great', which he regarded as sound both in biology and economics.

Until the latter half of the eighteenth century, the population of Britain increased very slowly, but then it rose sharply, like the rabbits after they invaded Australia, and the sparrows America.

The variation which enabled man to devour the coal measures, the capital of the earth, and thus increase swiftly in numbers, was 'that remarkable mutation, James Watt'.

He was opposed to free competition between human beings for means of subsistence, for as soon as social organization begins, 'competition is between societies and not between individuals'. He concluded that 'some elements of the doctrines vaguely described as socialism' are essential to social stability.

However, if Nature provided guidance on problems of distribution of goods, her teaching was even clearer as to the distribution of political power. 'Socialistic she may sometimes be, but democratic she is not.' Certain philosophers had asserted that all men are equal, to which the naturalist replied that it was not true. The statesman had said, 'Proceed, then, as if it were.'

How would the extension of political power to 'still lower grades of our population' work out? The gardener knows that if seedlings are given better conditions, they may develop into decent specimens, but 'they will not turn into better kinds'. The effect on human beings is seen in the newer countries. The emigrants prosper; all can develop; but what the gardener calls 'important novelties' are rare.

All progress in civilization depends on favourable mutational novelties: on the born discoverer. In Bateson's view, socialism and democracy are incompatible. The socialist community could succeed only if it had 'one mind, as the bees'.

Democracy regards class distinction as an evil, but biological observation taught that it was essential. The aim of social reform should not be to abolish classes, but to see that all individuals got into their proper class and stayed there.

In the Middle Ages, where the social system was stratified and stable, it was on the whole true that the righteous flourished. That is not so in our own society, which is changing very rapidly. Owing to this, many persons are not at their proper level, and are accordingly exceptionally discontented.

'As a boy in Cambridge I learned that if a man got a first-class he might be happy, if he got a second class he would be unhappy; if he got a third class, nothing but misery and a colonial life awaited him.' Bateson would have been interested to learn that in 1950 the suicide rate among Oxford undergraduates was eleven times that of young men of the same age in the general population, and seventeen times that of men of the same age in the Army.

But one learned in the course of life that it was possible to be happy on weekly wages and a pass degree. The social order was too unstable to last much longer, and to desire that it should would be callous. A social 'histolysis' was approaching, and no one could predict what would arise out of it. Those who wished that something better should come of it should be guided by 'the facts of natural physiology'.

Bateson delivered his lecture to an almost empty hall. It clashed with the 'Torpid' boat-races, which preoccupied the Oxford men.

He devoted the second of his British Association addresses in Australia to the implications of genetics for man. Precise knowledge of heredity was very new. Horse-breeders had realized only recently that the dam was as important as the sire.

The popular notions of the nature of man contained ancient and obscurantist survivals, and those of genetics were not much in advance of the ideas of the natives of Central Australia.

He attacked the conception of the equality of man, and Tom Paine's criticisms of the notion of hereditary political institutions. Paine said that a 'hereditary wise man' would be as absurd as a 'hereditary mathematician'. Bateson contended that one would be surprised if two mathematical parents did not have mathematical children.

Not even the patristic writers were farther from the truth than the Encyclopaedists, whose ideas were utilized as the basis for social reform. They went back to Nature without investigating what Nature is.

Genetic research would provide the means by which a nation might decide by what kind of beings it would be represented, and he thought it very probable that some nation would try the experiment. It would make 'awful mistakes'. He had been alarmed by Galton, who had suggested that 'Bohemians' were destined to perish. But for him the unbusiness-like artists, musicians, writers were the salt of the earth, and without them life would indeed be flat and wearisome. One should not forget that Beethoven's father was a drunkard and his mother consumptive.

He expressed the view that all important progress of civilization was due to rare genius. Men like Newton and Pasteur, who created new worlds, were inconceivably few. The mental differences between men were enormous; beside them their physical differences were insignificant. He could not imagine anything which would better 'teach humility, the only foundation of content', than the contemplation of the immeasurable superiority of genius over ordinary man.

We owe our escape from the conditions of the Old Stone Age to not more than a few thousand men of genius. Few 'contribute anything but their labour', and apart from imitation, 'their physiological composition is that of an earlier order of beings.' He discussed the efflorescence of genius in Ancient Greece. He quoted Galton's opinion that the Ancient Greeks surpassed the modern Europeans in intelligence, 'differing from us as much as we do from the negro'. But after a few generations, the display was ended.

What was the explanation? Bateson said that if he saw a fine stock of thoroughbred game fowls become, after a few years, a collection of unrecognizable mongrels, he would suggest that it was due to one or two birds of other breeds straying in.

The primitive Greek tribal organization had maintained a close endogamy, which forbade a man to marry a woman not of his kindred. This had produced a homogeneous population. In the sixth century it began to be broken down by foreign marriages, which were formally permitted by the reforms of Cleisthenes in 507 B.C. Pericles restored the old laws in 451 B.C., but they could not be enforced owing to the loss of men in the wars. Thus the decline of Greece appeared to be due to 'mongrelization'.

The decline that had occurred in Greece, and would almost certainly be repeated 'in the United States', was usually ascribed to the effects of political institutions and conditions of life. But conditions did not cause variations in the faculties of men.

Punnett says that Bateson found the great obstacle to progress lay in the nature of the ruling class. Contemporary democracy offered opportunities to what Bateson called the 'vocal' type of literary mind, which is not inclined to learn and make learning a basis of action. Church and State recruited 'vocal' leaders.

The British intellectual and professional classes (to which Bateson and Punnett belonged) produced, says Punnett, 'more scientists because they are more pure-bred for this type of mind'. But taxation fell on this class heavily, in view of its expensive training. It married late and limited its offspring. In comparison with the commercial and labouring classes, their numbers are decreasing, and they cannot be recruited from these classes because they are 'constitutionally different'.

They are gradually disappearing, 'Just as in a mixed population of cats the white animals would gradually disappear if a steady discrimination were exercised against them.'

Bateson expressed his views on *Science and Nationality* to the Yorkshire Science Association in 1919. He said that 'the life of one creature was the death of another.' The naturalist knew that every race held its place on the earth by 'power and power alone'; he did not 'cherish dreams of equal rights and universal peace.'

In war men committed crimes of which in normal times they scarcely knew the name. Truth was garbled and rewritten. We glorified ourselves and vilified the enemy, conduct formerly regarded as that of the cad. No nation admitted that it fought from an ignoble motive. Behind slogans of 'duty' and 'liberty' and the liberation of the oppressed were found commercial considerations and maps of mineral deposits.

He did not see any biological necessity for war. It arose from competition between nations, and he thought that this might be avoided by organization, by the creation of a World State, which was becoming technically possible through the advance of science.

Peace might be imposed by one dominant nation, as in Roman times. He believed that a dream of this kind was in the minds of an important section of the German people.

Men were not equal, nor were nations. 'There are proletariat races as there are proletariat families, and their redemption does not lie in statesmanship.' Was not Copernicus' mother a German, and Chopin's father from Lorraine?

The proposed League of Nations was a lawyer's conception, based on a democracy of nations. But 'democracy is the combination of the mediocre and inferior to restrain the more able.' It is a process of social decay, ending in proscriptions and lethargy. The Roman Empire died of such an 'ancient bolshevism'.

He suggested that the modern technique of propaganda, 'the greatest political invention of our age', might be used in order to persuade the people of the world to insist on peace, to create a public opinion against war. In his Galton Lecture on *Racial Problems*, also delivered in 1919, he remarked that the 'proclivities of the war-makers are infinitely more dangerous than those of the aberrant beings whom from time to time the law may dub as criminals.'

He shuddered at some of the opinions of eugenists. A German had expressed to him before the war of 1914 that the Latin races were destined to perish. Galton himself, whose Quaker grandfather had made a fortune out of the manufacture of small arms and was not without respect for material success, had suggested that futile people who could not earn their own living should be abolished.

He feared the types who would exercise selection. He did not find broadcloth and bank-balances an essential part of the ideal. He did not care for Thomas Bowdler, F.R.S., who prepared 'purified' versions of Shakespeare and Gibbon.

The intellectual middle class was in danger of being replaced by the thriftless majority, which 'a breeder might designate as culls'.

The Church had held up the development of mankind for a thousand years, and it might be 'held up indefinitely by the proletariat'. The proletariat were, however, deceived by the promises of power. The 'only foundation of contentment' is the humility which makes one satisfied with the place to which one is fitted.

Commercial success was rewarded far more highly than any other activity, and equality of political power had been extended to the lowest elements of the population.

But the potential force of the intellectual and professional class is enormous. If they organized themselves, they would be irresistible.

Some of Bateson's colleagues were even more extreme. Punnett quoted with approval Gobineau's assertion of the inequality of man.

Bateson's opinions on the implications of heredity for man are of the utmost gravity, from several points of view. The opinions in themselves are extraordinarily contradictory, erroneous and dangerous.

In the field of biology, in which he was learned, he held very simple clear-cut philosophical views. He accepted, virtually without qualification, Weismann's theory that environment has no effect on the germ-plasm. He conceived hereditary characters as virtually indestructible units. He argued as if there were relatively few of them and that they obeyed simple rules. His picture of the hereditary mechanism, even from his own point of view, was far too crude and oversimplified.

He saw that Mendelian heredity could throw no light on the origin of species, and he adopted the suggestion that species arose from 'mutations', chance unexplained events.

It is not surprising that Lysenko has described Bateson as an 'obscurantist'. His work in pure biology was marked by sharply held metaphysical conceptions of the definiteness of unit characters, of the dominance of nature over nurture, and the inability of the environment to modify the germ-plasm.

It is evident that in the last analysis the effects of environment must be inherited, because organisms are a product of environment. The earth has given birth to living things, somehow they have evolved out of the environment and are a product of it. They must, therefore, be responsive to its influence. It may be difficult to discover exactly and completely how this influence is exerted.

The science of heredity, which Bateson named genetics, became metaphysical, mechanical, and mathematical. Its results were expressed in concepts of hereditary units or genes, which resembled the atoms or molecules of physics and chemistry. Finally, the properties of heredity, and the refined effects of natural selection on these units were described more and more in mathematical terms.

Genetics drifted away from the category of life. In such a situation, a restatement of the fundamental conceptions of biology was inevitable.

Lysenko redefined heredity as 'the property of a living body to require definite conditions for its life and development and to respond in a definite way to various conditions.'

The rigidity of the inappropriate categories of chemistry and mathematics has disappeared, and is replaced by a flexible concept, capable of guiding a vast range of new development.

The genetic structure of the Mendelians, created especially by Bateson and Morgan on the basis of Weismann's germ-plasm theory, and developed in the Soviet Union by Bateson's pupil N. I. Vavilov, had become separated from fundamental conceptions of living organisms. It had tended to reduce them to chemistry and mathematics.

The agricultural population of the Soviet Union did not find Mendelian genetics an effective scientific aid in its direct concern with the biology of the cultivation of plants and animals. Lysenko has succeeded in finding a better set of conceptions for releasing the biological interests of the Soviet population. As a consequence, the people's activity in biology is perhaps ten times as great as before. Enormous projects for transforming the nature of the country, which require immense biological efforts, are being carried out.

Lysenko's biological approach, his capacity for interesting the people in biology and helping them to use it for changing the climate and characteristics of the country, are essential in these unique activities. He has succeeded in creating a biology with a meaning deeper and wider than that of Mendel-Morgan genetics.

Lysenko regards Darwinism as having expelled idealism and metaphysics from the explanation of the history of organisms. He therefore all the more fiercely criticized Weismann and Morgan for re-introducing them into biology in the guise of mythical immortal hereditary substances, which are themselves unchanging but direct the change of other living substances.

In the present epoch of struggle between two world-outlooks, the conflict between the materialist and idealist trends in biology is particularly sharp. The assertion of Mendel-Morganism that variation is random and indefinite raises a barrier to scientific foresight and experimental advance.

Michurin's teaching and work, which shows that the nature of living organisms can be guided in directions needed for practical purposes, is, in contrast, encouraging to the experimenter. It has changed the method of thinking on the problems of biology. Michurin himself, in his own life-time, produced more than 300 new strains of plants, all by strictly directed selection, which included systematic training.

In Lysenko's view, the surest way to a deeper insight into biology is through the scientific solution of practical problems concerning living organisms. Michurinism raises Darwinism to a higher level, for it converts it from a science which *explains* the history of living nature into one which is a creative *effective* means for systematically mastering it, and utilizing it for practical ends.

Lysenko has written that he was not led to study the species problem from 'mere curiosity or a fondness for abstract theorizing'. He was led to them by consideration of such problems as the conflict between crops and weeds, the blending of grasses for grass mixtures, and the swift afforestation of steppe areas.

Mendel-Morganism is built entirely on chance, and denies necessary relationships in living nature. Consequently, it condemns practical workers to fruitless waiting. It prevents planning. A science which does not give practical workers a clear perspective, and an indication of the way in which they should look, and hence confidence, does not deserve the name of science.

As Michurin said: 'We must not wait for favours from nature; our task is to wrest them from her'.

Lysenko regarded the fact that the Communist Party and the Soviet Government were helping Michurinism imposed an extra duty 'to arm the State farms and collective farms with an advanced theory. That is what the Soviet people expect of us.'

Their task was to place science still more effectively at the service of the people.

VII. A FRUSTRATED TALENT

Bateson was quite conscious of the depth of his own failure to relate genetics to the interests of the people and to secure support for its development. As we have seen from his social views, he failed to enter into social and political activity. He was the reverse of Lysenko, who, though by temperament very retiring, has learned how to take part in social affairs and thereby has successfully combined biological with social and political action.

The crudity of Bateson's views on the social implications of genetics, held almost with irresponsibility and levity, is due in part to the individualism of a British intellectual of the middle-class Liberal tradition. He felt at liberty to express views on social questions without serious preparation. He was not conscious of a responsibility to study

8. Academician T. D. Lysenko

face p. 304

them in theory and gain experience of them in practice, before he offered opinions on their biological significance. He took it for granted that one was permitted to be an amateur, and assumed the right of a member of a privileged class to express opinions without responsibility. He ended by expressing in the guise of biological truths the social prejudices of the class to which he belonged.

The crudities of Bateson's sociological opinions are characteristic of the views of many geneticists. They have very largely disqualified genetics as a guide on social questions, and indicate the existence of serious flaws in the fundamental assumptions of the pure science itself.

The defects of the general ideology of genetics, so vividly illustrated in Bateson, have deeply penetrated the science and affected the attitude and unconscious thought of many geneticists who would regard their own social views as entirely contrary to Bateson's. This was no doubt a contributory cause of the failure of N. I. Vavilov's development of Mendelian genetics in the Soviet Union. A science so heavily charged consciously and unconsciously with middle-class philosophical presuppositions was incompatible with a socialist society.

The extraordinarily controversial character of genetics arises in part from the gravity of the social issues involved. In part, also, it is a reflection of an immature science. The history of science shows that when many scientists of great ability differ violently among themselves, it is usually found that their science is lacking in some necessary fundamental scientific concepts. This was seen with regard to the theory of combination in chemistry in the eighteenth century, and in the theory of infinity in mathematics in the nineteenth century.

Mendelian genetics, even if its internal logical and mathematical coherence are accepted, is a narrow and artificial structure compared with the phenomenon of life. The future science of heredity must be an altogether wider and subtler thing, and that is what Lysenko has started to create. He has cleared the ground.

Before the vast complexity of life, any set of qualifications is limited, and requires fundamental revision from time to time. As Dembowski has remarked, sciences themselves exhibit the processes of growth and decay, and Mendelian genetics has passed into a condition of abstract detachment and proliferation of details which characterizes the period of senility which precedes the disintegration and rebirth of a science.

The Mendelian geneticists assert that Lysenko is ignorant of Mendelian genetics. Even if this were true, it should not disturb Mendelians. One of the most remarkable features of Mendel himself was his ignorance of other branches of biology. His answers to examination questions for a diploma to teach natural history have been preserved, and they show that he did not know the elements of the subject. In fact, his answers were regarded as a sign of mental defect, and only after pressure through the Church was he granted permission to teach children natural history, and then only under the age of nine.

Nemeč has expressed the opinion that the motive of Mendel's research was to disprove Darwin's theory of evolution, in the interests of Christian theology. He made his experiments on variation to discover whether species could have originated in the manner suggested by Darwin. He considered that he had disproved Darwin's theory of the origin of species, and hence his theory of evolution, and had thereby performed a great service to the Church.

Bateson in fact promulgated metaphysical presuppositions of theology in a non-theological form, as nineteenth-century middle-class rationalists promulgated Christian ethics in non-Christian terminology.

From time to time fresh starts are necessary. Mendel by-passed systematic zoology; Bateson by-passed comparative embryology; and Lysenko has by-passed Mendel-Morgan genetics.

When such changes occur some good things are temporarily lost, but this is outweighed by the much greater primary gains. The mass-utilization of biology which has begun in the Soviet Union is necessary for the survival of the human race. William Vogt has described the world-wide wastage of resources that is occurring under the spur of capitalist exploitation. His remedy is the reduction of the size of human populations, an erroneous answer derived from a fallacious biologism.

The answer of a socialist society is not the extinction of peoples, but the development of biological knowledge in a form that can be applied to human survival. The conservation of resources cannot be accomplished by capitalism based on the principle of exploitation, and the science that has been selected and fostered by such a system.

Bateson's scientific thought was affected by individual characteristics as well as by class-ideology. He had an abrupt and controversial temperament. The subject selects the man, as well as the man the subject. He was selected by a subject to which these qualities were fitted. He was naturally inclined to discontinuity, and therefore apt to recognize it.

His qualities appeared to some as the model of absolute honesty and fearlessness inspired by a burning passion for truth. To others, these same qualities seemed to show a lack of self-control and an inability to enter into other people's points of view, and cooperate with them.

Punnett said that Bateson was never deterred by other men's opinions. He liked knowledge to be 'hard and clear', the phrase he used in describing Doncaster's quality of mind, but which also described his own. It was admitted by all that his unmoderated outspokenness gave offence.

Bateson had an outspoken scientific talent.

J. B. S. Haldane has said that his contributions to science were as great as those of Vesalius.

Why were his experimental results not of the first importance? His main genius was not experimental; he was primarily a propagandist. At times he wrote like H. G. Wells, and his slashing, antithetic style has influenced certain of his successors.

He was unable to combine genetics with agriculture because Cambridge was out of tune with agriculture. It was difficult to foster co-operation between Cambridge science and agricultural practice. British agriculture was under a strong leisure-class influence. Cambridge students of agriculture were generally the sons of landowners and were not farmers. Cambridge agricultural science did not grow out of the agricultural community.

British agriculture is that of an imperialist industrial country. The main genius of British science was reflected in industrial and imperial interests, in physics arising out of the steam-engine and electricity, in biology arising out of imperial exploration and exploitation. The expedition which took Darwin round the world, and provided him with the data for the theory of evolution, was organized for the charting of harbours for British ships and the collection of specimens of raw materials for British trade. Darwin was reinforced in his introduction of the principle of natural selection by the example of competition in capitalist society.

The class-outlook of the Soviet people demanded a mass-changing of nature, and it has produced the biology suitable for that purpose. The class-outlook of capitalism has produced Mendelian genetics. Its individualism concentrated attention on units and segregation, and did not lead to bold views on the changing of nature for the betterment of all mankind. It failed to explain evolution, so it was necessary to look outside Mendelian genetics for a principle more profound than any within it—the cause of evolution.

Bateson had a strong artistic sense which gave him a passionate interest in nature as a spectacle. He reflected with strong emotion on the wonders of his gardens, with their marvellous variety of shape and colour in plants and flowers. He was specially interested in the variations of patterns of plants, which he named meristic variations. He believed that mathematicians might be able to discover something of the laws of their geometry.

At the beginning of his *Materials for the Study of Variation*, he wrote: 'In the old times the facts of Nature were beautiful in themselves and needed not the rouge of speculation to quicken their charm, but that was long ago, before Modern Science was born.' This passage reveals an anti-scientific strain.

It appears that Bateson might have been satisfied with the Garden of Eden before the discovery of Knowledge. He might have been content to enjoy Nature as a spectacle only. As Singer has remarked, there was a good deal of the *Naturphilosoph* in Bateson.

Punnett has written that Bateson regarded his experiments with plants and animals as a means by which man might clarify his conceptions of the place of himself in the scheme of things.

The faculty which enabled Bateson to enjoy the aesthetics of nature also made him an amateur of art. He collected drawings by old masters, Japanese colour prints, and Chinese works of art. He invested his savings in these rather than in other forms of property.

In his depression after the death of his second son in 1922, his old scientific interests were unable to distract him, but he was revived by his appointment as a Trustee of the British Museum, which gave him opportunity for the wider exercise of his aesthetic faculties.

Bateson had an intense curiosity to see the Soviet Union and visit N. I. Vavilov. He was invited to attend the Bicentenary of the Academy of Sciences of the Soviet Union in 1925. He wrote an account of this journey in *Nature*.

He said that the courtesy and hospitality with which he and the other two British delegates were received could not have been exceeded. Their baggage was passed in and out of the country unexamined.

The buildings of Moscow and Leningrad were dingy, and the British scientists had a novel feeling of being 'conspicuously well-dressed'. They noticed that Soviet students had patched trousers and artisans wore rags suggestive of the 'casual ward of a workhouse'. Among the statesmen who addressed them on science and what it might do for the people were Kalinin, Krassin and Lunacharsky.

It became very clear that 'the revolutionary government is perfectly sincere in its determination to promote and foster science on a very large scale.'

He visited the institutes of zoology and botany directed by Philiptschenko and Dogiel, the cytological school of Nawaschin, Koltsov's Institute of Experimental Biology, in which Serebrovsky directed a department of genetics, Prianischnikov's school of agricultural chemistry, Kohts' Darwin Museum, which contained valuable material 'on the incidence of variation', and N. I. Vavilov's Institute of Applied Botany and Plant Breeding.

In one of the Museums he saw a recently discovered skeleton of the giant *Indricotherium*, about twelve feet high. He viewed it with mixed feelings, for it had been discovered at Chalkar Tengis, and a photograph of the site showed that he must have passed within a few yards of it during his expedition of 1886 and missed it.

He wrote that he and his British colleagues 'left with no clear conception of the principles or practice of communism . . . of liberty we saw no sign.'

Bateson and his science were the prisoners of his social ideology, and he was unable even to begin to try to understand the new world he was visiting. He frankly regretted having missed so much during his visit of 1886; if he had lived sufficiently long, perhaps he would have come to regret that he had also missed very much in 1925.

Under the direction of Vavilov, his scientific conceptions were nevertheless given a long and extremely generous trial, but their failure and supersession were inevitable.

A few months after his return to England he had a final heart-attack, and died on February 8th, 1926, at the age of sixty-four.

Bateson published about 150 papers and five books, including his major *Materials for the Study of Variation.*

In spite of his shortage of means in his earlier years, his estate was sworn as £25,435.

Bateson was a scientist of great ability, who was unable to overcome the obstructions of the ideology of the society in which he lived. Not only was his scientific ability deflected by the ideology outside himself, it was distorted also from within, owing to his conscious and unconscious personal acceptance of a large part of this ideology. His life and work provide a vivid illustration of the frustration of a great scientific talent by a declining imperialist social system.

BIBLIOGRAPHY

William Bateson, F.R.S., Naturalist. His Essays and Addresses together with a Short Account of his Life, by Beatrice Bateson (1928).

Materials for the Study of Variation, by W. Bateson (1894).

Scientific Papers of William Bateson. Edited by R. C. Punnett, vols. 1 and 2 (1928).

Mendel's Principles of Heredity, by W. Bateson (1902 and 1909).

Problems of Genetics, by W. Bateson (1913).

The Causes of Evolution, by J. B. S. Haldane (1932).

New Paths in Genetics, by J. B. S. Haldane (1941).

Evolution, the Modern Synthesis, by Julian Huxley (1942).

Soviet Biology, by T. D. Lysenko (1949).

The Situation in Biological Science, Proceedings of the Lenin Academy of Agricultural Sciences of the U.S.S.R. (1949).

A Short History of Biology, by C. Singer (1931).

Letters from the Steppe. Written in the Years 1886–87, by W. Bateson. Edited with an Introduction by Beatrice Bateson (1928).

Dictionary of National Biography, **68** (1922–30).

William Bateson, R. C. Punnett, *Edinburgh Review* (July, 1926).

'Early Days of Genetics,' R. C. Punnett, *Heredity,* vol. 4, pt. I (April, 1950).

The Elements of Genetics, by C. D. Darlington and K. Mather (1949).

Animal Genetics, The Science of Animal Breeding, by F. A. E. Crew (1925).

Obituary Notices: *Proc. Roy. Soc. B.*, **101** (1927); *Times* (February 9th, 1926).

Obituary Notices of L. Doncaster, by W. Bateson: *Proc. Roy. Soc.*, **92** (1921); *Nature* (June, 1920).

SELECT BIBLIOGRAPHY

The reader may find these volumes valuable for further reading, in addition to those recommended at the end of each chapter.

The Atom, by G. P. Thomson. 1942.
Atomic Physics, by Max Born. 1951.
The Structure of the Atom, by E. N. da C. Andrade. 1927.
The Conduction of Electricity through Gases, by K. G. Emeleus. 1951.
General Astronomy, by H. Spencer Jones. 1951.
Fundamentals of Biochemistry in Relation to Human Physiology, by T. R. Parsons. 1939.
An Introduction to Comparative Biochemistry, by E. Baldwin. 1948.
Genetics in the 20th Century, edited by L. C. Dunn. 1951.
Soviet Genetics, by A. G. Morton. 1951.

INDEX

Abbott, Dr., 206f
Abel, F. A., 205, 208
Academic Assistance Council, 84, 86, 91
Adamkiewicz reaction, 218, 225
Adams, W. S., 158
Adams, J. C., 36, 150, 197
adsorption equation, 11
Aikin, J., 251
Aitken, 17, 27
Allibone, T. E., 84
alpha particles, in photographic films, 77
  rebound of, 72
  scattering of, 72
ancestral heredity, law of, 270, 274f, 279
Anderson, C. D., 86
Appleton, E. V., 29, 37, 79
Aston, F. W., 13, 29, 34, 79f, 160
astrology, 122
atom, 20
  disintegration of, 64, 78, 86
  of mass three, 80
  nuclear, 43, 74, 76
  nuclear theory of, 91
  photograph of disintegration of, 81
  radius of, 73
  structure of, 10, 19, 73
atomic energy, 66, 179
  problems of exploitation of, 90
  release of, 88
  sun's, 67

Babinet, 118
Bach, J. S., 123
Bacon, Francis, 244
Baer, K. E. von., 253, 261
Balanoglossus, 253ff, 259ff
Baldwin, E. H. E., 180
Balfour, A. J., 36, 180, 253
Balfour, F. M., 253, 255f
Ball, R., 56, 150
Bankhead, Tallulah, 95
Barker, T., 4
Barkla, C. G., 20, 29, 75
Barrow, I., 69
Bartlett case, the, 209
Bateman, H., 75
Bateson, Anna, 249
Bateson, Mary, 249
Bateson, Mrs. B., 281
Bateson, W.:
  ancestry, 248f
  attitude to chromosome theory, 284, 287f, 291, 294f
  Chair at Cambridge, 283
  character, 288, 295ff, 306
  class-outlook, 289, 302, 304, 308f
  death of first and second sons, 284, 293, 295, 308
  discovers linkage, 285
  on eugenics, 301
  finance, 281, 289, 309
  John Innes Institution, 283
  literary and artistic taste, 252
  marriage, 280
  on Mendelism and mankind, 290, 296, 302
  more popular in U.S.A., 282
  proposes word 'genetics', 282
  relations with Weldon, 270–280
  search for endowments, 281
  starts on *Balanoglossus*, 254
  style, 307
  visits Kazakstan, 256
  visits U.S.S.R., 284, 308
Bateson, W. H., 248f
Becker, 85
Becquerel, H., 2, 54, 59f, 65
Bergson, H., 189
Berkeley, Bishop, 135
Bernard, C., 225
Bethe, H., 160
Bickerton, A. W., 52f
Biffen, R., 286
biogen, 224
biometry, 271, 280
Birkbeck, G., 245
Blackett, P. M. S., 37, 79, 81, 83, 86
Bohr, Niels, 32, 44f, 74ff, 81f, 84, 103, 109f
  quantum atomic theory of, 75, 110f

Boltwood, R. A., 70
Boltzman, L., 99
bombardier beetle, 219
Born, M., 2, 12, 25, 29, 35
Bothe, W., 85
Boveri, T., 284
Bragg, W. H., 29, 36
Bragg, W. L., 29, 75
Braithwaite, R. B., 184ff
breeding, 278, 289
Broad, C. D., 243
Broglie, L. de, 13, 81
Broglie, M. de, 36
Brook, F. W., 224
Brooks, W. K., 254, 271, 293
Browning, O., 56
Bruno, Giordano, 129
Bryan, W. J., 294
Bumstead, H. A., 25, 35
Butler, Montagu, 38f
Butler, Samuel, 248, 290f
butterfly pigments, 199, 213, 219f, 265

Callendar, H. L., 27, 29, 62
Carnegie, A., 281
Carroll, Lewis, 170
cathode rays, 8f, 14f
Caudwell, C., 190
causality, 186
Cavendish dinner, 26
Cavendish, H., 13
Cavendish Laboratory, 2, 8, 20ff, 36, 40, 57, 78f, 98, 144, 148
  an imperial institution, 23
  finance of, 29, 40, 87
  tea, 25
Cavendish Physical Society, 16, 24
Chadwick, J., 79ff, 85ff
chain reaction, 88
Chamberlain, N., 39
Chandrasekhar, 195
chimaeras, 287
Christie, W., 142
chromosomes, 284, 291, 294
chromosphere, 178f
Clark, A., 178
Clifford, W. K., 171f
cloud expansion chamber, 28, 80f, 134
cloud formation, 17
Cockcroft, J. D., 83, 85, 88
Cole, S. W., 199, 215, 225
Condon, E. U., 82
Conway, A. W., 148
Cook, C. H. H., 49
Cornforth, M., 184
corpuscles, 18
cosmic rays, suggestion of existence of, 28
Cottingham, E. T., 164
Cotton, A., 83
Cox, J., 62
Crommelin, A. C. D. C., 164
Crookes, W., 8f, 11ff, 35, 70
Crowther, J. A., 19, 21, 26
Curie, Marie, 2, 54, 61, 64f
Curie, P., 2, 30, 54, 61, 65

Dale, H. H., 198, 202, 218
Dalton, J., 4, 140
Darbishire, 278
Darwin, C., 94f, 242, 255, 262ff, 266f, 274, 277, 290f, 293f, 306ff
Darwin, C. G., 74f
Darwin, F., 281
Darwin, G. H., 94, 100, 102, 104, 113f, 136, 149
Darwin, Miss, 281
Davidson, C. R., 148, 164
Debye, P., 110
Dee, P. I., 86
Department of Scientific and Industrial Research, 37, 85f
Descartes, R., 174
dietetics, 228
Dingle, H., 194
Dirac, P. A. M., 45, 81f, 86, 163, 185, 191, 194
Dirac's 'holes', 81, 86
discontinuity, 174
Dogiel, 308
Dokuchaiev, 289
Doncaster, L., 281f, 284, 286, 306
Durham, Beatrice (Mrs. Bateson), 280
Durham, H., 206
dynamic equilibrium, 198, 236
Dyson, F. W., 164

Eagle, 232
earth-moon system, 119
eclipse expeditions, 164f
Eddington, A. H., 140
Eddington, A. S., 29, 36, 104, 136, 203f
  ancestry, 140
  Chair at Cambridge, 149
  discovers mass-luminosity law, 158
  eclipse expedition, 164
  at Greenwich, 148f
  like Icarus, 153, 161
  internal constitution of stars, 144ff
  number-theory of nature, 191ff

Eddington A. S.—*continued*
on philosophy, 167ff, 181ff
pulsating stars, 144
Quaker, 140, 153, 195
on relativity, 161ff, 165, 181, 191
religion, 183, 190
social class, 142
stellar movements, 149
style, 172, 175, 180, 189
'world-building', 162
Edison, T. A., 19, 23
Einstein, A., 2, 9, 13, 43, 67, 84, 86, 94, 107, 110, 112, 133, 139, 161. 163ff, 170, 192
electromagnetic detector, 23
field, 43, 91
electron, 1, 80f, 86
discovery of, 17
must be prime material, 18
proof of existence of, 19, 59
size of charge in, 29
electronics, 40
Eliot, T. S., 180
Ellinger, 225
Elliott, K. A. C., 232
Ellis, C. D., 79, 83, 157
Elster, J., 28, 70
emanation, 64
Encyclopaedists, 299
energy, conservation of, 99
equipartition of, 94, 126
Engels, F., 4, 46, 190
entropy, 185
ether, 111
eugenists, 301
Evans, A., 154
Everett, E., 31
evolution, 269, 293
mechanism of, 278
Ewing, A., 243
Exhibition of 1851, Commission of, 22
Eykman, 229

Faraday, M., 5, 11f, 33, 41, 43, 51, 57, 135, 137
Feather, N., 85
Fessenkov, 138
Fischer, E., 233
Fitzgerald, G. F., 10, 106
Fitzgerald-Lorentz contraction, 166, 169
Flamand, 225
Fleming, J. A., 51
Fletcher, W. M., 199, 202, 216, 231f, 241
Foerster, 153
force, notion of, 170
Ford, W. J., 48
Forsyth, A. R., 8
Foster, M., 214, 234f, 253, 256, 292
Frankland, P. F., 208
Franklin, Benjamin, 68
Frazer, J. G., 35
freedom of the will, 186
Friedmann, A., 163
Friends, Society of, 141
Funk, 230

Galileo Galilei, 43, 114, 122, 130, 170, 184, 199
Galton, F., 270f, 273f, 297, 300, 302
games and physics, 31
Gamow, G., 82
Garnett, W., 10
Garrod, A. E., 212, 222
gases, electric discharge in, 14f
electrification of, 18
Geiger, H., 69ff, 74ff
counter, 70
Geitel, H., 28, 70
genetics, 267, 282, 303
Gibbs, J. W., 11
Gladstone, J. H., 213
Gladstone, W. E., 41, 249, 298
Glazebrook, R. T., 10, 16
glutathione, 200, 231
Godlewski, E., 63
Goldhaber, M., 86
Goldstein, E., 8f, 13, 20, 57
gravitation, 171
gravitational condensation, 94
Greeks, ancient, 300
Greenwich Observatory, 148, 150
Gregory, R. P., 282, 284, 287
Grijns, 229
Gurney, R. W., 82
Gurvitsch, 242

Haeckel, E., 253
Hahn, O., 63, 87
Halban, F., 88
Haldane, J. B. S., 280, 294
Haldane, J. S., 238f
Hamilton, W. R., 163
Hardy, G. H., 91, 98, 189
Hardy, W. B., 215
Harkins, W. D., 80
Harmer, S. F., 259
Harris, L. J., 244
heavy hydrogen, 79
Heisenberg, W., 81, 85, 131, 185
Helmholtz, H., 8f, 33, 35
Henderson, W. C., 26

Henry, J., 51
Herman, R. A., 147f
Herschel, W., 114, 123
Hertz, H., 2, 14, 16f, 43, 51, 57f
Hertzsprung, E., 151, 158
Hicks, G. D., 187
Hittorf, W., 8f, 14
Hock, Suzanne (Lady Jeans), 105
Hofmeister, W., 233
Holism, 238
Hope, W. B., 223
Hopkins, F., 205
Hopkins, F. G., 265f, 284f
  achievement, 197ff
  ancestry, 204f
  appointment at Cambridge, 214
  biochemical philosopher, 233ff
  butterfly pigments, 199, 219, 220
  chemistry of muscular contraction, 199, 216, 230
  crystallization of proteins, 199, 224
  dynamic equilibrium, 198, 201, 236
  first professor of biochemistry, 216
  glutathione, 200, 231
  literary taste, 203
  marriage, 213
  Nobel Prize, 201, 229
  on protoplasm, 235
  research organizer, 200, 218
  on science writing, 247
  socialist, 241
  style, 202
  tryptophane, 199, 217
  urine analysis, 221
  vitamins, 199, 226, 229, 241
Hopkins, Gerard Manley, 204
Hoyle, F., 138
Hunter, 232
Hurst, C. C., 278f, 281, 286
Huxley, A., 245
Huxley, J., 294
Huxley, T. H., 207, 242, 256, 260

ionization, 177
isotopes, first separation of, 21

Jacobi, C. G. J., 24, 102, 114
Jeans, J. H., 36, 157, 164, 175
  ancestry, 95ff
  annihilation of matter, 94, 112
  attitude towards popularization, 122, 130, 135f
  Cosmogony, 95
  equipartition, 94, 126
  first marriage, 104
Jeans, J. H.—*continued*
  fortune, 104, 133
  gifts and limitations, 112
  origin of solar system, 118
  passion for music, 105, 123
  principal temperature, 100
  professor at Princeton, 104
  rotating compressible fluids, 94, 101, 116
  rotational instability, 101, 117
  second marriage, 105
  stellar fission, 103, 116
  style, 93, 100, 117, 123, 130
  tidal theory, 118, 138
Jeans, W. T., 95ff
Jex-Blake, Dr., 249, 251
Joad, C. M., 246
John Innes Horticultural Institution, 283, 292
Joliot-Curie, F., 32, 64, 85f, 87f
Joliot-Curie, Irène, 64, 85f, 87
Jolley, W. J., 215
Joseph, H. W. B., 186
Joule, J. P., 4

Kalinin, M. I., 308
Kant, E., 115, 184
Kapitza, P. L., 81, 83f, 86f
Kaufmann, W., 10, 16
Kay, W., 69, 77
Kepler, J., 124
Killby, Miss H. (Mrs. Onslow), 281
Kirchhoff, G., 4, 106
Knight, T. A., 276
Kohts, 308
Koltsov, N. K., 308
Kowarski, L., 88
Krassin, L. B., 308

lactic acid, 230
Lakin, Miss, 251f
Lamarck's theory, 295
Lamb, H., 72, 141
Lane, J. H., 145
Langevin, P., 15, 26, 36, 44, 77
Langmuir, I., 236
Lankester, R., 241, 260, 262, 272
Laplace, S. P., 114f
Larmor, J., 7, 106
Laxton, 276
Leader, Miss V. R., 229
Leavitt, Miss H., 124
Le Chatelier's principle, 11
Lemaître, G., 163, 291
Lenard, P., 34, 74

Leonardo da Vinci, 195
Liapounoff, A., 94, 113
Liebig, J. von, 234
Lindemann, F. A. (Lord Cherwell), 110, 157, 167
linkage, 285
Littlejohn, W. S., 49
Lock, R. H., 282, 284
Lockyer, N., 67, 145f, 151
Lodge, O., 29, 50f, 292
Loeb, J., 68
Lorentz, H. A., 106
Lucretius, 170
Lunacharsky, A. V., 308
Lyell, C., 242
Lysenko, T. D., 289, 302ff
Lyttelton, A., 8
Lyttleton, R. A., 138

McClelland, J. A., 26
Macdonald, W., 66
MacDonald, J. Ramsay, 90, 213
Mach, E., 235f
Maclaurin, J. S., 53f, 101, 114
McLennan, J. C., 29
McVittie, G. C., 163
magnetic detector, 51f, 55f, 58
  field, powerful, 81
  viscosity, 51f
Malthus, T. R., 290, 298
Marconi, G., 47
Marris, W. S., 49, 53
Marryat, Miss, 282
Marsden, E., 48, 50, 53, 69, 72, 76f
Marx, Karl, 54
mass-luminosity law, 158
  spectrograph, 21, 34, 80
matter, annihilation of, 94, 112
Maxwell, J. Clerk, 2, 6ff, 22, 26, 28, 40f, 43, 58, 93, 98, 105, 111f, 139, 146, 163, 180, 197
Mayer, 19
Mellanby, E., 217f, 241f
Mendel, G., 269, 272f, 289
Mendelism, 303
mesons, 28, 193
metal, electrical conductivity of, 19
Michelson-Morley experiment, 105, 133, 161
Michurin, I. V., 304
Millikan, R. A., 29
Milne, E. A., 93, 100, 137, 178, 189, 194
minerals, age of, 67
Minkowski, H., 162, 166, 170
Mitchell, Charlotte Tiffany (Lady Jeans), 104
Mond Laboratory, 87
Morgan, E. J., 232
Morgan, T. H., 287, 294f, 303
Morier, R., 256
Moseley, H., 260
Moseley, H. G. J., 75f, 260
muscular contraction, chemistry of, 199, 216, 230

Nagaoka, H., 74
natural selection, 271
nature, transformation of, 307
Nawaschin, 308
nebulae, shape of, 117
  stability of, 101
nebular hypothesis, 94, 115
Needham, Dorothy M., 214f, 217
Needham, J., 214f
Nemeč, B., 306
Nernst, W., 110
neutron, 79f
Newall, H. F., 157
Newton, A. P., 260, 282f
Newton, Isaac, 6f, 21, 40, 43, 69, 74, 98, 112, 114, 124, 130, 137, 162f, 170, 180, 190, 197, 275, 300
  First Query, 173
Newton, Mary (Lady Rutherford), 53, 61, 65
New Zealand, history of, 46
Nicholas, Grand Duke, 257
Nicholson, J. W., 109, 141
nuclear fission, 63
  physics, 74
nucleus, 73

Occhialini, G. P. S., 86
Oliphant, M. L. E., 88
organicism, 238
origin of species, 295
Ostwald, W., 235f
Owens College, 2
oxidations, 200, 232

Paget, Rose (Lady Thomson), 24
Pailhade, R. de, 231
Paine, Tom, 299
Papalexi, N., 29
Pasteur, L., 211, 234, 300
Pavlov, I. P., 242
peace, 301
Pearson, K., 271ff, 274, 279
Pekelharing, 229

periodic table, 19
Perkin, A. G., 206
Perkin, W. H., 206
Perrin, J., 17, 160, 236
Petavel, J. E., 141
Philiptschenko, 308
piezo-quartz vibration, 77
Pinkus, S. N., 224
Pitt, Newton, 211
Planck, Max, 2, 13, 43, 71, 74, 106ff, 131, 139, 164, 190
Planck's constant, 110, 132
  law, 109
planetary systems, 95
planets, formation of, 119
platinum resistance thermometer, 27
Plücker, J., 14
Plummer, H. C., 191, 193
Poincaré, H., 102f, 106, 109, 113f, 129, 168
pointer-readings, 183f, 190
Poulton, E. B., 279
Powell, C. F., 28, 77
Prianischnikov, 308
principle of least action, 173
proteins, crystallization of, 224
proton, 80
protoplasm, 224
Prout's hypothesis, 19, 80
Przibram, K., 29
Punnett, R. C., 277f, 279, 282, 284ff, 288, 300ff, 306, 308
Pye, W. G., 30

Quantum Theory, 71, 94, 107ff, 121, 131

radar, 88
radiation pressure, 146, 155
radiations, characteristic, 75
radiative equilibrium, 155
radioactivity, 60
  artificial, 80
  spontaneous transformation theory of, 66
Ramsay, W., 58
Rayleigh, third Lord, 8, 10, 22, 26f, 32f, 35f, 39f, 58, 69, 94, 106, 109, 197, 253
Rayleigh, fourth Lord, 18, 22, 24, 26, 29, 32
Rayleigh-Jeans formula, 106
Regener, E., 71
Reid, A., 296
Reinganum, 77
relativity, theory of, 67, 84, 94, 121, 133, 161, 175
research discussion group, 24
  method and organization of, 21, 41, 79
  planning of, 90
  students, 15, 53, 55
  team work, 24
Reynolds, O., 3f, 10, 33, 112
Richardson, O. W., 19, 29
Righi, A., 32
Ritchie, A. D., 184, 190
Robb, A. A., 26
Robertson, H. P., 163
Robinson, R., 287
Roche, E., 114
Rømer, O., 8
Röntgen, W. K., 2, 15, 30, 146
Ropmann, F., 230
Rosenblum, S., 83
Rosenhain, W., 29
Routh, E. J., 6, 98, 147
Royal Society, 104
Royds, T., 72
Ruhmkorff, H. D., 14
Russell, A. S., 69
Russell, Bertrand, 167, 180, 240, 246
Russell, H. N., 151, 158, 160
Rutherford, E., 15f, 23, 25f, 29, 31f, 36, 38, 40, 43, 109, 135ff, 139, 141, 157, 167, 197f, 285
  aims in life, 44
  ancestry, 47f
  arrival in Cambridge, 55f
  Cavendish professor, 78
  Chair at McGill, 62
  discovers alpha particle, 61, 66
  discovers emanation, 64
  discovers nucleus, 73
  disintegrates atom, 78
  ideas, 44
  Manchester period, 69–78
  marriage, 65
  misapplication of his discoveries, 89
  Nobel Prize, 68, 72
  prejudices, 71, 84f
  qualities, 89ff
  research method, 79
  spontaneous transformation theory, 66
  style, 52
Rutherford, G., 47
Rutherford, J., 47f
Rutherford, Lady, 31

Sampson, R. A., 146

saturation current, 59
Saunders, Miss E. R., 271, 278, 281, 285
Schäfer, E. A., 222f
Schrödinger, E., 81, 185
Schuster, A., 9f, 14, 34f, 68f, 75, 141
Schwarzschild, K., 146, 154, 157
science, social origin of, 184
  writing, 201, 247
scientific Calvinism, 293
  laws, 185
  workers, organization of, 91
Scientific Workers, Association of, 244
  National Union of, 1
scintillation method, 70f
Searle, G. F. C., 8, 16, 24, 29
Sedgwick, A., 254, 260, 267, 282f
sex-linked inheritance, 286
Shakespeare, W., 45
Shapley, H., 124, 159
Sherrington, C. S., 238, 240, 241
Shimizu, 80f
showers, 86
Sidgwick, H., 253
Sidgwick, Mrs., 36
Simpson, G. C., 141
Singer, C., 308
Sitter, W. de, 161f
Skobeltzyn, D., 83
Smoluchovski, S. T., 29
Smuts, J. C., 238
Soddy, F. W., 21, 43, 65ff, 70
solar system, origin of, 118, 120
Sollas, Miss, 281f
Soviet Union, 285, 303
space, magnitude of, 176
space-time, 166
Spencer, H., 281
Staples-Browne, R., 281
Starling, E. H., 212
stars, Cepheid, 124, 154, 159
  energy of, 127
  hydrogen in, 160
  interior of, 177
  mass of, 159, 177
  motion of, 118
  pulsating, 144, 154
  temperature of, 177
Steel, A. G., 8
stellar energy, source of, 146, 160
  fission, 103, 116
Stephenson, Marjory, 217, 241
Stevens, Jessie (Lady Hopkins), 213
Stevenson, R. L., 238
Stevenson, T., 209f, 213
Stewart, Balfour, 3ff, 12, 48
Stokes, G., 22, 33, 98, 197
Strømgren, B., 160
submarine detection, 77
Swift, Dean, 169

Tartter, 221
Taylor, G. I., 29
Tchebycheff, 113
thermionic emission, 19, 148
thermodynamics, second law of, 99, 173, 185
Thompson, Martha, 48
Thompson, W. H., 36
Thomson, G. P., 2, 25, 29, 40
Thomson, J. J., 1, 55, 57f, 59f, 63, 69, 74, 78f, 80, 98, 157, 197
  achievement, 1
  ancestry, 1ff
  Cavendish professor, 10
  class origin, 40f
  discovers electron, 17
  fortune, 34, 38
  health, 31
  isotope analysis, 21
  literary taste, 39
  marriage, 24
  Mastership of Trinity College, 38f
  moving charge, mass of, 9
  Nobel Prize, 27, 39
  physical chemistry, 11
  political ideas, 41
  research organizer, 21
Thomson, Mrs. E. S., 3
Thomson, Mrs. R. E. (Lady Thomson), 55, 61
Thomson, W. (Lord Kelvin), 11, 19, 29, 32f, 40f, 51, 98, 106, 114, 145
Threlfall, R., 23, 37f
Tizard, H. T., 38
Todhunter, I., 7
Townsend, J. S., 15, 17f, 26, 29, 55
tubes of force, 13

uncertainty principle, 131f
universe, age of, 128
  expanding, 163
urine analysis, 221

vacuum pumps, improvement of, 16
variations, 256, 261, 271, 292
Vavilov, N. I., 285, 303, 305, 308
Vavilov, S. I., 74
Vesalius, 307
Virchow, R., 266
vitalism, 238

vitamins, 216, 229, 230, 241
Vogt, W., 306
Vries, de, 270, 277

Wade, J., 210
Wallace, A. R., 262, 279
Walpole, H., 276
Walton, E. T. S., 84f
Washbourne, 211
Watson Watt, R. A., 37, 40
Watt, James, 298
Weismann, A., 267, 290, 295, 303
Weldon, W. F. R., 253f, 260, 270ff, 273ff, 279
Wheeler, 103
Wheldale, Miss M. (Mrs. Onslow), 281f, 287
Wiechert, E., 16
Wieland, H., 200, 219ff
Wien, W., 20, 71
Willcock, E. G., 226
Wilson, C. T. R., 16f, 27, 29
Wilson, H. A., 26, 29
world-building, 162, 187f

X-ray analysis, 85
  diffraction, 75
X-rays, 58
  analysis of crystal structure by, 29
  effect on gases, 58
  scattering of, 29, 157

Yukawa, H., 193

Zaviska, F., 29
Zdanov, 77
Zeleny, J., 26

For Product Safety Concerns and Information please contact our EU representative GPSR@taylorandfrancis.com
Taylor & Francis Verlag GmbH, Kaufingerstraße 24, 80331 München, Germany

www.ingramcontent.com/pod-product-compliance
Ingram Content Group UK Ltd.
Pitfield, Milton Keynes, MK11 3LW, UK
UKHW020145200926
475702UK00005B/933

* 9 7 8 0 4 1 5 4 7 4 5 3 5 *